Student Poster Sessions 2014

Core Programming Area at the 2014 AIChE Annual Meeting

Atlanta, Georgia, USA
16-21 November 2014

ISBN: 978-1-5108-1273-4

Printed from e-media with permission by:

Curran Associates, Inc.
57 Morehouse Lane
Red Hook, NY 12571

Some format issues inherent in the e-media version may also appear in this print version.

Printed by Curran Associates, Inc. (2015)

For permission requests, please contact AIChE
at the address below.

AIChE
120 Wall Street, FL 23
New York, NY 10005-4020

Phone: (800) 242-4363
Fax: (203) 775-5177

www.aiche.org

Additional copies of this publication are available from:

Curran Associates, Inc.
57 Morehouse Lane
Red Hook, NY 12571 USA
Phone: 845-758-0400
Fax: 845-758-2634
Email: curran@proceedings.com
Web: www.proceedings.com

TABLE OF CONTENTS

Robust Bifunctional Catalysts via Grafting Polyoxometalates to Mesoporous Silica with Intercalated Noble Metals

Frida Damián fridamian9@gmail.com
Yazhou Ji yaji@mymail.mines.edu
Ryan Richards rrichard@mines.edu

Abstract

The synthesis of a bifunctional catalyst with high thermal stability, catalytic performance and recyclable is a matter of great interest for green chemistry research. By having a low deactivation rate, such a catalyst could potentially be used in biomass upgrading reactions. In this project, we focused on the synthesis and characterization of a bifunctional catalyst with polyoxometalates (POMs) grafted on the surface, acting as an acid catalyst, and metal nanoparticles incorporated on the mesoporous silica wall as a hydrogenation catalyst. The first surface modification anchored transition metals, Au, Pd and Pt, inside the walls of the SBA-15mesoporous network. The second grafted the POMs, tungstosilicic acid hydrate. Catalytic performance and recyclability were tested with a benchmark reaction, the oxidation of n-hexadecane. The results showed a high catalytic activity and low deactivation rates over a total of three cycles. Further, characterization methods including transmission electron microscopy, physisorption and infrared spectroscopy, proved the successful synthesis of the bifunctional robust catalyst. Future testing of the catalysts on more complex reactions is being pursued.

Mechanistic Examination of Acid-Base Bifunctional Aminosilica Catalysts Through Hammett Analysis

Virginia E. Collier, George I. Lindy, Eric G. Moschetta, and Christopher W. Jones*

*cjones@chbe.gatech.edu

School of Chemical & Biomolecular Engineering
Georgia Institute of Technology, 311 Ferst Drive NW, Atlanta, GA 30332

Bifunctional catalysts, also known as cooperative catalysts, use two species that operate cooperatively to increase the catalyst's activity as compared to the activities of the independent species. In cases where the two components are an acid and a base, the active species must be kept spatially separated such that they do not negate the other's effects while still allowing them to favorably interact. The benefit of tailoring the spatial interactions between catalytic species on a solid support is that the resulting interactions between both catalytic species and the reactants augment the activity in a fashion that may not otherwise be possible in solution. The heterogeneous nature of the bifunctional material facilitates separation from the reaction mixture and the ability to synthesize metal-free, recyclable bifunctional materials makes it environmentally friendly. Mesoporous silica is a popular choice for a catalyst support as the surface silanols can act as weak Brønsted acids for an acid-base bifunctional material. For this study, trialkoxysilanes bearing amine groups were selected as the basic species and grafted onto the silica surface. Acids and bases are common components of heterogeneous cooperative catalysts because anchoring these species on the same support allows the material to have both acidic and basic character without the catalytic species deactivating each other as they would in solution.

The effects of the silica surface amine-silanol ratio on the rates of the aldol and nitroaldol (Henry) condensations were examined. These reactions were selected due to their application in industry for organic synthesis on a larger scale. Six catalysts were produced with amine loadings ranging from 0.20 mmol N/g silica to 1.9 mmol N/g silica. Aldol and nitroaldol reactions were run on a 10 mol% base catalyst basis for one mole of reactant. The aldol reaction displayed a lower conversion than the nitroaldol reaction and showed more sensitivity to amine density as high reactant conversion was obtained within a much narrower range of amine loadings for the same total reaction time. Nitroaldol activity was highest for a 1.0 mmol N/g silica loading with an optimum loading range of 0.6 to 1.2 mmol N/g silica, while the aldol reaction activity peaked at a 1.0 mmol N/g silica loading and had a narrower optimum loading range of 0.7 to 1.0 mmol N/g. Hammett analysis for each reaction was performed using a 0.5 mmol N/g catalyst loading to further examine the mechanism of each reaction. The aldol reaction data indicated a non-ionic species was formed during reaction whereas the nitroaldol results suggest formation of a cation or near-cation species, potentially by nitromethane's ability to form a resonance structure, while the acetone used in the aldol reaction cannot form such a resonance state. The rate of reaction for the aldol and nitroaldol reactions was found to be enhanced by electron withdrawing groups in the para position.

Catalyst Preparation by Addition of Metal Ions During the Synthesis of Silica Aerogels

Silica aerogels have been the subject of much study and speculation over the past several decades. First produced in the 1930s, silica aerogels are lightweight, transparent materials consisting of more than 90% air. Silica aerogels exhibit several unique and unusual properties, including high specific surface area, high porosity, transparency, low thermal conductivity, low density, and high optical transmission. Due to the rare qualities they possess, aerogels have numerous scientific and commercial applications. These applications include thermal insulation, acoustic barriers, and catalytic support. Researchers strive to discover new applications of silica aerogels, as well as improve existing applications. An application of particular interest is that of catalytic support.

There are three principal steps involved in the synthesis of silica aerogels: preparing the gel, aging the gel, and drying the gel. The primary factors to take into consideration for gel preparation are starting materials and the hydrolysis procedure. The starting materials involved in silica aerogel synthesis include a source of silicon, a solvent, a catalyst, and water. In general, silicon alkoxides are used as the source of silica, and tetramethoxysilane (TMOS) and tetraethoxysilane (TEOS) are the preferred precursors. When these substances undergo hydrolysis, silicon dioxide, or silica, is produced. In addition, methanol is produced as a byproduct from TMOS, and ethanol is produced as a byproduct from TEOS. TEOS was chosen as the silicon derivative for this project. Alcohols, particularly methanol and ethanol, are popular solvents because they reduce the hydrolysis rate during the synthesis. Ethanol was chosen because ethanol is also a byproduct of TEOS hydrolysis. The three accepted gel preparation procedures are acid catalysis, base catalysis, and two-step catalysis. In both acid and base catalysis, the catalyst is combined with the silicon derivative, solvent, and water to create a sol. The two-step catalysis procedure involves the addition of an acidic catalyst and a basic catalyst at different times. First, the acidic catalyst is combined with the silica derivative and the solvent. Then, a mixture of a basic catalyst and water is added to create the sol. The sol is then left to age. The two-step process has been proven to increase the rate of condensation, reduce the gelation time, and induce a broader distribution of pores in the gel. For these reasons, the two-step process was chosen with citric acid and ammonium hydroxide.

The second step in silica aerogel synthesis is aging. During the aging process, the sol is submerged in a solvent for a designated amount of time. This time period can range from several hours to several days. This leads to a stronger, more rigid silica network. Once again, ethanol was chosen as the solvent because it is a byproduct of the silicon derivative, TEOS. In this work, the gels were aged in solutions containing cerium and platinum salts.

There are several methods used to dry silica aerogels. The most common drying method involves supercritical carbon dioxide. First, the gel and solvent, usually an alcohol, are placed in an autoclave. Then supercritical carbon dioxide is pumped in at constant temperature until the pressure is above the critical point. The solvent is then replaced by the supercritical carbon dioxide. Finally, the pressure is lowered at constant temperature, and once ambient pressure is reached, the autoclave is cooled to room temperature. This method was chosen because drying methods implementing the use of carbon dioxide have proven to result in crack-free gels.

The three-step synthesis is the most basic form of synthesizing silica aerogels. This research will focus on how the addition of metal ions affects the synthesis of a silica aerogel. Specifically, it will determine the effect of metal ions on the rate of gelation, surface area, and atomic content of the gel. This research focuses on how the addition of metal ions during the synthesis of a silica aerogel affects its surface characteristics and catalytic activity. In particular, effects of metal ions on the rate of gelation, surface area, and atomic content of the gel are addressed. This project focuses primarily on the metals cerium and platinum. Cerium ions are good oxygen promoters for oxidation and reduction reactions, and platinum is an active catalyst for hydrogen enrichment reactions. A high dispersion of metal ions on the surface of silica aerogels will enhance their catalytic activity. Silica aerogels have been prepared, aged in ethanol with and without platinum and cerium salts, and dried in supercritical carbon dioxide. The specific surface areas of these gels were all in excess of 500 m^2/g, suggesting that the presence of the salts did not adversely affect the formation of the gel network. An elemental analysis of these gels as well as catalytic property testing using water-gas shift reaction will be completed to further this research.

Propylene (C_3=) Metathesis by Supported WO_x/SiO_2 Catalysts: Structure-Reactivity Relationship and Reaction Kinetics

Edgar Macuil, Soe Lwin Israel E. Wachs

Department of Chemical Engineering, Lehigh University, Bethlehem, PA 18015, USA

There currently is a shortage of propylene that will continue fro the next decade. To meet this propylene shortfall, many chemical companies are turning to metathesis of ethylene and 2-butene that yields two propylene molecules. The commercial olefin metathesis catalyst is supported WO_x/SiO_2, but only limited research has previously been performed on this catalyst system in spite of its industrial importance of this catalyst system, there has been very little research performed on supported WO_x/SiO_2 in prior years. The objective of this research is to determine the structure-activity relationship of supported WO_x/SiO_2 catalysts for olefin metathesis.

The supported WO_x/SiO_2 catalysts were prepared by the incipient wetness impregnation method followed by drying and calcination in molecular air. Characterization with *in situ* Raman spectroscopy revealed that the tungsten oxide phase was present as both isolated surface WO_4 sites and WO_3 nanoparticles on the SiO_2 support, with the latter only present at high tungsten oxide loading (> 8% WO_x). The reactivity and kinetics of the supported tungsten oxide phase were determined with temperature programmed surface reaction (TPSR) spectroscopy and steady-state olefin metathesis studies. The supported WO_3/SiO_2 catalyst was initially activated by exposure to propylene at 500°C. The best performing catalyst for propylene metathesis was supported 8% WO_3/SiO_2 demonstrating that the surface WO_4 sites, and not the WO_3 nanopartcles , are the catalytic active sites. The propylene metathesis kinetics were found to follow 1st–order dependence on the propylene partial pressure suggesting that the rate-determining-step is the propylene adsorption step. The E_{act} = ??? kJ/mol with a pre-exponential value of ???/s. This new information can be used to develop better catalysts and more accurately design the chemical reactors.

Effects of Temperature on Activation Energy and Pre-Exponential Factor in C-C Bond Forming Reactions Using a Cooperative Aminosilica Catalyst

George I. Lindy, Virginia E. Collier, Eric G. Moschetta, and Christopher W. Jones*

School of Chemical & Biomolecular Engineering

Georgia Institute of Technology, 311 Ferst Drive NW, Atlanta, GA 30332

gilindy@gatech.edu; vcollier3@gatech.edu; eric.moschetta@chbe.gatech.edu

*cjones@chbe.gatech.edu

In the organic synthesis of chemical products, the formation of covalent carbon-carbon bonds is consistently an essential process in the production of important molecules. Overcoming the activation energy of these various reactions often requires a catalyst consisting of a transition metal. These metals are expensive and can deactivate and/or deteriorate over time. Silica based catalysts are easily synthesized, and do not require the rare elements needed for some metal-based catalysts. These silica materials contain surface silanols that can act as weak acids in chemical reactions. These silanols react with trialkoxysilanes bearing functional groups such as amines and carboxylic acids to impart different catalytic effects. Incorporating basic amines on these silica surfaces allows for cooperative catalytic behavior, in which the material has acidic and basic properties that can participate in the catalytic process without deactivating each other. In addition to their catalytic nature for organic synthesis, amine-functionalized silicas also have promising results in CO_2 capture, making them an interesting topic for study. In this study, effects of temperature in aldol and nitro-aldol condensations facilitated by amine-loaded, silica based catalysts is observed.

In this study, we chose 4-nitrobenzaldehyde as the substrate because the strong electron-withdrawing group in the para position increased the catalytic activity for each reaction. The aldol and nitro-aldol condensations use acetone and nitromethane, respectively, as solvents and reactants. The active catalyst is an MCM-41 silica catalyst with (4-aminobutyl)triethoxysilane grafted on the surface. Physical and chemical characterization of the catalyst is confirmed using physisorption, XRD, and TGA. A HDMS capped version of the catalyst is used as a control to ensure that cooperative catalytic activity is, in fact, from the grafted amine groups and silanols. We performed the aldol and nitroaldol condensations on the aminosilica catalysts at four different temperatures to determine the initial activity at each temperature. The kinetic results for each trial allow for the activation energy and pre-exponential factor to be calculated for the two reactions. Our results indicated that the aldol condensation has a higher activation energy than the nitro-aldol, and that both activation energies are lowered when the silanols are uncapped. These results suggest that silica based catalysts, upon further research, could have a large impact on the chemical industry involving organic synthesis.

395547 The Adsorption of PEM Degradation Products on Pt Catalyst

Abstract:

ABSTRACT

The polymer electrolyte fuel cell is a promising alternative technology for automotive applications, using H_2 gas for power. After operation, the polymer electrolyte membrane in the fuel cell can degrade and produce organic contaminants that can decrease the fuel cell performance. Electrochemical quartz crystal microbalance (EQCM) technique was used to determine the effects of organic compounds Da-Naf and Da-3M, representing degradation products from a Nafion®membrane and a 3M™ membrane, respectively. The effects of Da-Naf and Da-3M concentration on Pt catalyst surface coverage were studied. By using EQCM with cyclic voltammetry, a richer set of data was obtained than by using cyclic voltammetry alone. The results suggest that both Da-Naf and Da-3M adsorb onto Pt catalyst sites, with Da-3M occupying more sites than Da-Naf, possibly due to a difference in chemical structure. Neither Da-Naf nor Da-3M adsorbed irreversibly to the Pt electrode. Due to the fact that these contaminants can be produced within the fuel cell and their apparent poisoning potential, the poisoning mechanism of both Da-Naf and Da-3M should be studied further.

397408 Ethane and Methane Dehydrogenation over Pt and PtSn Alloys

Abstract:

The effect of platinum tin alloy structure and composition on the kinetics and thermodynamics of dehydrogenation and coke formation pathways during light alkane dehydrogenation have been studied using density functional theory. Ethylene is in high demand as a valuable chemical precursor and is among the most produced organic compounds worldwide. Ethane is a cheap, abundant chemical, but has less practical value. Therefore, studying catalysts for ethane dehydrogenation to its corresponding olefin is relevant to modern demands and greatly increases the value of this particular alkane. Supported Pt catalysts are known to be active for ethane dehydrogenation, but the high temperatures required by these endothermic reactions leads to significant coke formation and deactivation. Alloying Pt with Sn and other main group elements has been shown to decrease the amount of coke formed and lead to more stable catalysts. We apply periodic density functional theory to compare the potential energy surfaces from ethane and methane dehydrogenation. This will allow us to better appreciate the effect of catalyst composition (e.g. Pt/Sn ratio) and surface geometry on both reaction energetics and coke formation. As compared to pure Pt, C-H bond scission is more difficult on the alloys, and dihydrogen desorption is more facile.

397605 Uncertainty Quantification of a Theoretical Study of the (reverse) Water-Gas Shift Reaction over Pt (111)

Abstract:

Uncertainty quantification of a theoretical study of the (reverse) water-gas shift reaction over Pt (111)

Donald Mitchell

The City College of New York

Advisors: Eric Walker, Dr. Andreas Heyden

The water-gas-shift (WGS), $CO + H_2O \leftrightarrow CO_2 + H_2$, and its reverse reaction are key processes in the purification of hydrogen, synthesis of methanol and production of long chain hydrocarbons. In this work density functional theory is employed to study the reverse and forward water-gas shift reaction on a Pt(111) catalyst. A Pt catalyst was modeled as a four-layer slab with a 4x3 surface unit cell using the VASP software with Perdew, Burke, Ernzerhof functional. Transition state searches for surface reactions were conducted with nudged elastic band and dimer methods. A microkinetic model was developed to generate model turnover frequencies at relevant temperature and pressure conditions. Elementary reaction rate constants of the microkinetic model were computed using harmonic transition state theory. Model turnover frequency results are compared to experiments at a variety of conditions. The model qualitatively agrees with experiments although the model consistently underestimates the experiments. Finally, due to the inexact nature of density functional theory and the stiff equations in the microkinetic model, uncertainty quantification is necessary to report reliable turnover frequency results. Therefore, uncertainty quantification is displayed for selected model results. Future work to improve the model includes adding lateral interaction effects to binding energies of molecules adsorbed to the catalyst.

Photocatalytic Reactor Design and Synthesis of Selected Metal–Organic Frameworks for Carbon Dioxide Reduction

Utilization of Solar Energy in photocatalytic reactions for reducing carbon dioxide to fuels and useful chemicals (i.e. CO, CH_4, CH_3OH) has caught a great deal of attention in the past decades. As the chemical reduction of CO_2 is energetically difficult due to its remarkable thermodynamic stability, this process requires a significant amount of energy transfer. Therefore, Metal–organic frameworks (MOFs), a new class of crystalline molecular solids built from linking organic ligands with metal or metal-cluster connecting points, have recently emerged as a versatile platform for developing single-site solid catalysts. MOFs have been used to drive a range of reactions, including Lewis acid/base catalyzed reactions, redox reactions, asymmetric reactions, and photocatalysis. MOF catalysts are easily separated from the reaction mixtures for reuse, and yet their molecular nature introduces unprecedented chemical diversity and tunability to drive a large scope of catalytic reactions. In this study, a photocatalytic reactor was designed to meet the objective of handling the reaction proposed. While designing the reactor, different parameters were considered in order to reach the optimal conversion rate. Moreover, selected MOFs were synthesized and examined for specific criteria (i.e. water stability, semi-conductivity, etc…).

Thermochemical CO_2 and H_2O Splitting via Chemical-Looping with Cerium and Cobalt Mixed Oxides for Oxygen Generation.

Ari Fischer, Chunxiang Zhu, George M. Bollas
[*] Department of Chemical & Biomolecular Engineering, University of Connecticut, Storrs, 191 Auditorium Road, Unit 3222, Storrs, CT, 06269-3222, USA. Email:ari.fischer@uconn.edu

Reliable oxygen supply for NASA exploratory manned space missions is of strategic importance. The best method to provide the O_2 required for manned missions is to recycle the oxygen contained in CO_2 and H_2O produced by the astronauts in the space vessel. To accomplish this goal, this projects bridges elements of materials science and reactor and process design in an effort to break the thermodynamic equilibria of the unfavorable splitting of CO_2 and H_2O, at a minimum energy penalty and system size. This presentation will discuss progress made in enhancing the oxygen production from CO_2 and H_2O by utilizing, in a sequential manner, the oxidation and reduction properties of CeO_2 and Co_3O_4, respectively. Co_3O_4 has been shown to release oxygen via a process known as Oxygen Uncoupling at temperatures 800-900°C.[2] This work will explore the feasibility of pairing the thermal oxygen uncoupling (reduction) of Co_3O_4, with a regeneration step of oxygen transport from CeO_2, which in turn is oxidized by CO_2 in a separate process step. Oxygen uncoupling operates on the principle that for certain metal oxides (e.g., Co_3O_4) it is thermodynamically feasible to uncouple gaseous oxygen, while being exothermally reduced in an inert atmosphere. At the same time, certain metals (e.g., CeO) show favorable kinetics for oxidation by CO_2 at elevated temperatures. It is also well known that metals alloys allow for the transport of oxygen between the metals comprising them. Thus, one can envision a process, in which a Ce-based oxygen carrier performs CO_2 splitting in an endothermic process, transports the captured oxygen to a Co-based oxygen carrier, which then uncouples it to gaseous oxygen in an exothermic process. The objective of this work is to explore experimentally such mixed metal oxygen carriers and show the thermodynamic and kinetic feasibility envelopes for an oxygen production process.

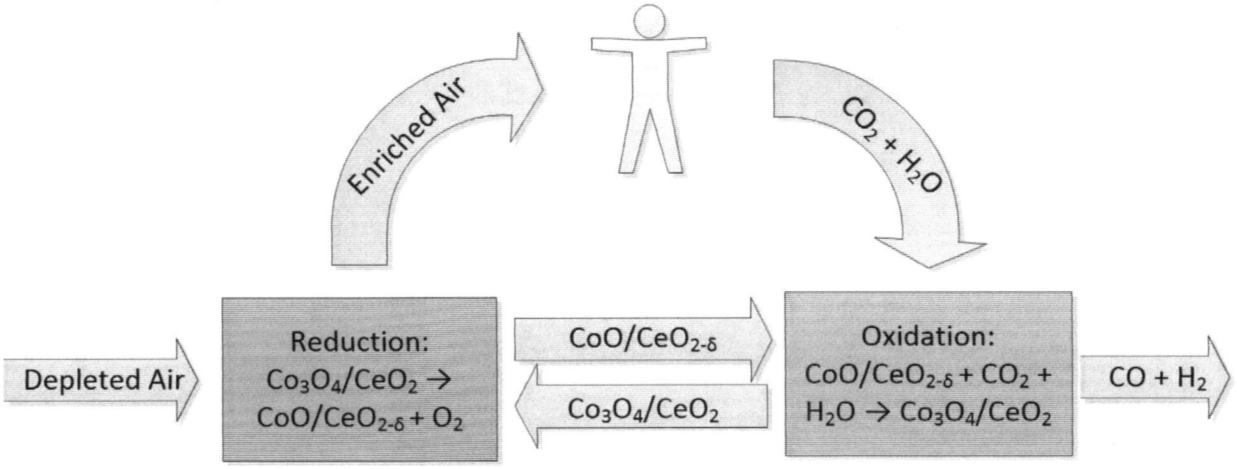

Figure 1: Basic concept of CO_2 and H_2O splitting via chemical-looping with cerium and cobalt oxides.

High temperature cycling of CeO_2 and Co_3O_4 mixed metal oxides can achieve thermochemical splitting of H_2O and CO_2. CO_2 and H_2O are stable molecules; for example H_2O requires prohibitive temperatures of about 4100 K before direct thermolysis is thermodynamically feasible.[1] However, metal redox cycles using cerium oxide (CeO_2) have been proven capable of CO_2 and H_2O reduction at temperatures as low as 900°C, as described by Scheffe et al.[1] This cycle was tested by Chueh et al.[3] under realistic solar concentrating conditions over 500 cycles to yield syngas that reflected in composition the CO_2 and H_2O feed ratio[1]. Figure 1 shows a basic diagram for the oxidation and reduction reactions. The reduction phase occurs at high temperature via nonstoichiometric oxygen release (the cerium oxide lattice structure is retained despite oxygen vacancies) in the cerium and by the cobalt oxide uncoupling reaction. In the oxidation phase, $CeO_{2-\delta}$ is oxidized by CO_2 and H_2O to produce CeO_2 and syngas. Oxygen transport from the ceria oxidizes CoO to form Co_3O_4. For pure ceria cycles, δ is limited to 0.25 and therefore the extent of oxygen generation is small. By pairing CeO_2 and Co_3O_4 oxides, the CO_2 and H_2O splitting properties of ceria and the oxygen uncoupling properties of cobalt oxide can increase the amount of oxygen released. In this study, CeO_2 and Co_3O_4 mixed oxides were prepared using a co-precipitation method. These oxygen storage materials were characterized using XRD, SEM, and TGA. A fixed-bed reactor was used to study high temperature oxygen release in an inert gas and CO_2 reduction over the oxygen depleted material.

Acknowledgments

This work is supported by the Fall 2013 Connecticut Space Grant Consortium Undergraduate Research Fellowship.

References:

1. Scheffe JR, Steinfeld A. Oxygen exchange materials for solar thermochemical splitting of H2O and CO2: a review. *Mater. Today* 2014;00(00):1-8.

2. Adanez J, Abad A, Garcia-Labiano F, Gayan P, de Diego LF. Progress in Chemical-Looping Combustion and Reforming technologies. *Prog. Energy Combust. Sci.* 2012;38(2):215-282. doi:10.1016/j.pecs.2011.09.001.

3. Chueh W, Haile S. A thermochemical study of ceria: exploiting an old material for new modes of energy conversion and CO2 mitigation. *Philos. Trans. A. Math. Phys. Engr. Sci.*, vol. 368, no. 1923, pp. 3269-94, Jul. 2010.

Role of C_2H_4 during CH_4 Conversion to Benzene by Supported MoOx/ZSM-5 Catalysts

Chris Walker, Yadan Tang and Israel E. Wachs
Department of Chemical and Biochemical Engineering
Lehigh University
Bethlehem, PA 18015 USA

The recent discovery of significant reserves of shale gas has triggered interest in gas-to-liquid technology (GTL) and especially a one-step process to convert methane to liquid hydrocarbon fuels. Methane dehydroaromatization (DHA) to benzene by supported MoO_x/ZSM-5 catalysts has generated much interest. Despite extensive studies on this catalytic reaction system, debate remains about the reaction mechanism, catalytic functions of Mo and ZSM-5 support, and the rate-limiting step. This work aims to provide fundamental insights into these issues *via* CH_4 and C_2H_4 temperature programmed surface reaction (TPSR) spectroscopy.

The CH_4-TPSR measurements revealed that initiation of benzene formation requires ~700 °C and only with MoO_3/ZSM-5 catalysts, while Mo-free HZSM-5 does not activate CH_4. C_2H_4 is observed as a reaction intermediate during CH_4 conversion to benzene. Corresponding C_2H_4-TPSR experiments demonstrated that C_2H_4 also produces benzene, but at much lower temperatures (~500 °C) and. These observations lead to the conclusions that the

- higher reaction temperature of CH_4-TPSR demonstrates that methane C-H bond cleavage is the rate-limiting-step,
- activation of CH_4 requires Mo sites,
- conversion of C_2H_4 to C_6H_6 does not require Mo sites and is catalyzed by Brønsted acid sites of the ZSM-5 support, and
- Mo sites increase the yield of benzene and decreases the activation temperature.

The presence of surface Mo sites on the ZSM-5 support decreases the amount of Brønsted acid sites and the moderate concentration of Brønsted acid sites for supported MoO_3/ZSM-5 catalysts might suppress severe coking in comparison to Mo-free HZSM-5.

397950 The Synthesis and Oxygen Transport Properties of La2NiO4 Nanoparticles

Abstract:

This work highlights a novel approach for synthesizing $La_2NiO_{4+\delta}$ nanoparticles with controlled morphology through the use of reverse micro-emulsion systems. $La_2NiO_{4+\delta}$ is a mixed ionic and electronic conducting (MIEC) oxide that exhibits excellent oxygen transport properties due to oxygen hyperstoichiometry in the lattice. The oxygen exchange/transport properties play an important role in the catalytic activity of $La_2NiO_{4+\delta}$ toward reactions involving oxygen. In this contribution, we show that the oxygen exchange properties of these materials can be tuned by controlling the size and shape of the oxide nanoparticles via a reserve micro-emulsion synthesis approach. We show that the ratio of the surfactant (CTAB) to water (W_0) plays a key role in controlling the aspect ratio and the distribution of surface facets of the nanostructures. Variations in W_0 lead to changes in the geometry of the particles from nanorods to polyhedra. In addition, the choice of precipitating agent directly influences the size of nanoparticles. Thermal gravimetric analysis (TGA) of the synthesized $La_2NiO_{4+\delta}$ shows anisotropic behavior of oxygen diffusion, which is directly linked to the degree of oxygen hyperstoichiometry (δ) and distribution of surface facets. This method has provided opportunities for optimizing the morphology of $La_2NiO_{4+\delta}$ nanoparticles, which has implications in catalytic processes that involve redox reactions with oxygen where mixed ionic and electronic conductivity is required.

398972 Characterizing Coke Deposition on ZSM-5 during Catalytic Fast Pyrolysis

Abstract:

Catalytic fast pyrolysis of biomass has gained interest as a pathway for producing bio-oil as a sustainable, cleaner alternative to traditional fossil fuels. One of the challenges facing the implementation of catalytic fast pyrolysis is the formation of carbonaceous coke during subsequent rounds of catalysis which leads to rapid deactivation. This coke may be combusted off, but this requires large amounts of energy. The precise mechanism of coke formation is not well understood, but by examining the coke formed under various conditions, we hope to understand more about how catalytic fast pyrolysis may be optimized to produce more hydrocarbons with less coking. For this research, a quartz annular flow horizontal packed bed reactor was used to pyrolyze ten quarts boats containing 50.25±0.25 mg biomass. The boats were pyrolyzed successively; the produced vapors were upgraded over 0.5 g HZSM-5 catalyst and then analyzed via molecular beam mass spectrometer. The overall biomass to catalyst ratio was 1:1 for each experiment. Eight experiments were conducted: four different biomasses were pyrolyzed and upgraded in the presence of hydrogen, then repeated in the absence of hydrogen. Biomasses sampled include microcrystalline cellulose, larch wood xylan, pine shavings, and red oak lignin. Following pyrolysis experiments, mass spectrometer results were analyzed for product formation and coked catalyst samples were analyzed for weight and acidity using thermogravimetric analysis and temperature programmed ammonia desorption. The catalyzed pyrolysis runs carried out in hydrogen were expected to have less coke formation, as that the effective hydrogen index would be increased. However, the zeolites in ZSM-5 were not able to activate the hydrogen, so addition of hydrogen had little effect on coke formation. Catalytic fast pyrolysis of xylan had the lowest overall coke formation, but produced low to moderate yield of hydrocarbons, similar to that of lignin. Lignin and Xylan consist of shorter polymers or amorphous resins and must build up a hydrocarbon pool within the pores of the catalyst before upgraded vapors are produced. Catalytic fast pyrolysis of cellulose is capable of producing the greatest yield of hydrocarbon product, but also has the greatest deactivation due to coke formation.

397417 Optimization of the Synthesis of Koser's Reagent Using Electron Deficient Arenes

Abstract:

Koser's reagent ([hydroxyl(tosyloxy)iodo]benzene), and derivatives thereof, can be used for many different kinds of reactions, such as the synthesis of diaryl iodane tosylates which are useful aryl electrophiles. The preparation of Koser's reagent depends on the nucleophilicity of the iodine atom in the aryl iodide precursors. Accordingly, derivatives of Koser's reagent with electron deficient arenes are more difficult to synthesize than electron rich arenes along with extended reaction times (multiple hours) and expensive 2,2,2-trifluoroethanol as a co-solvent are required for synthetically useful yields (57 – 83%). A greener, cost effective and efficient reaction would ideally be conducted in short reaction time and without the need for the fluorinated co-solvent, while maintaining high chemical yields. Experiments were conducted using microwave and conventional heating at varying temperature, time, and reaction concentration to determine which variable had the greatest influence on reaction yield using the highly electron-deficient aryl iodides 1-iodo-4-trifluromethylbenzene and 1-iodo-4-nitrobenzene. It was found that the most influential variable was reaction concentration. The koser's reagent using 1-iodo-4-trifluromethyl-benzene was prepared in 85% yield in only 2 minutes at 40°C in dichloromethane at 0.67 M in aryl iodide. Using the 1-iodo-4-nitro-benzene, the reaction was optimized at 40°C for 30 minutes in dichloromethane at 0.67 M for 88% yield. Future experiments will test the breadth of this effect across a diverse range electron deficient arenes.

A Systematic Study of Cu, Fe Based Non-Precious Metal Catalysts for Use as PEM Fuel Cell Cathodes

Aston Thompson[a], Gabriel Goenaga[a], Nelly Cantillo[a] and Thomas Zawodzinski Jr.[a,b]

[a] Chemical and Biomolecular Department, University of Tennessee – Knoxville
[b] Physical Chemistry of Materials Group, Oak Ridge National Laboratory

State of the art proton exchange membrane fuel cells (PEMFCs) utilize platinum and platinum alloy metals to catalyze the reactions taking place on both the anode and cathode electrodes. Since platinum group metals (PGMs) are an expensive and limited resource, replacing the Pt with a non-PGM catalyst is a primary research goal in the PEMFC area. Cu, Fe and Cu-Fe alloys supported on a carbon black were investigated as the cathode catalyst material for PEMFC. Samples were thermally activated for one hour under inert gas atmosphere at temperatures ranging from 600°C to 1000°C. Electrochemical characterization using rotating ring disk electrode (RRDE) experiments was carried out to determine the oxygen reduction reaction (ORR) catalytic activity and basic mechanistic aspects of these materials. In addition, single cell experiments were conducted to determine the cathode performance and stability under cell operating conditions.

High Stability, High Activity Pt/ITO Fuel Cell Catalysts

Abbey E. Wangstrom, Shuai Zhao, William E. Mustain
University of Connecticut

Abstract:

In this work, Tin-doped Indium Oxide (ITO) has been studied as a stable non-carbon catalyst support material for both the Proton Exchange Membrane Fuel Cell (PEMFC) and the Anion Exchange Membrane Fuel Cell (AEMFC). ITO was prepared with various tin doping and electrochemically tested in a three electrode electrochemical cell. Linear sweep voltammetry was used investigate the activity of the Pt/ITO catalysts for the oxygen reduction reaction in both acid and alkaline media, which exceeded that of commercial Pt/C. The stability of raw ITO and Pt/ITO was tested by cyclic voltammetry, and showed excellent durability compared to Pt/C.

Electrochemical Evaluation of Novel Binders for Alkaline Fuel Cell Applications

Samantha Medina[1], Gabriel Goenaga[2], Samuel St. John[2], Shane Foister[2] and Thomas A.
Zawodzinski Jr.[2,3]
[1] Material Science Department, The University of Tennesse, Knoxville, TN
[2] Chemical and Biomolecular Department, The University of Tennesse, Knoxville, TN
[3] Physical Chemistry Materials Group, Oak Ridge National Laboratory, Oak Ridge, TN

Low temperature fuel cells, such as proton exchange membrane fuel cells (PEMFC) and alkaline fuel cells (AFC), are of great interest because they offer the possibility of being used in portable applications. PEMFCs commonly use Pt as the catalyst for both the hydrogen oxidation reaction (HOR) and oxygen reduction reaction (ORR), as well as Nafion, a cationconducting (H^+) polymer, as the membrane and catalyst layer ionomer and binder. Both, Pt and Nafion, are expensive materials that increase the production cost of PEMFCs. One advantage AFCs have over PEMFCs is that non-precious metal catalysts can replace Pt for the ORR and HOR, thereby reducing the overall cost associated with the fuel cell. However, there is not a standard anion-conducting (OH^-) polymer that can be used as the membrane or catalyst layer ionomer and binder. Instead, Nafion or sodium exchanged Nafion is commonly used. In our lab we have synthesized novel low cost anion-conducting binders. These binders were electrochemically evaluated using rotating ring disk electrode (RRDE) experiments in a standard three-electrode electrochemical cell in 0.1M KOH with a precious metal catalyst (PMC) and a non-precious metal catalyst (NPMC) working electrode. We found that the performance of our binders surpassed that of Nafion when tested with a platinum catalyst, increasing the half way potential $E_{1/2}$ from 0.8 V with Nafion binder to 0.82 V with one of our binders.

398753 Engineering the Nano/Microstructure of Solid Oxide Fuel Cell Cathodes

Abstract:

Solid Oxide Fuel Cells (SOFCs)are high-temperatureelectrochemical devices that have the capacity to convert the inherent chemical energy of fuels (i.e, H_2, CO, hydrocarbons) into electrical energy. SOFCs are composed ofan anode where the fuel is electro-oxidized, an electrolyte that transports oxygen ions, and a cathode thatelectrochemically reduces gas-phase O_2 from air to oxygen ions, O^{-2}.In order to achieve structural stability of SOFCs at elevated, operating temperatures, the ion-conducting electrolyte oxide material is in corporated in both the anode and cathode compartments of the cell. The commonly utilized approaches for synthesizing button-cell SOFCs involve ball milling of the electrode electrocatalysts with the ion-conducting electrolyte oxide (i.e., yttria stabilized zirconia(YSZ)), and formation of the electrodes based on dry-pressing or tape casting the mixture into thin films. While these methods have led to structurally stable SOFCs, they provide limited flexibility on the incorporation of different electrocatalysts (only the ones that arestable and compatiblewith YSZ at high sintering temperatures of larger than 1000C that can be used), and lead to electrocatalyst structures that are fairly large in size (micron) with low catalytically active surface area.In this contributing, we described a method for preparing SOFC cathodes thatallow for flexibility over the cathode electrocatalyst and preservation of its nanostructure. This method relies on a controlled synthesis ofthecathode electrocatalysts with a desired nanostructure and its incorporation in a microporous scaffold of the ion-conducting oxide.We show that the developed cathode structures lead to an enhanced triple phase boundary (interface between the electrocatalyst and the ion-conducting electrolyte) and enhanced electrochemical performance.

398832 A Study of Kinetics for a Fuel Cell Contamination Model

Abstract:

The oxygen electrode of a low temperature fuel cell was studied using a spinning electrode, painted with a catalyst ink, submersed in an electrolyte solution and reacted via cyclic voltammetry scans. Potentiostatic measurements were used to determine the reaction kinetics. The experiments used various parameters , which included pH of the electrolyte solution, catalyst concentration, and contamination levels. The goal is to collect data that will help us to obtain parameters that will predict performance loss in the fuel cell due to contamination.

Catalyst Study and Characterization for a Polysulfide Bromine Redox Flow Battery

Chris Bruneau[a], Yu-Jia Bai[a], Gabriel Goenaga[a] and Thomas Zawodzinski Jr.[a,b]

[a] Chemical and Biomolecular Department, University of Tennessee – Knoxville
[b] Physical Chemistry of Materials Group, Oak Ridge National Laboratory

This research aims to understand the effect of different catalysts for the sulfur redox reaction on the negative electrode of a polysulfide bromine (PSB) battery. To test the catalyst activity, the performance of the battery is evaluated by charging and discharging polarization curves. With pure high surface area carbon paper electrodes, the discharging polarization curve exhibits very fast kinetics. However, the charging polarization curve shows a significant kinetic loss. Our research shows that deposition of a metal catalyst on the high surface area carbon paper yields an appreciable improvement on charging kinetics. This results in symmetrical polarization curves for charging and discharging, suggesting that our catalysts were effective in improving reversibility of the PSB. In addition, metal-oxide catalysts were also tested. Results show that the metal oxide catalyst yields even better kinetics for this system. The addition of these catalysts significantly lowered the overpotentials in the kinetic region of the polarization curves.

Fatty acid methyl ester biodiesel fuels produced from soybean oil with honeycomb monolithic catalysts

Priscilla Lang[1], Shamim Begum[1], Jonathan Mbah[1], Nader Vahdat[1], Kyung C. Kwon[1]

([1]) Department of Chemical Engineering, Tuskegee University, Tuskegee, AL 36088

ABSTARCT

Soybean oil as a representative of liquid vegetable oils is converted to fatty acid methyl esters (FAME) in the presence of methanol and formulated honeycomb monolithic catalysts via a transesterification process. Monolithic catalysts were formulated by impregnating honeycomb γ-alumina-wash-coated monolithic catalyst supports with various active metals. Reversible conversion of soybean oil to FAME was in favor of decreased reaction temperatures with monolithic catalysts, but catalytic activities of monolithic catalysts decrease more steeply with decreased temperatures, as reaction duration increases. Deactivated catalysts can be regenerated by heating them in atmosphere. Multi-layers of various active metals impregnated onto a honeycomb monolithic catalyst support stabilize the monolithic catalyst and increase catalytic activities of the monolithic catalyst with increased reaction duration. Order of loading catalytically active metals on monolithic catalyst supports affects stabilities and catalytic activities of honeycomb monolithic catalysts. Calcium oxide appears to stabilize catalytic activities of formulated monolithic catalysts. Repeatedly fluctuated conversion of soybean oil indicates that formulated monolithic catalysts are deactivated and regenerated repeatedly by themselves.

396520 Residence Time Distribution Studies on a Rotary Calciner with Flights

Abstract:

Rotary calcination is a thermal treatment process poorly understood, yet commonly used in catalyst manufacturing. Problems arise particularly upon scale-up of the process from the laboratory and pilot plant scales to the manufacturing scale. Developing such fundamental understanding of rotary calcination can improve product quality and cut energy and material costs. The two important indices for continuous calcination are: (1) in the axial direction, the residence time of the particles inside the calciner, and (2) in the radial direction, the characteristic time of calcination. This research seeks to provide a methodology for scale-up through understanding of the effects of calciner geometry, operating conditions and material properties on particle residence time and temperature distributions. To optimize calciner performance, the particle residence time must exceed the time required for heating and calcination. For uniform treatment, the particles must also exhibit low axial dispersion. A combination of discrete element method (DEM) simulations and experiments are used to explore the influence of these competing timescales on scale-up. This poster focuses on the results of recent residence time and dispersion studies conducted on a pilot plant scale calciner. In the experiments, the influence of operating conditions and material properties on the mean residence time, hold up and axial dispersion in a pilot plant rotary calciner. It was observed that increasing the feed rate did not significantly affect the mean residence time, while increasing the calciner incline and rotary speed decreased mean residence time.

Catalytic Biomass Pyrolysis: The Micropore Effect

Luzmary Sabino[1], Christopher J. Keturakis[1] and Israel E. Wachs[1*]

[1]*Operando Molecular Spectroscopy and Catalysis Laboratory, Chemical Engineering Department, Lehigh University, Bethlehem, PA 18015 USA*

*iew0@lehigh.edu

Biofuel production via catalytic pyrolysis of biomass has been studied as an alternative biofuel option in order to end the dependence on fossil fuels and develop a sustainable fuel. The biomass pyrolysis literature has not examined the dynamic changes of the catalyst or biomass as the pyrolysis reaction is taking place during pyrolysis. Previous studies in our lab using *in situ* and *operando* optical spectroscopy have provided molecular detail about catalytic active sites and reaction mechanism during biomass pyrolysis using HZSM-5 and supported Al_2O_3/SiO_2 catalysts. The molecular spectroscopy studies on catalytic biomass pyrolysis have shown both similar and differing product formation between H-ZSM-5 and supported Al_2O_3/SiO_2 catalysts although both catalysts possess surface AlO_x active sites. The objective of this investigation is to examine the effect of microporosity on the formation of products during pyrolysis since the literature claims that a "hydrocarbon pool" in the micropores is responsible for reactivity/selectivity. Four catalysts were chosen for this investigation, Silicalite-1 (Al-free ZSM-5 crystal structure, micropores, no acidity), fumed SiO_2 (no micropores or acidity), HZSM-5 (acidity, micropores), and supported Al_2O_3/SiO_2 (acidity, no micropores) catalysts.

The pore volume, pore size distribution, and surface area of the model catalysts were determined from the Brunauer-Emmett-Teller (BET) surface area analysis and the Barrett-Joyner-Halenda (BJH) and the Density Functional Theory (DFT) model pore size and volume analysis. The molecular structures of the catalysts were analyzed under dehydrated conditions using *in situ* Raman and IR spectroscopy. *In situ* temperature programmed IR (TP-IR) and GC-MS analysis were used to monitor the formation of reactive species and products during pyrolysis of cellulose, lignin, and xylan. These three biomass model compounds were selected because they represent the three major components of biomass. Finally, *operando* Raman Spectroscopy-Mass Spectrometry simultaneous monitored, in real time, product formation, catalyst structure and surface intermediate formation during catalytic biomass pyrolysis. The results showed that the microporosity of the catalysts during biomass pyrolysis affect the formation of several pyrolysis products, such as Levoglucosan (m/z 60), Furfural (m/z 96), Benzene (78), Toluene/Xylene (91), and Diethyl Phthalate (150).

398447 Synthesis and Characterization of Palladium-Iron Nanoparticle Catalyst for the Conversion of Biomass

Abstract:

The increasing demand and dwindling reserve of fossil fuels and the associated environmental issues have motivated industry and society to search for reliable, renewable and clean energy. Accordingly, the conversion from non-food source lignocellulosic biomass into biofuels and valuable chemicals has attracted increasing research interest. In the conversion of biomass, the hydrodeoxygenation (HDO) reaction is a critical step and therefore requires high-performance catalyst to optimize its yield and selectivity.

In this work, we explored the controlled synthesis of Palladium-Iron (Pd-Fe) nanoparticles catalyst for the HDO reaction of phenol and cresol. We demonstrated synthesis of bimetallic Pd-Fe nanoparticles via a simultaneous chemical reduction approach. Specifically, we reduced iron chloride ($FeCl_3$) and tetraaminepalladium nitrate ($Pd(NH_3)_4(NO_3)_2$) using sodium borohydride (NaBH4) as a reducing agent in the presence of polyacrylic acid (PAA) and polyvinyl pyrrolidone (PVP) as capping agents. The X-ray diffraction (XRD) pattern indicated an FCC structure with lattice constant larger than that of bulk Pd. The transmission electron microscopy (TEM) indicated a multiply twinned bimetallic structure and the average crystallite size from XRD was of 4.6 nm as calculated using the Scherrer equation. Energy dispersive x-ray spectroscopy (EDS) showed narrow composition distribution with Fe:Pd ratio between 2.5 and 3.5.

This work developed a feasible and controlled approach to synthesize Pd-Fe nanoparticles that will provide us better understanding of the function of Pd-Fe nanoparticles in the HDO reaction and basis for further studies into designing more active and selective catalysts to optimize the conversion of biomass into biofuel and valuable chemicals.

Andrew Fox

Analysis of Catalytic Conditions for Biochemically Produced p-Toluic Acid

In this research, various reaction parameters were altered to determine the ideal environment for the catalytic production of p-toluic acid from renewable biochemical. The method in development used a high pressure/high temperature reactor to convert coumalic acid, in the presence of propene and a dehydrogenating catalyst, into the desired product. Various catalysts and reaction conditions were tested to determine the parameters producing p-toluic acid with the highest conversion of coumalic acid, conversion to toluic acid, and selectivity of para form of toluic acid.

Current production of polymers in the industry is almost entirely dependent on the use of petrochemicals, a nonrenewable resource. In previous research, a method for developing one such polymer, polyethylene terephthalate (PET), from renewable biological sources has been developed; this research, however, was purely experimental and did not pursue optimization of the process. On aspect of this procedure, the conversion of coumalic acid to p-toluic acid, has been shown to be technically possible but not efficient enough for full scale operations. This project investigated the ideal conditions and catalysts that could lead to usable production of p-toluic acid.

The research conducted consisted of three main procedural stages: sample preparation, reaction development, and sample analysis. The samples consisted of the coumalic acid reactant dissolved in a solution of toluene at a ratio of 20 milligram of reactant per milliliter of solvent. 10 mg of active metal catalyst, either palladium, platinum, or ruthenium, on an activated carbon base was added to each sample. The samples were then pressurized with nitrogen gas to remove contaminants and the nitrogen gas replaced with propene gas for the reaction. Each sample was heated up to either 140, 160, or 180 °C and left to react for either two, four, or eight hours. The samples were then analyzed using Ultra Performance Liquid Chromatography (UPLC) and Gas Chromatography-Mass Spectrometry (GCMS). The UPLC was used to determine the concentration of products, intermediates, and reactants in the samples and the GCMS was used to determine the products, intermediates, and reactants present in each sample that could not be determined via UPLC. The results showed that coumalic acid was converted almost uniformly at high percentages. The yield of toluic acid was found to be the highest when palladium was used as a catalyst, followed closely by platinum, and with ruthenium producing a mere third of the other two; it also increased as time and temperature of the reaction increased. Selectivity was found to be highest at low temperatures and shorter run times and had higher trends when palladium or platinum catalysts were used.

While the experimentation produced interesting trends, the individual results were significantly lower than that of literature values and previous experimental work. The source of this error could arise from the sample preparation, analytic method, or the analytic equipment, but further studies would need to be conducted to pinpoint the error source. Future research could also investigate the kinetics of the reaction, as the current understanding of the reaction mechanism is limited and the most prevalent reaction pathway has yet to be determined. In addition, further optimization of the yield and selectivity with alternate catalysts, conditions, and solvents could be tested.

Interactions between SO_2 and Fly Ash Components for Oxycombustion Conditions

Robert Alexander MacDonald
West Virginia University

Advisor: Dr. Bihter Padak

Under air combustion, the flue gas from a coal power plant is primarily a mixture of water, nitrogen, and carbon dioxide. Oxy-combustion that utilizes combustion under oxygen and carbon dioxide (CO_2) results in a highly concentrated stream of CO_2 that can be sequestered before it enters the atmosphere. This is due to the flue gas lacking nitrogen, and the ease of condensing water from the gas stream. In oxy-combustion the lack of nitrogen causes oxygen to have a higher partial pressure, which oxidizes more sulfur dioxide (SO_2) to sulfur trioxide (SO_3) than in air combustion. The SO_3 reacts with water to form sulfuric acid causing corrosion. The higher amount of SO_3 results in higher sulfur retention on the fly ash, but the mechanism of sulfur binding is not well understood.

In this study, calcium oxide (CaO) and magnesium oxide's (MgO) ability to adsorb SO_x was examined under different temperatures and exposure times to SO_2 in air. The samples were characterized using Fourier transform infrared spectroscopy, x-ray diffraction, and a scanning electron microscope. The results so far have indicated that MgO has a lesser binding ability relative to CaO, this is partially due to the lower binding energy between MgO and SO_2.

398410 Palladium Theory of Aqueous-Phase Heck Alkynylations

Abstract:

The influence of water on the catalysis of biphasic Heck alkynylation, a family of palladium-catalyzed carbon-carbon bond formations that enable materials, natural products, and pharmaceuticals syntheses, was investigated. Kinetic theory derived from Hatta moduli and pseudo-stationary-state approximations discovered the biphasic cross-coupling to occur within thin films of immiscible aqueous-organic interfaces. Remarkably, water reduced Gibbs activation barriers compared to values previously estimated by density functional theory calculations (DFT) of purely organic syntheses. Quantum tunneling in proton transfer mechanisms partly accounts for the change. Water also influenced the rate-determining steps of the catalytic cycles, and it potentially switched the cross-coupling's mechanism from deprotonation, previously thought to govern the reaction, to carbopalladation. Carbopalladation theory identified the ~35% of the Pd wasted during synthesis was ~10% greater than the amount predicted by deprotonation. Our discoveries enabled E-factor predictions that could someday help reduce chemical wastes generated during materials, natural products, and pharmaceutical manufactures.

Enhancing Rate of Selective Hydrogenation Catalysts by Modulation of Pd Crystallite Diameter

Nathan Prisco[1], Simon Pang[2], Carolyn Schoenbaum[2], J. Will Medlin[2]

[1]University of Texas at Austin, John J. Mcketta, Jr. Department of Chemical Engineering, 200E Dean Keeton St. Stop C0400 Austin, TX 78712-1589
[2]University of Colorado Boulder, Department of Chemical and Biological Engineering, 596 UCB Boulder, CO 80309-0596

Background and Motivation

Introduction

Currently there is much interest in developing processes for the production of advanced biofuels and chemicals. Unlike conventional petroleum refining, reactions of biosourced molecules are very complex and often difficult to control. For this reason, designing solid catalysts for bio-based processes requires great consideration in order to obtain products in appreciable yields. One such example is the hydrogenation/hydrodeoxygenation of furfural which is an important chemical intermediate produced from hemicellulose. Possible products from this reaction include 2-methylfuran, which has great potential to be used in gasoline fuel blends, and furan, which has little to no potential to be used in gasoline fuel blends.[1]

Production 2-methylfuran from biomass:

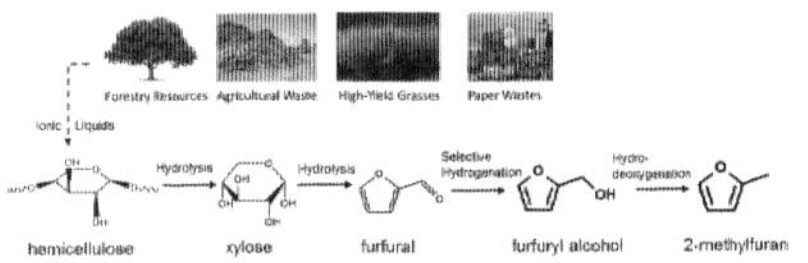

[1] Christensen, et al. (2011). Energy Fuels, 25, 4723-4733.
[2] Wyman, et al. Trends in Biotech. Volume 25 (4), 153-157.

Catalyst Design

Hypothesis

Previous work has indicated that when octadecanethiol (C_{18}) self-assembled monolayers (SAM's) are coated onto supported palladium crystallites, the SAM blocks adsorption of furfural to terrace sites but still allows access to corner sites. When terrace sites are suppressed the hydrogenation of furfural favors the formation of 2-methylfuran. Therefore, in addition to applying a SAM, the activity of terrace sites may also be further suppressed by reducing the crystallite diameter of uncoated Pd crystallites.

Driving Site Specific Interactions via Self-Assembled Monolayers

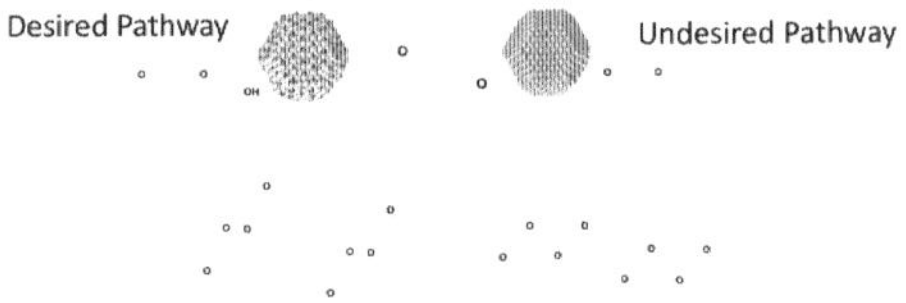

Graphic Adapted from: [1] Pang, et. al. (2013). Nat. Commun. 4, 2248.

How does catalytic site distribution influence reactivity and selectivity?

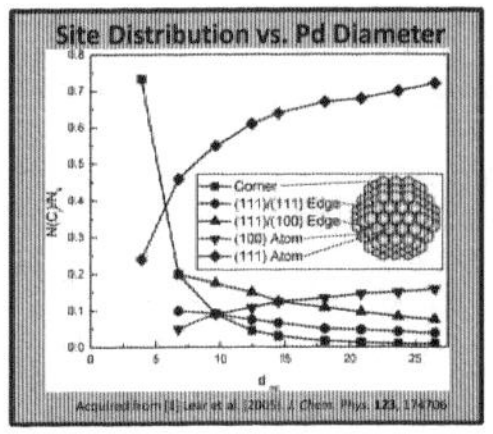

Adapted from [1] Lear et al. (2005). J. Chem. Phys. 123, 174706

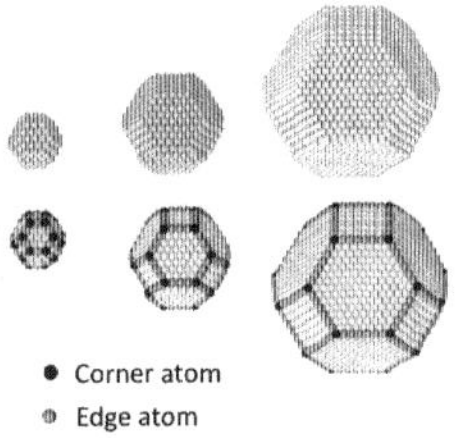

Materials and Characterization

Synthesis and Materials

Precursor:

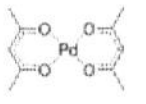

Palladium (II) acetylacetonate
or
Pd(acac)$_2$

Support:

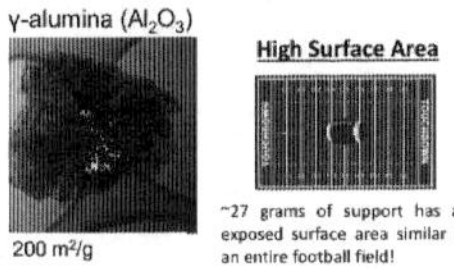

~27 grams of support has an exposed surface area similar to an entire football field!

Surface Reaction[1]:

$$Al_S^{3+} + Pd(acac)_2 \rightarrow [Al(acac)_x]_S^{(3-x)+} + Pd_S^{2+}$$

- Driven by a Lewis Acid exchange on the alumina surface
- The amount of available Al^{3+} reactive sites may be tailored with pre-treatments such as calcination

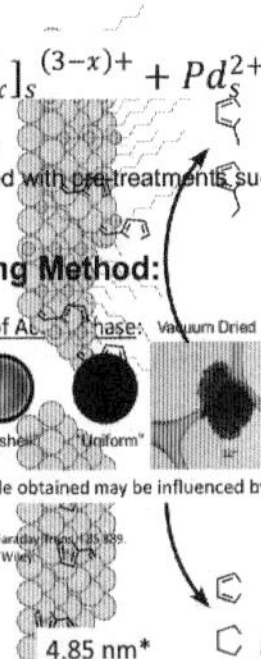

Impregnation Techniques:

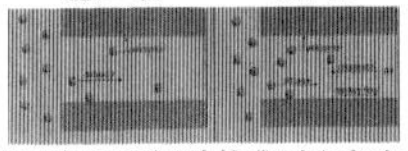

|Pore Diffusion Dominates| |Capillary Action Dominates|
Graphic adapted from de Jong[2]

Drying Method:

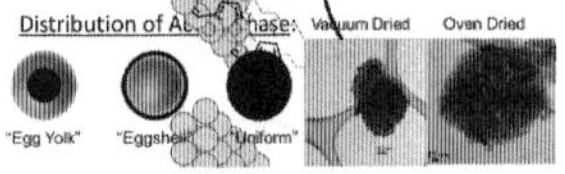

-Active phase profile obtained may be influenced by drying procedure[2]

[1] Van Veen, G. Jonkers and W.H. Hesselink. (1989). J. Chem. Soc. Faraday Trans. 85, 389.
[2] de Jong, K. P. (Ed.). (2009). Synthesis of Solid Catalysts (2nd ed.). Wiley.

Prepared Catalysts

1.63 nm* 3.81 nm* 4.45 nm* 4.85 nm* 8.45 nm*

Impregnation	%Pd	SA (m²/g$_{cat}$)	<Diameter> (nm)	S/Pd
Dry†	~0.10†	0.308±0.032	1.63±0.17+	N/A+
Wet	0.61±0.02	0.804±0.003	3.81±0.02	0.22±0.05
Wet	1.00±0.05	1.129±0.079	4.45±0.32	0.13±0.03
Wet	2.10±0.01	2.170±0.156	4.85±0.35	0.03±0.04
Wet	2.03±0.04	1.209±0.002	8.45±0.20	0.04±0.04

*ICP elemental analysis is pending

Diffuse Reflection Infared Fourier Transform Spectroscopy (DRIFTS):

Linear CO (corner site) @ ~2077 cm⁻¹
Bridged CO (edge site) @ ~1924 cm⁻¹
Three-fold CO (terrace site) @ ~1838 cm⁻¹

3.81 nm* CO DRIFTS

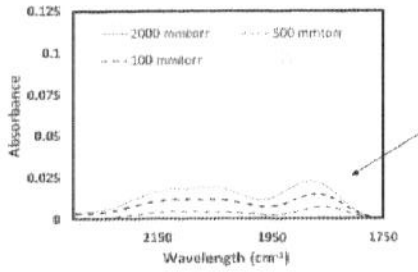

Larger Particles have much more three-fold CO bonding

4.85 nm* CO DRIFTS

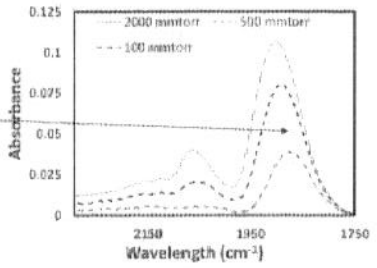

Reference peak wavelengths obtained from [1] Lear et al. (2005). J. Chem. Phys. 123, 174706

Reaction Kinetics

Uncoated Catalysts

2-Methylfuran Selectivity of Selected Catalysts

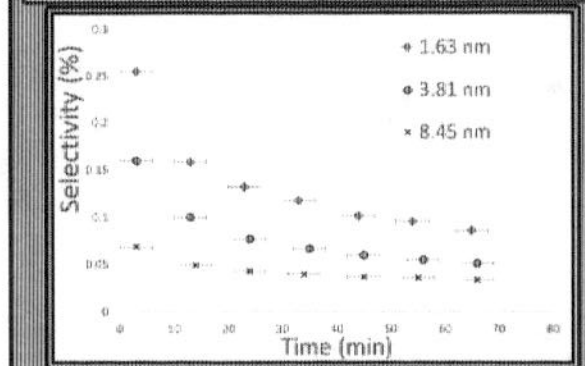

Furfural Conversion of Selected Catalysts

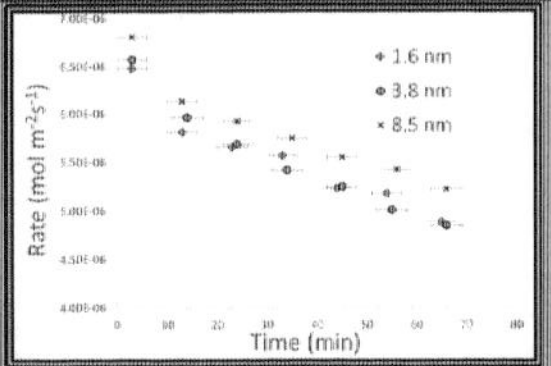

Figure 4 & 5. Selectivity and reaction rate data for the gas phase hydrogenation of furfural at 190°C in a PFR with a flow of 20 sccm of H_2, 59 sccm of He, and 5.95±0.40E-07 mol s⁻¹. An attempt was made to keep the active metal surface area constant for each trial such that 27.9±0.1 mg, 6.35±0.10 mg, and 8.92±0.10 mg of catalyst samples corresponding to 1.63 nm*, 3.81 nm*, and 8.45 nm* respectively. The total amount of solids was diluted to 25.0 mg if necessary using alumina.

Smaller particle uncoated catalysts have a higher 2-methylfuran selectivity. This is because the smaller diameter catalysts have a large proportion of edge and corner sites relative to terrace sites.

SAM Coated Catalysts

2-Methylfuran Selectivity of Selected Catalysts

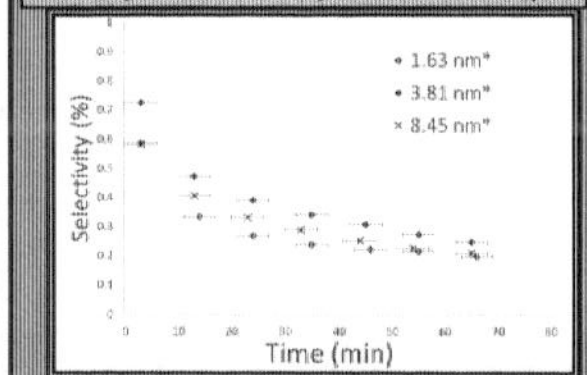

Furfural Conversion of Selected Catalysts

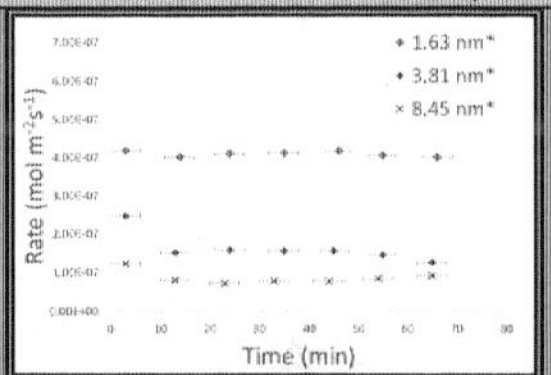

Figure 6 & 7. Selectivity and reaction rate data for the gas phase hydrogenation of furfural at 190°C in a PFR with a flow of 20 sccm of H_2, 59 sccm of He, and 5.95±0.40E-07 mol s⁻¹. An attempt was made to keep the active metal surface area constant for each trial such that 35.5±0.1 mg, 18.71±0.10 mg, and 26.9±0.1 mg of catalyst samples corresponding to 1.63 nm*, 3.81 nm*, and 8.45 nm* respectively. The total amount of solids was diluted to 25.0 mg if necessary using alumina.

Smaller particle SAM coated catalysts have a higher furfural consumption per uncoated Pd surface area. This observation suggests that fewer active sites are obstructed by application of the SAM coating onto smaller diameter particles. The observation supports previous work which suggests the SAM blocks access to terrace sites.

Conclusions

- This work demonstrates that finely dispersed catalytic particles with a minimal crystallite diameter are best suited for selective conversion of furfural.
- This work broadly demonstrates the importance of considering site specific interactions in catalyst design.

Acknowledgements

- Simon Pang
- April Corpuz
- Dr. Hans Funke
- Medlin Group
- Audrey Fisher & the C2B2 Program
- National Science Foundation
- Philips 66 Center for Energy Innovation

TiO_2 as a Catalyst in Thin Film Coatings of Glass Beads for Water Purification Purposes

Johnson Scholar: Sydney Quinton-Cox
Department of Bioengineering, Oregon State University, Corvallis, OR 97331, USA

Graduate Mentor: Simon Fowler
Department of Physics, Portland State University, Portland, OR 97207, USA

Professor: Dr. Jiao
Department of Engineering, Portland State University, Portland, OR 97207, USA

Abstract

The purpose of finding a standard process for coating 1 mm borosilicate glass beads in TiO_2 is creating an effective alternative for making clean water through optimizing the photo degradation of the sun. Creating a thin film TiO_2 coating that can optimize the photo catalytic degradation of organic compounds in water purification systems is essential because this type of photocatalyst is capable of degrading, "dyes, pesticides, herbicides, pharmaceuticals, cosmetics, phenolic compounds, toxins, and more" (Lazar, Varghese, and Nair 2012). The goal was to determine which surfactants degrade organic compounds most effectively and then make a coating that was less than 1 micron thick. Previous studies concluded that in order to make a coating that does not crack it must be thinner than 1 micron. However, creating a uniform coating that is less than 1 micron proved to be a difficult process.

The first group of synthesis experiments was done with several different types of surfactants. The effectiveness of each coating was tested by how quickly it could degrade methylene blue dye when in contact with solar light. The results showed that PAA (poly acrylic acid) and TEOS (tetraethylorthosilicate) are the most effective combination for degrading organic compounds. After the most effective surfactants were determined, the next step was to design a recipe and process that would coat the beads uniformly without producing cracking and flaking. A systematic approach was used by narrowing down the options by what worked the best at each stage of the drying process for producing a coating that will not crack or flake.

The development of the most effective TiO_2 coating for water purification purposes was accomplished by testing different surfactants, drying techniques, and baking temperatures to sinter the beads. The coatings were examined with scanning electron microscopes, Raman spectroscopy, and tested using an organic compound, methylene blue dye, for degradation results. The most effective process was drying the beads with a tumbling machine and then the beads were baked at 600°C for 8 hours to preserve the 1 mm borosilicate glass beads' shape.

The degradation tests found the TiO_2 coatings are approximately 13 times more effective at degrading organic compounds than sunlight alone. The next step in developing this technology is to do a full scale test with all the tubes in the reactor filled with TiO_2 coated glass beads.

In the future, the information learned during the internship will be applied to creating a strong crystalline scaffold structure without glass, so that only the catalytic material is occupying space in the water purification reactor.

Hydrogen Evolution Reaction Catalyst for Solid-State Alkaline Water Electrolysis

Yanxin Li

Illinois Institute of Technology

As global warming appears to have greater impact on our life, the demand of renewable, environmentally friendly, affordable and reliable alternatives to the traditional fossil-based fuel is expanding. Among the many possibilities, hydrogen economy have the potential to revolutionize the energy production due to its clean nature and its non-harmful combustion product—water, as well as its abundant source in nature. The hydrogen economy starts with hydrogen production by water electrolysis; Hydrogen Evolution Reaction (HER), which serves as the reduction step in water electrolysis is one of the important steps towards efficient and clean hydrogen generation. However, to date, large-scale industrial production of hydrogen in alkaline electrolyzers is constrained by low reaction kinetics even in the presence of platinum catalysts, otherwise known for its very good activity in acid environments. Therefore, in this research, substitute electrocatalysts will be synthesized and compared to Pt catalyst for HER activity.

The mechanism of water electrolysis involves two steps: the initial formation of hydrogen intermediates ($H^+[H_2O]+e^- \square H_{ad}^-[+OH^-]$) and followed by the recombination step ($2H_{ad} \square H_2$) or an electron transfer step ($H^+[H_2O]+e^-+ H_{ad} \square H_{ad}^-[+OH^-]$) involving water. This later step suggests the presence of a bifunctional catalyst with water splitting ability, in this case $Ni(OH)_2$/metal, will improve the overall reaction (Danilovic, 2012). Based on previous results obtained in ring disk electrode (RDE) by Danilovic et al., 2012, we will evaluate the performance in an solid-state alkaline water electrolyzer of platinum, iridium and nickel (supported on carbon) catalysts mixed with $Ni(OH)_2$

Metal supported on carbon catalysts of Ni and Ir are expected to have lower activity than Pt but improved reaction kinetics should be observed when metal nanoparticles (Pt, Ni or Ir) are in presence of $Ni(OH)_2$ which act as hydrophilic clusters. Platinum metal group catalysts could be rewarding alternatives if higher effectiveness in catalyzing hydrogen evolution reaction is steadily preserved. In other words, it might provide an accessible path to the clean hydrogen energy.

Reference

[1] **N. Danilovic, Ram Subbaraman**. “Enhancing the Alkaline Hydrogen Evolution Reaction Activity through the Bifunctionality of $Ni(OH)_2$/Metal Catalyst”. *Angew. Chem.* Int. Ed. 51, 1-5 (2012)

Techno–economic analysis of bio-based production of Hydroxy-Methyl-Furfural (HMF)

Jingyao Wang[1], Zhaojia Lin[1], Marianthi Ierapetritou[1]

Deparment of Chemical and Biochemical Engineering, Rutgers – The State University of New Jersey

This work discusses the design and simulation and techno-economic analysis for the production of hydroxymethylfurfural (HMF) from fructose. Petroleum is the main raw material for the production of fuels and chemicals. However, due to exhausting resources and emerging environmental issues from oil industry, alternative production paths have been investigated. Biomass sources are among the alternatives with the potential to produce both chemicals and fuels.[1] Particularly, the biobased building block chemicals, with multiple functional groups have the potential to be transformed into new families of useful chemicals.[2] Among these high-value chemicals, HMF has attracted a lot of interest since it is an important chemical that can be converted into many crucial furan derivatives such as alkoxymethylfuran, 2,5-furandicarboxylic acid, 5-hydroxymethlfuroic acid, bishydroxymethylfuran, 2,5-dimethylfuran etc. Besides those, HMF could also be used for the production of some useful nonfuranic compounds such as levulinic acid, adipic acid and caprolactam etc.[3]

Existing work in the literature for the biobased production of HMF can be categorized into three main types: I) Fructose degrades in aqueous solutions at low conversion but high HMF selectivity;[4] II) HMF formation is carried out in an organic co-solvent systems;[5] III) HMF is formed and concentrated by reactive extraction[3,6] or reactive adsorption.[6] The project will exclusively focus on the first alternative, which is to produce HMF from fructose in aqueous solutions. The formation of HMF in aqueous phase includes the following reactions: the acid-catalyzed dehydration of fructose to HMF, formic acid (FA), and humins, followed by HMF further dehydration to levulinic acid (LA), FA and humins. LA which serves as precursor to various additives and HMF are the more valuable products during the process while the undesired side-reactions cause significant loss of sugar and limit the profitable production of HMF. In comparison to the other two alternative reaction paths II and III, approach (I) avoids using excessive organic solvents and NaCl required to perform the HMF extraction process proposed in approaches II and III.[7] Moreover, HMF selectivity is higher and less sensitive to temperature when operating at low conversion (<15%) in aqueous solution while selectivity decreases rapidly and becomes much more sensitive at high conversion. This indicates that a low conversion reactor coupled with separation process and a recycle stream might be the most economical alternative.[4]

The production of HMF in aqueous solution is with high selectivity whereas as mentioned it is still challenged by low conversion which requires efficient separation and recycle of a large amount of fructose remaining in the aqueous phase to make this an economically favorable process. Several separation methods have been proposed including extraction, chromatography, distillation and crystallization. However, those may not make the aqueous process economically competitive due to insufficient separation of fructose from HMF, the high utility cost, the equipment cost, and the loss in raw materials due to side reactions. Instead, this work focuses on adsorption method that has the advantage of achieving separation of the relatively small amount of HMF from aqueous solution with less impurity from fructose. Lower operation temperature in adsorption column also reduces the raw material waste due to side reactions [8].

The reaction model is built and analyzed using Matlab. Then the aqueous reaction process is incorporated into Plug Flow Reactor Model followed by the adsorption purification, which are designed and simulated using Aspen Plus. Aspen Economic Analyzer is then used to perform the economic evaluation for the HMF production process. Sensitivity analysis is utilized to evaluate the effects of reaction conversion and selectivity to the optimal production of HMF.

REFERENCES

1. JJ, B., Feedstocks for the future - biorefinery production of chemicals from renewable carbon. *Clean Soil Air Water.* **2008,** *36(8)*, 641–647.

2. (a) Werpy, T.; Petersen, G. *Top Value Added Chemicals from Biomass: Volume I -- Results of Screening for Potential Candidates from Sugars and Synthesis Gas*; DOE/GO-102004-1992; TRN: US200427%%671 United States10.2172/15008859TRN: US200427%%671Tue Feb 12 21:52:38 EST 2008NRELEnglish; 2004; p Medium: ED; Size: 76 pp. pages; (b) Werpy.T, P. G. *Top Value Added Chemicals from Biomass*; Pacific Northwest National Laboratory: August 2004; p 1.

3. (a) Robert-Jan van Putten, J. C. v. d. W., Ed de Jong,, Carolus B. Rasrendra,⊥ Hero J. Heeres,; Vries, a. J. G. d., Hydroxymethylfurfural, A Versatile Platform Chemical Made from Renewable Resources. *Chemical Reviews* **2013,** *113(3)*, 1499-1597; (b) van Putten, R.-J.; van der Waal, J. C.; de Jong, E.; Rasrendra, C. B.; Heeres, H. J.; de Vries, J. G., Hydroxymethylfurfural, A Versatile Platform Chemical Made from Renewable Resources. *Chemical Reviews* **2013,** *113* (3), 1499-1597.
4. Swift, T. D., Kinetics of Homogeneous Brønsted Acid- Catalyzed Fructose Dehydration and HMF Rehydration: A Combined Experimental and Computational Study. *ACS Catalysis* **2014,** *4*.
5. Román-Leshkov, Y.; Chheda, J. N.; Dumesic, J. A., Phase Modifiers Promote Efficient Production of Hydroxymethylfurfural from Fructose. *Science* **2006,** *312* (5782), 1933-1937.
6. Swift, T. D.; Bagia, C.; Nikolakis, V.; Vlachos, D. G.; Peklaris, G.; Dornath, P.; Fan, W., Reactive adsorption for the selective dehydration of sugars to furans: Modeling and experiments. *AIChE Journal* **2013,** *59* (9), 3378-3390.
7. Kazi, F. K.; Patel, A. D.; Serrano-Ruiz, J. C.; Dumesic, J. A.; Anex, R. P., Techno-economic analysis of dimethylfuran (DMF) and hydroxymethylfurfural (HMF) production from pure fructose in catalytic processes. *Chemical Engineering Journal* **2011,** *169* (1–3), 329-338.
8. Geier, D. F.; Soper, J. G., Method for purifying hydroxymethylfurfural using non-functional polymeric resins. Google Patents: 2011.

398203 Development of an Ipad Application for Distillation Analysis

Abstract:

The Chemical Engineering Application was formulated to revolutionize the information presented in a typical separations and distillations theory class through textbooks into an interactive collection of calculations, equations, and graphs. The information includes substance volatility, Raoult's Law using Antoine's Equation, two substance multi-phase composition graphs, and the McCabe-Thiele equation. The time saved by using the application to complete calculations and graphing can be spent analyzing patterns and trends based on changing variables. The student will be allowed the time to focus on the theory and study of distillation analysis, thus building a stronger foundation in data-driven learning. Ultimately, the creation of the application will allow the student to immediately visualize local and global behavior of the underlying function through an interactive and social learning environment.

A Multiphysics Simulation of a Simplified Tri-reforming Reactor

Charles Suleskey, Ryan A. Kent, Prof. Babu Joseph, Prof. John Kuhn

For the past decade, the United States has been producing approximately 250 million tons of municipal solid waste per year (US EPA), over 50 percent of which was discarded into landfills. The natural breakdown of these wastes leads to the production of about 100 million metric tons of methane and a nearly equivalent amount of carbon dioxide, both potent greenhouse gases. The most typical method for dealing with this gas release is to flare the methane (as methane has a greenhouse potential roughly 20 times that of carbon dioxide, CO_2 is often not processed), however this fails to capture the energy stored in the methane, thus doing little to offset the carbon footprint of the waste. Various methods and processes of reclaiming the energy produced by waste decomposition have been developed. This research project is developing a combination of four of these processes in one system: a tri-reformer, using three reactions to produce syngas, and a Fischer-Tropsch synthesis reactor, which converts the syngas into hydrocarbon fuels. This presentation is a simulation and analysis of the tri-reformer stage, in which a temperature and flow profile while be made to model steady-state operation using Comsol. The information produced by this model can then be used to better project reactor sizing and heating in practical applications.

Modeling, simulation and optimization of polybenzimidazole (PBI) hollow fiber-based membrane systems for water-gas shift reaction applications

Andrew Radcliffe and Fernando V. Lima
Submitted to Topical Area: Computing and Process Control

Hydrocarbons such as coal, natural gas and biomass serve as the principal source of H_2, which could see use as a feedstock in myriad chemical and clean energy production processes. H_2 production from hydrocarbons generates CO_2 emissions which must be mitigated to reduce their impact on climate change. In production of electricity from coal, integrated gasification combined cycle (IGCC) plants with carbon capture generate syngas that is subjected to the water-gas shift (WGS) reaction in a conventional reactor subsequently followed by CO_2 removal by physical absorption with Selexol. Implementation of membrane reactors (MRs) performing the WGS reaction provides advantages over the current method, granting increased CO conversion, facilitated CO_2 removal/sequestration as well as process intensification, i.e., the reduction of the number of pieces of process equipment. Membranes for these reactors are typically composed of high cost dense metal alloy or ceramic. Polymeric membranes possessing suitable gas separation properties offer greatly reduced costs but with some performance drawbacks such as temperature limitations and reduced H_2 selectivity/permeability.

In this presentation, the suitability of a polybenzimidazole hollow fiber membrane for use in an MR, a single-stage reactor/separator and a multi-stage reactor/separator system is assessed. Performance of the material is determined by modeling and simulation of these reactor systems in MATLAB; the models are subsequently employed to solve optimization problems with multiple constraints to more efficiently exploit membrane area by considering alternative membrane placements around the reactor length. Simulation and optimization results show that the PBI hollow fiber membrane's performance satisfied constraints on CO conversion, H_2 productivity, CO_2 capture and effluent mole fractions for the stand-alone MR and multi-stage reactor/separator systems. Solutions were obtained to the proposed optimization problems relating to these two reactor systems with the membrane area being successfully reduced by up to 15%, corresponding to significant capital investment savings for industrial-scale application.

398933 Optimization of a Two Step Batch-Cooling Crystallization Process

Abstract:

A population balance equation describing unseeded batch cooling crystallization processes was simplified with the method of moments, and the processes were simulated using kinetic models incorporating crystal growth, primary nucleation, and secondary nucleation. These models were used to predict the development of concentration and crystal size distribution (CSD) over time. A multi-objective optimization was solved in order to maximize mean crystal size as well as to decrease variance of the CSD. It is found that a two-step cooling strategy, which holds the temperature constant in the middle of the cooling to induce primary nucleation, can improve the process, in comparison with a linear cooling profile. Nucleation and growth kinetics were analyzed to discuss the reason for this advantage in the two-step cooling system.

Modeling Flow Characteristics Within a Perfusion Bioreactor Using Computational Fluid Dynamics

M. J. Robeson[1], J. Gandhi[2], E. Berson, Ph.D.[1], E. M. Brey, Ph.D.[2]
[1] University of Louisville, [2] Illinois Institute of Technology

Introduction: One means by which to grow tissues outside the body is through the use of a perfusion bioreactor (PBR), consisting of a circular pump system that constantly perfuses a nutrient-rich growth media through a cell-seeded porous scaffold material. Recently, a PBR consisting of polymer beads packed in a cylindrical column has been used to promote growth and proliferation of the cells. The goal of this study was to use computational fluid dynamics software to quantify the ratio of advective to diffusive transport of oxygen within polymer beads across a range of bead diameters and permeabilities, to determine in which regime bulk flow within the beads plays a significant role in nutrient delivery.

Materials and Methods: Spherical alginate beads were created by crosslinking alginate solution drop-wise in 1.1% calcium chloride solution, then imaged under a microscope and sized with a custom MATLAB script. Once size ranges were determined, randomly packed bead columns were generated in SolidWorks using gravity simulation. The assemblies were then imported into ANSYS Workbench14.0, where fluid flow was simulated using the FLUENT solver. The reactor was modeled with a growth chamber 6.4 mm in diameter, and with bead diameters of 2.75, 2.23, and 1.91 mm, corresponding to the three sizes currently used in lab trials. A range of permeability values from $1.2x10^{-12}$ to $1x10^{-8}$ cm^2 were tested, with $1.2x10^{12}$ cm^2 corresponding to the alginate beads currently used in lab trials.

Results and Discussion: Twenty-five mm high columns of each of the three bead sizes were first simulated at 3 mL/min flow rate and $1.2x10^{12}$ cm^2 permeability. Average velocity in the interior of the beads was computed and converted to an average Peclet number, using 2.08 $x10^{-9}$ m^2/s as the diffusion coefficient of oxygen in alginate. It was determined that the largest beads (2.75mm) produced the highest average Peclet number, at 0.023, with 0.55% of computational cells within the beads experiencing Pe>1 (advection dominant). The large diameter beads were then simulated with varied permeability values, $1.2x10^{-12}$ to $1x10^{-8}$ cm^2, to determine a critical permeability value where species transport becomes dominated by advection. At a permeability value of $5.7x10^{-9}$ cm^2, the average Peclet number was 0.99, with 41.1% of computational cells experiencing Pe>1, marking the transition to advection-dominated oxygen transport.

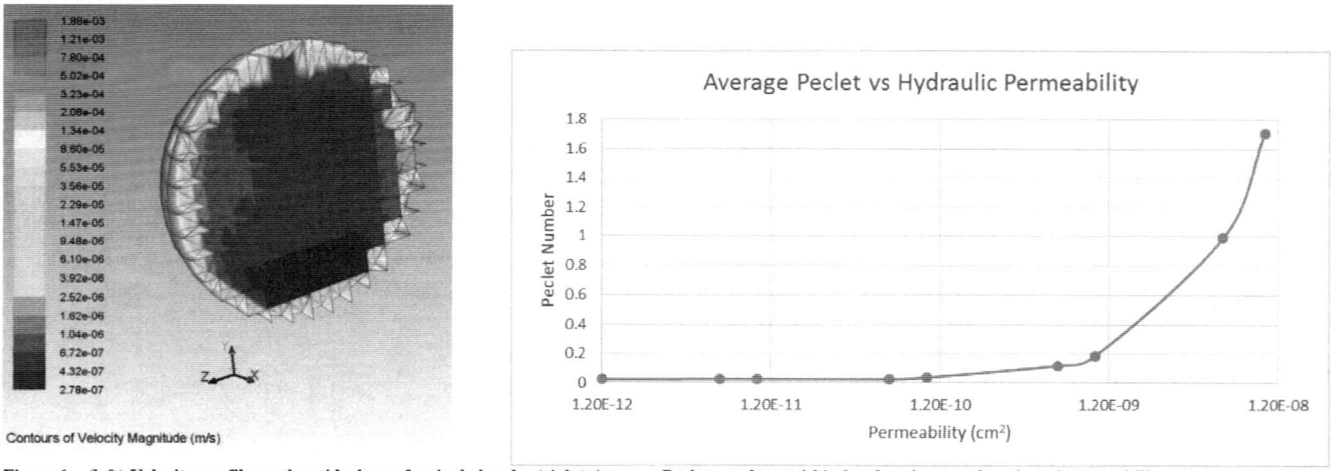

Figure 1 – (left) Velocity profile on the mid-plane of a single bead (right) Average Peclet numbers within bead region as a function of permeability

Conclusions: With the conditions currently utilized in lab trials, diffusive transport predominates except in very small portions of the beads. This study determined the permeability at which species transport becomes predominantly advective. Plans for future trials include using different or modified hydrogels with higher permeability, as well as building a more complex mass transfer model within FLUENT.

Acknowledgements: This research was conducted through funding by the NSF (Grant No. EEC-1157041)

398919 Predicting Three Phase Boundary Density and Interfacial Areas through Mechanistic Modeling

Abstract:

With the world's ever climbing energy needs and our current source of energy, fossil fuels, slowing eating away at the ozone layer; the world needs a source of clean renewable energy. One of the best options for this clean renewable energy is fuel cells. In this research we look at one type of fuel cell: the Solid Oxide Fuel Cell (SOFC). SOFCs are electrochemical devices characterized by their solid, ceramic electrolytes that directly oxidize a fuel to produce electricity with low emissions and high efficiency. For the fuel cell to be functional the electrodes must have porosity for fuel transport through the cell, oxygen ion conductivity to transport the oxidizing agent to the fuel, electrical conductivity to transport electrons from the reaction, and three phase boundary (TPB) sites to enable chemical reaction. TPB sites occur at the intersections of electrical conductor particles, ceramic particles and empty space so that the oxidation reaction may occur at a site with the electrons, oxidizing agent and fuel all present. Recent work has shown that electrode fabrication by infiltration of electronic conductor/oxidation catalyst particles into an oxygen ion rich porous backbone provides excellent performance while decreasing the amount of conductor material necessary as compared with other fabrication methods. Infiltrated electrodes also allow for flexibility in composition and structure of the electrodes. This fabrication method, however, has many design parameters whose impact on the resulting cell's performance is not fully understood. Experimentally exploring these parameters is very time consuming, thus a modeling effort can help to guide the desired experiments.

In this work, we show a model for the design process of a SOFC and the corresponding determination of both total and active TPB density as well as electrical conductivity for the formulated electrodes. The model is based on the underlying physics of the electrode fabrication process. The fabrication process begins with the slurry formation and film creation, followed by pore former combustion, ceramic particle sintering and conducting particle infiltration. Highlighted results include (1) that among all the model inputs the scaffold:infiltrate size ratio has the greatest impact on TPB density (both active and total), (2) increasing scaffold:infiltrate size ratio decreases TPB density, and (3) TPB density shows a maximum when varying porosity. Each highlighted result is explained through comparison with the interfacial surface areas of the three phases involved in the TPB density. Composite results including both TPB density and effective conductivity are also presented.

397281 Tribological Study of Polyethylene Glycol Monolayers on Silica Substrate

Abstract:

Lubricating micro- and nano-electromechanical systems (MEMS and NEMS) has become increasingly challenging as devices are continuously miniaturized. Decreasing the dimensions of these systems causes their surface area to volume ratios to increase significantly, making frictional forces between moving parts dominant. Solid-state lubrication via self-assembled monolayers presents a promising alternative to traditional macro-scale lubricants, which are too bulky to use in such confined systems.[1,2] Previous studies have shown that self-assembled monolayers containing alkylsilanes and mixed alkylsilane and perfluoroalkylsilane chains reduce friction coefficients compared to equal length systems. Although, hydrocarbon-based monolayers have been widely studied and shown to improve frictional properties and device life, mixed monolayers impart the beneficial properties of perfluoroalkylsilanes to the film including low critical surface tension and high thermal and mechanical stability. Furthermore, a mixed length configuration with longer alkylsilane chains relative to the perfluoroalkylsilane chains leads to a liquid-like layer formed by the taller chains that fold over and align with the shearing direction.[3]

In this study, we investigate polyethylene glycol (PEG) monolayers using molecular dynamics simulations as a potential solid-state lubricant for NEMS. The presence of oxygen atoms within the chains render these chains more flexible than their alkyl- and perfluoroalkyl-counterparts. Thus, even sparsely functionalizing a surface leads to a uniform film. We have investigated the tribological properties of such films across a range of surface coverages and chain lengths. Preliminary results show the effects of surface coverage on friction coefficients is diminished as chain length is increased. This suggests that equivalent friction coefficients could be achieved whilst reducing separation distances by using longer chains on sparser surfaces.

1. Booth, B. D., Vilt, S. G., Lewis, J. Ben, Rivera, J. L., Buehler, E. A., McCabe, C., & Jennings, G. K. (2011). Tribological Durability of Silane Monolayers on Silicon. Langmuir, 27(10), 5909–5917. doi:10.1021/la104778q

2. Lorenz, C. D., Chandross, M., Grest, G. S., Stevens, M. J., & Webb, E. B. (2005). Tribological properties of alkylsilane self-assembled monolayers. Langmuir, 21(25), 11744–8. doi:10.1021/la051741m

3. Lewis, J. Ben, Vilt, S. G., Rivera, J. L., Jennings, G. K., & McCabe, C. (2012). Frictional properties of mixed fluorocarbon/hydrocarbon silane monolayers: a simulation study. Langmuir, 28(40), 14218–26. doi:10.1021/la3024315

Simulated Annealing Polymerization

A General Algorithm for Preparing Atomistic Model Structures of Amorphous Polymers

Michael Wurmstein, Sriramvignesh Mani, and Rajesh Khare

Abstract

The quest for new polymers with specialized properties surges onward to fill the inherent industrial demand for materials with novel applications. The combinatorial approach of experimentally synthesizing and characterizing new polymers becomes increasingly expensive and time consuming for any project requiring an appreciable library of polymer systems. However, molecular modeling can provide insights into the relationship between the structure and properties of polymeric systems, issuing guidance for further investigation. Such an approach, for example, can be used to design polyacrylate copolymer membranes for efficient separation of alcohol-water mixtures for biofuels production. Previously, molecular models of linear chains and fully cured cross-linked epoxy have been created through polymerizing a mixture of monomers in simulations. For copolymer systems of interest here, however, each type of reacting monomer pair has a different reaction probability, thus rendering usage of simple polymerization algorithms based on uniform reaction probabilities inappropriate. The long term goal of this work is to create a general algorithm that accounts for specific chemical reactivity considerations while preparing model structures of amorphous polymers.

Problem Definition

- Steps to construct the monomer:
  1. Pack the desired proportions of monomers into a simulation box.
  2. Equilibrate molecules in a molecular dynamics simulation to liquid state intermolecular distances.
  3. Form new bonds between spatially close monomers.
- Energy of the new bonds is minimized by minimizing total bond length.
- The problem is thus reminiscent of the classic Travelling Salesman Problem.
- This optimization is completed using simulated annealing.

Optimization Algorithm

- Total bond length is minimized with simulated annealing, as follows:
  1. A random initial configuration is taken.
  2. A starting "Temperature" is assumed.
  3. Stochastic transport and reversal moves are proposed.
  4. Each change is evaluated by Metropolis Criterion.

$$P = e^{\frac{-\Delta f}{T}}$$

Where P is the probability that a move is accepted;

Δf is the change in the total bond length due to the move; and

T is the current temperature, a parameter for controlling the likelihood of uphill moves.

  5. Temperature is reduced after a specified number of moves.
  6. The simulation ends when total bond length is approximately constant.

Simulation Design

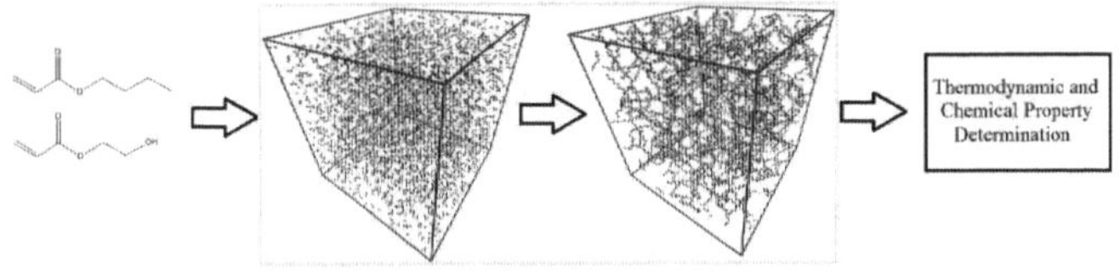

- •Monomers are input by the user with specified reactive sites.
- •Monomers are replicated and equilibrated in a molecular dynamics simulation.
- •New bonds are formed and relaxed to minimize energy.
- •Thermodynamic and chemical calculations are carried out.

Improvement Scheme

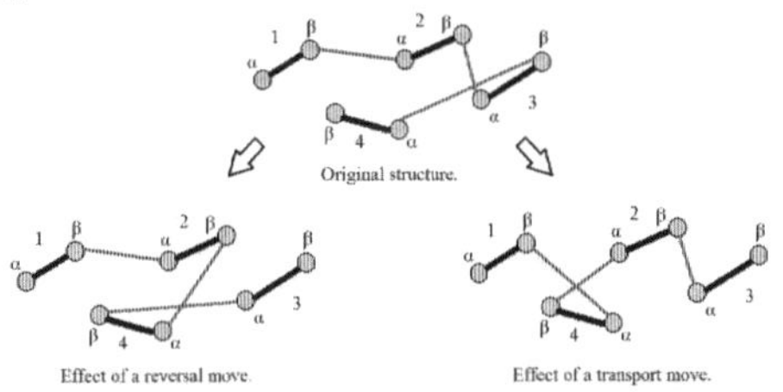

- Monomers are represented as alpha-beta pairs.
- Alpha atoms may only bond to beta atoms.
- Each iteration, a random transport or reversal move is evaluated.

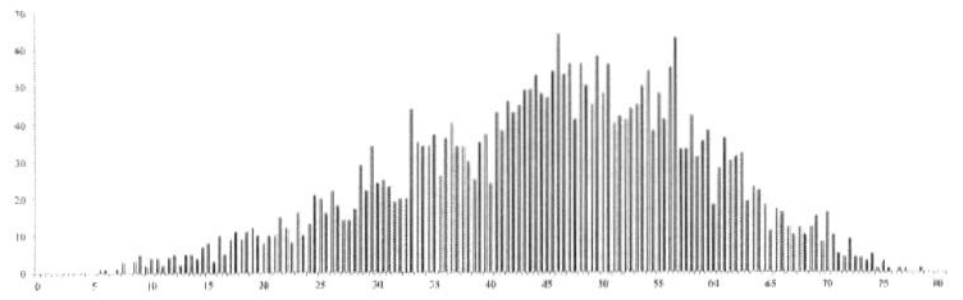

Figure 1. Bond length distribution of a poly-butyl acrylate before optimization.

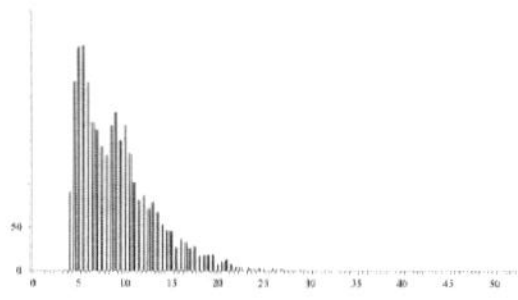

Figure 2. Bond length distribution of a poly-butyl acrylate after optimization.

Results

- •Polymers may be formed with controlled interspecies and intraspecies reactions.
- •Models of the following structures have been formed:
  1. Homopolymers of acrylates.
  2. Copolymers of acrylates.
  3. Cross-linked systems of acrylates with pentaerythritol tetraacrylate.
- •Other possible models include polystyrenes.
- •Figure 1 and Figure 2 show representative histograms of bond lengths before and after optimization.

Conclusions

- A general algorithm was developed for creating different polymer structures by reacting a mixture of monomers.
- The algorithm was used to form poly butyl acrylate (below left) and a copolymer of butyl acrylate and 2-hydroxy ethyl acrylate (below right).

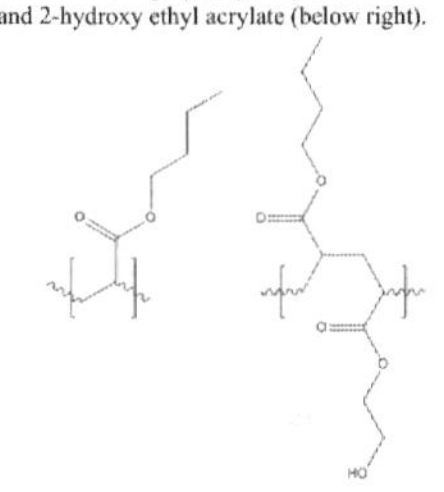

- Improvements are underway to allow user-specification of the conversion of a species or set of species.

References:

Khare, R., Paulaitis, M. E., and Lustig, S. R. "Generation of Glass Structures for Molecular Simulations of Polymers Containing Large Monomer Units: Application to Polystyrene," Macromolecules, Vol. 26, No. 26, pp. 7203-7209. 1993.

Kirkpatrick, S., Gelatt, C.D., and Vecchi, M.P. "Optimization by Simulated Annealing," *Science*, New Series, Vol. 220, No. 4598, pp. 671-680. May 23, 1983.

Acknowledgements:

Funding: This work is supported by the National Science Foundation under Grant: NSF CMMI-1335082.

Computational resources: 1. Texas Advanced Computing Center (TACC), The University of Texas at Austin. 2. High Performance Computing Center (HPCC), Texas Tech University.

Protracted Colored Noise Dynamics Applied to Molecular Dynamics Simulations of Block Copolymers

Daniel Nicoloso
dannicoloso@gmail.com
Clifford L. Henderson
School of Chemical and Biomolecular Engineering
Georgia Institute of Technology

Mean field models and Monte Carlo simulations have been used to study polymeric systems in recent years. While these methods provide great computational expedience, they include less accurate potentials and require intelligently selected modes of motion, which can adversely affect outcomes. Conversely, molecular dynamics (which uses Newton's second law to calculate molecular movements) combined with realistic potentials provide more accurate movements and states, though computational efficiency is sacrificed. Using coarse graining and parallel computing greatly increases the time and length scales that can be accessed in MD simulations of polymeric systems, but some systems, especially systems with large energy barriers or with long chains, still require unrealistic simulation times. One example of such a system is a lamellar forming BCP with a defect. Although the defect density is very small at equilibrium, the activation energy can be large, causing a drastic increase in computation time.

The purpose of this study is to apply protracted colored noise dynamics (PCND) to such systems with the goal of decreasing the simulation time required to reach equilibrium. PCND is a technique that applies time correlated stochastic forces in MD simulations. In linear polymer systems these forces can be made to act along the backbone of the chain. These forces broaden the energy distribution of the system, increasing the diffusivity of the molecules, and allowing energy barriers to be passed over more easily. Of course, the distribution of these forces must be carefully chosen so that the equilibrium state is not excessively perturbed. In this study, PCND was applied to BCP simulations from two initial states: mixed state, and defective lamellar state. The results of the mixed state simulations were used to fine tune PCND parameters to avoid excessive perturbation of system behavior. The defective state simulations were used to determine how much PCND would reduce computational time. The results have been promising, showing an order of magnitude reduction in computational time required to reach equilibrium for defective state simulations.

396064 Contact Angles and Hysteresis on Nanoscale Textured Surfaces Via Atomistic Molecular Dynamics Simulations

Abstract:

We quantify the effects of patterned, physical surface texture on nanoscale contact angles at differing levels of attraction between fluid and surface atoms. We also examine the effects of surface texture on contact angle hysteresis for fluids with differing terminal center-of-mass velocities. In order to accurately measure these quantities, a two-surface, circular fit approach to calculating static and dynamic contact angles on arbitrarily textured nanoscale surfaces via atomistic molecular dynamics simulations is developed in detail. We pay particular attention to the molecular dynamics simulation details, system characteristics, analysis techniques, regression type, and analysis parameters required to minimize bias and maximize statistical efficiency. Under these conditions, dynamic contact angle hysteresis, unlike the static contact angle, is not affected by the introduction of surface texture. Additionally, we observe a linear relationship between advancing and receding dynamic contact angles and contact angle hysteresis on both smooth and textured surfaces.

UNIVERSITY OF SOUTH CAROLINA

College of Engineering and Computing

Molecular Modeling of ssDNA—Salt Interactions via Mean Field Theory and Molecular Dynamics Simulations

Katherine Driscoll[1] and Mark Uline[2]

[1]Department of Physics and Astronomy, Department of Mathematics; [2]Department of Chemical Engineering, University of South Carolina

1. Biotechnology and ssDNA

•Many kinds of biotechnology frequently incorporate the phenomena of DNA binding to various ligands. To this end, single-stranded DNA (ssDNA) is frequently used in biosensors and microarrays.

•ssDNA biosensors have been predicted to become an important new feature of medical diagnostics as they allow investigators to detect abnormalities in genetic sequences and to assess how DNA sequences interact with surrounding molecules[1,2].

• These devices immobilize one end of the aptamers (short strands of naturally occurring nucleotides) to a surface, allowing the rest of the strand to interact with target molecules[3,4].

•These systems are immersed in ionic solutions to mimic the complex salt environments of cells.

•XPS studies and earlier mean field studies showed a tradeoff between the protonation fraction and ion incorporation into ssDNA structure[5,6].

2. Biosensor Setup

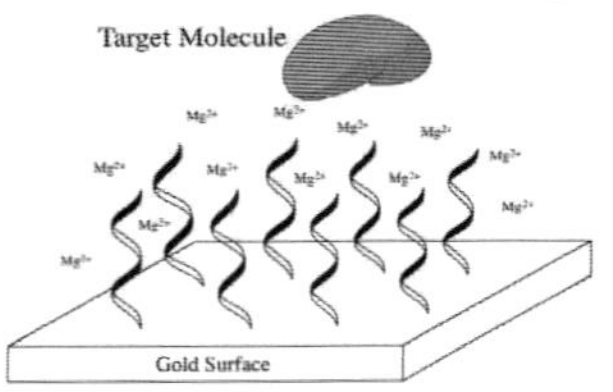

Figure 1. ssDNA aptamers are attached to a gold surface and interact with target molecules and ions in solution.

- Phosphate groups on ssDNA backbone induce repulsive electrostatic forces that cause the molecule to stretch out in a linear fashion.
- DNA can relieve the strain in its backbone by incorporating protons and cations from solution.
- Preliminary studies found that increasing surface coverage increased the protonation fraction and that increasing salt concentration increased cation uptake[6].

3. Volume Fraction and Planar Surface with Na^+

Figure 2 *(right)*. MFT calculation of the volume fraction confirmed findings that Mg^{2+} compacts ssDNA structure more. The volume fraction is shown as a function of distance from the wall with Na^+ (black) and Mg^{2+} (red).

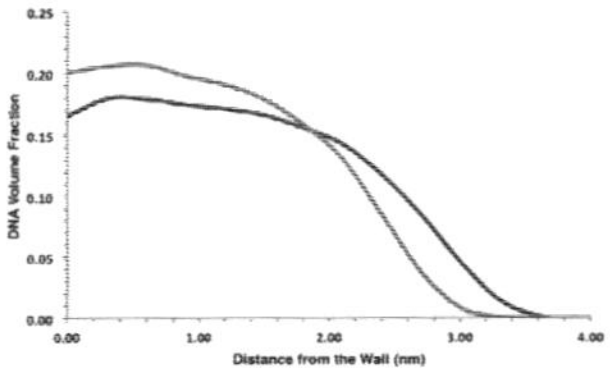

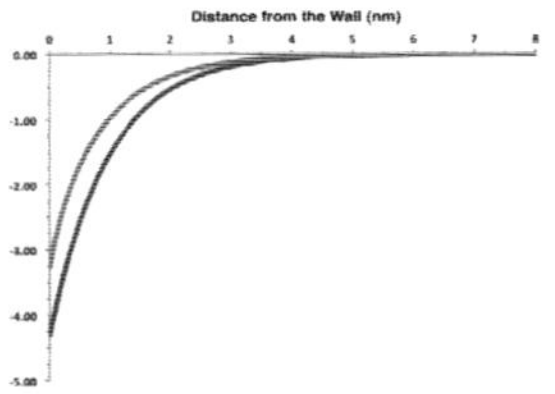

Figure 3 *(left)*. MFT calculation was set up for a simple system of ssDNA 21-mers bound to a gold surface and immersed in a sodium chloride (NaCl) solution with a surface charge density of -0.5 *e/nm*^{2}. Analytical solution (red), linear approximation (blue).

4. Curved Surface with Na^+ and Mg^{2+} Considered Separately

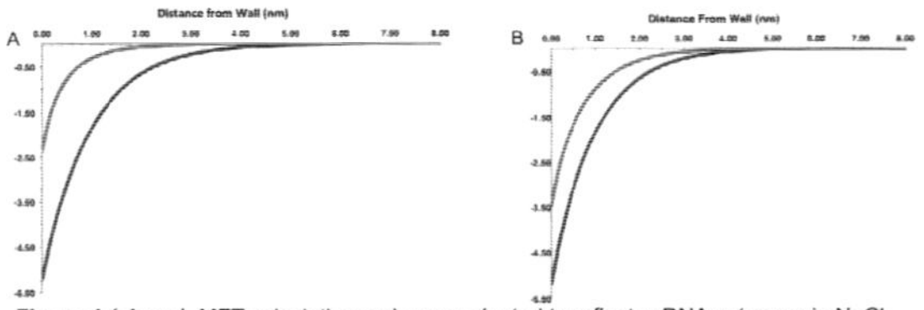

Figure 4 *(above)*. MFT calculation code was adapted to reflect ssDNA aptamers in NaCl solution grafted to a curved surface as they are in biosensors. *A* and *B* were both run with a salt concentration of 0.1 M and a surface charge density of -0.6 *e/nm*^{2}. The radius of the surface in A is 5 nm and the radius in B is 50 nm, showing that as radius increases, the results resemble those of a planar surface. Analytical solution (red), linear approximation (blue).

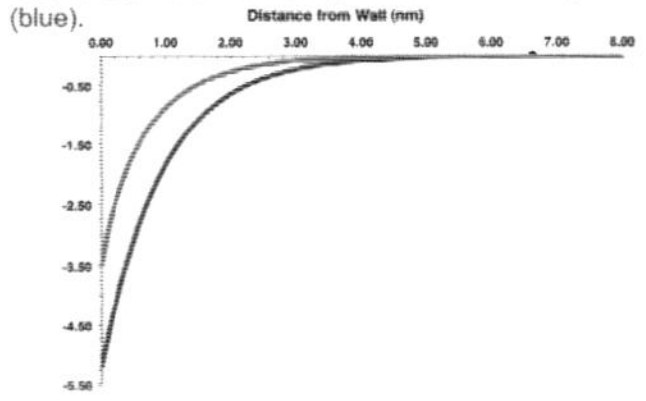

Figure 5 *(left)*. The MFT curved surface code was further adapted by replacing the NaCl solution with a $MgCl_2$ solution. Although this code does not yet contain the proper charge correlation term to account for divalent ions in solution, the results shown are similar to those for the ssDNA immersed in NaCl solution. Analytical solution (red), linear approximation (blue).

6. Planar Surface with Na^+ and Mg^{2+} Combined

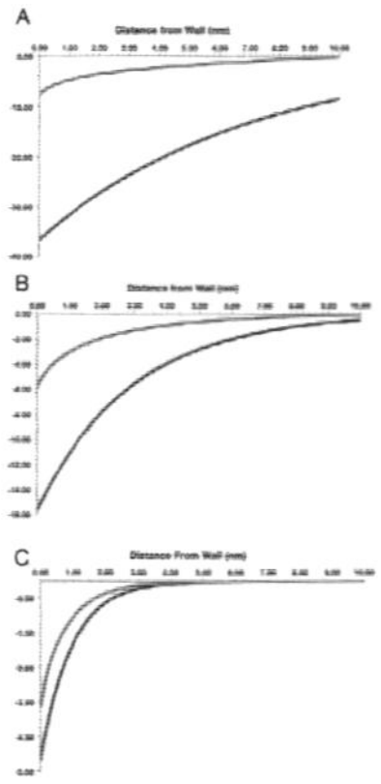

Figure 6. Mean field theory calculations of using a planar surface and a solution made of both ions. In A, the concentrations of Na^+ and Mg^{2+} were 0.001M and 0.001 M, respectively. In B, the concentrations of Na^+ and Mg^{2+} were 0.01M and 0.001 M, respectively. In C, the concentrations of Na^+ and Mg^{2+} were 0.1M and 0.001 M, respectively. Analytical solution (red), linear approximation (blue).

In reactions involving complex molecules, such as ssDNA, the thermodynamic dissociation constant must be multiplied by a correction term in order to yield the actual dissociation constant.

$$K_d = K_d^{\circ} \exp\left(\frac{-\psi(z)e}{k_b T}\right) C$$

7. Conclusions & Future Work

- We have confirmed that magnesium ions are more effective in condensing ssDNA structure than sodium ions due to their divalent properties.
- Several models of ssDNA 21-mers composed of simply thymine nucleotides immersed in various salt solutions have been successfully constructed and are ready for the incorporation of the charge correlation term from MD simulations.
- With the incorporation of this correction term into the MFT calculations, more complex systems with varied nucleotide composition and various salt concentrations will be conducted and tested .

8. References

[1] Piehler J, Brecht A, Gauglitz G, Zerlin M, Maul C, Thiericke R, Grabley S (1997) Anal Biochem 249:94-102
[2] Sassolas A, Leca-Bouvier B, Blum L (2008) Chem Rev 108:109-139
[3] Holland J, Geiger F (2012) J Phys Chem 116:6302-6310
[4] Hoffman T, Canziani G, Jia L, Rucker J, Doms R (2000) PNAS 97:11215-11220
[5] Ray S, Cohen H, Naaman R, Rabin Y (2005) JACS 127:17138-17139
[6] Uline M, Rabin Y, Szleifer I (2011) Langmuir 27:4679-4689

9. Acknowledgments

I would like to thank Dr. Mark Uline for the opportunity to work with him and for all of his guidance over the past year and a half. This work is supported by a Magellan Undergraduate Research grant from the Office of Undergraduate Research at the University of South Carolina.

398526 Partitioning of Fluoroalcohols and Their Environmental Implications

Abstract:

Fluoroalcohols are a class of alcohols that have fluorine atoms in place of the hydrogens on the alkyl chain and are found in many industrial products produced throughout the world. Fluoroalcohols can be oxidized in the environment to form perfluorooctanoic acid (PFOA), a persistent organic pollutant. Therefore, it is important to understand how fluorination affects the bioavailability of these compounds, especially in places where bioaccumulation is becoming a problem, such as the Arctic. The octanol-water partition coefficient is an accurate and widely-accepted way to compare the bioavailability of various chemicals, and can be found both experimentally and through simulation.

Molecular simulations using the OPLS-AA force field parameters and the molecular dynamics simulation engine NAMD (ver. 2.9) were performed on fluoroalcohols of varying chain length from perfluoromethanol to perfluorooctanol. Free energy perturbation was used to determine the free energies of solvation (in 1-octanol) and hydration. Once known, these free energies were used to determine the octanol-water partition coefficients. Simulations showed that fluoroalcohols tend to solvate better in 1-octanol and hydrate worse in water. The calculated octanol-water partition coefficient for fluoroalcohols tended to increase with chain length and number of fluorine atoms, which by the guidelines of AC-BAP (Artic Contamination and Bioaccumulation Potential) means that heavier fluoroalcohols are likely to become persistent contaminants and are more likely to be oxidized into persistent organic pollutants.

398425 Experimental and in silico Study of the Thermophysical Properties of Ionic Liquids with Less Common Imidazoium Cations

Abstract:

Although there are at least 1 million ionic liquids possible, a great portion of the ILs that have been experimentally synthesized and studied are based on relatively simple 1-*R*-3-methylimidazolium cations. Here, we present results for ILs that have additional functional group(s) at the 2, 4 and/or 5 positions of the imidazolium ring. Research in such ILs requires characterization of each species in terms of thermophysical properties associated with mathematical models. Given the vast number of possible structures and potential compounds, computational methods are required to predict these properties and benchmark against experimental data. Our work focused on modeling a number of new cation structures in COSMOTherm software and the actual synthesize of selected species as a means of validating predicted IL density. Excellent agreement was observed between calculated and predicted densities, which were then correlated to the IL fractional free volume (FFV) as calculated by COSMOTherm. Increased substitution of the imidazolium ring was observed to increase FFV more than isomeric 1-R-3-methylimidazolium cations. As FFV has been previously shown to influence the solubility of CO_2 in ILs, these results suggest there are still very large computational and experimental spaces to be studied in the search to identify "optimal" ILs for gas separation applications.

Conformational Analysis of Lacritin Simulations in Solution

Alexander Yang, Michael Shirts
Department of Chemical Engineering, University of Virginia

Abstract

Lacritin is a protein that has been found to increase basal tearing in rabbits, indicating potential as a remedy to dry eye. To assess lacritin's stability and understand its mechanism of action, molecular simulations in salt solution of the entire lacritin molecule and lacritin fragments were performed.

Experimental studies with lacritin have narrowed down the sites of interest to the last 25 residues. Additional experiments have demonstrated a binding site between the amphipathic alpha helix in the C-terminus of lacritin with the N-termnus of receptor syndecan-1.

Utilizing lacritin as a protein drug, however, poses various problems, including bulk synthesis as well as high costs. Current efforts include determining peptide fragments that include the amphipathic alpha helix of the lacritin molecule and exhibit the same structural behavior as the entire protein.

Molecular simulations are used to study the dynamic behavior of lacritin and its peptide fragments. Significant stability in the alpha helicity in the c-terminus of lacritin is observed from two residues within the C-terminus hydrogen bonding with residues located in the amphipathic alpha helix. Significant stability in the alpha helicity in the c-terminus of lacritin is exhibited from two residues interacting with the amphipathic alpha helix.

The addition of salt reduces the helicity of residues 101-113 in laboratory salinities (100 mM). However, the presence of salt increases local folding.

398301 Simulation of Mechanical Properties: Strain Rate Dependence of Epoxies

Abstract:

In recent years, many have sought to understand the modes of fracture in epoxy systems to further improve their applications in aerospace technology as adhesives and composites. The fracture patterns of epoxies and other thermosets differ from those of their thermoplastic counterparts due to the inherent differences in how the materials are formed. Thermoplastic materials exposed to tension exhibit hydrogen bond cleavage as polymer chains slide past one another, whereas thermoset materials fracture under tension due to broken covalent bonds. Thermosets are valued particularly for their toughness and their properties as part of polymer matrix composites (PMCs). Epoxies in particular have been studied extensively both experimentally and computationally, and this project looks to combine the two strategies to understand how epoxies fracture when exposed to differing strain conditions.

Simulations completed thus far have studied diglycidyl ether of bisphenol A (DGEBA) with 4,4'-diaminodiphenylmethane (MDA) without diluents present. The DGEBA monomers have varying molecular weights, accomplished by varying the chain length of the prepolymers. Multiple distinct configurations of the epoxies of interest were constructed as 70 Å cubic systems composed of stoichiometric ratios of DGEBA and MDA. A computer program cross-linked each configuration to a minimum 88% degree of cure, calculated as the ratio of "reacted" end groups to those present at the beginning of the simulation. This method uses an isothermal-isobaric ensemble to relax the system between molecular dynamics cross-link steps and assumes equal reactivity of primary and secondary amine groups.

Five independent cross-linked configurations for each system were created and subjected to uniaxial strain using LAMMPS (Large-scale Atomic/Molecular Massively Parallel Simulator), a molecular dynamics simulation program. Stress-strain relationships for each individual system were examined for strain rates varying from 10^6 s^{-1} to 10^{11} s^{-1}. Properties examined during the project included Young's modulus and strength at yield, which enabled analysis of the effects of molecular weight (chain length) and strain rate on epoxy fracture. The stress-strain curves will be used as inputs for a continuum-level theory in order to calculate the stress integral ahead of the crack tip, ultimately allowing the calculation of the J_{IC} integral.

Experimental results will be used to validate the computational method. The first goal will be to develop a process to construct highly cross-linked epoxy systems using DGEBA resins of varying molecular weight with both 4,4'-diaminodiphenyl sulfone (DDS) and 3,3' DDS. Thermogravimetric analysis (TGA) will be used to determine the degradation points of the uncured and cured samples. Differential scanning calorimetry (DSC) will be used to provide the degree of cure for each epoxy system created. Further plans for experiments include combination of materials characterizations including varieties of spectroscopy, microscopy and mechanical testing to understand morphology (amorphous/crystalline), chemical composition and bonding mechanism within matrix, at fiber/matrix interfaces/interphases, and micro-failure features (like micro-voids, matrix and interfacial cracks).

398442 Molecular Simulation of Bismuth Telluride Exfoliation in an Ionic Liquid Solvent

Abstract:

Nanosheets of bismuth telluride (Bi_2Te_3) are of great research interest because of their potential usefulness in diverse material applications, such as thermoelectrics, heterogeneous catalysis, and superconductors. One proposed method of their fabrication is by ionic liquid (IL) assisted exfoliation. ILs display unique and tunable solvation properties, and they are considered to be green alternatives to conventional solvents. In this work, molecular simulations are used to probe the mechanism of exfoliation of nanosheets of Bi_2Te_3 in the ionic liquid [BMIM]Cl. We study the thermodynamics, kinetics, and structure of the exfoliation process over a wide temperature range. We observe spontaneous exfoliation without artificial forces, allowing a kinetics-based estimate of the apparent activation energy of exfoliation. We also estimate the surface free energy of Bi_2Te_3, and the results are comparable to experiment. The structure of the IL solvation layers is found to be strongly ordered, with density oscillations near the interface alternating in charge. This is consistent with the observed stability of experimental dispersions of Bi_2Te_3, as well as previous simulation results. We find that surface free energy is independent of the size of the Bi_2Te_3 model used, which implies that the size of the system studied is adequate to describe experimental scale systems and to predict effective exfoliation conditions. The methods used are extensible to other ILs, and our current and future results should accelerate the experimental search for an effective exfoliation process for the manufacture of Bi_2Te_3 nanosheets.

Understanding the Early Stages of Metal Nanoparticle Growth Using Density Functional Theory Calculations

Olabobola Shobayo[1,2], Giannis Mpourmpakis[2]
[1]Department of Chemical, Biochemical, and Environmental Engineering, University of Maryland Baltimore County
[2]Department of Chemical and Petroleum Engineering, University of Pittsburgh

Metal nanoparticles are unique in their diversity of technological applications, ranging from active catalysts for energy generation to drug delivery carriers in medicine. Since nanoparticle physicochemical properties are directly related to their structural characteristics, our aim is to gain an understanding of the nanoparticle growth mechanisms. We focus on the early stages of metal nanoparticle growth, investigating the metal thiolate coordination polymer growth of $(CuSCH_3)_n$ and $(AuSCH_3)_n$, with n=1-8, by using Density Functional Theory calculations. We determined that after the formation of the trimer, for both Au and Cu complexes, the overall stability of the complexes remain the same. When the Bimetallic complexes were investigated, we saw that the complex stability is a linear function of both Au and Cu monometallic stability, indicating that the metals can be well mixed. Solvent effects slightly decrease the absolute values of the reported complex stability values (BE/n) and the trends remain the same as in the gas phase.

Computational Study of Copper as a Catalyst for Gaseous Carbon Dioxide

Ashaen Patel and Dr. Líney Árnadóttir
School of Chemical, Biological, and Environmental Engineering
Oregon State University
Corvallis, OR 97331

According to the World Bank oil drilling operations currently flare or release approximately 150 billion cubic meters of natural gas from oil wells annually. This is equivalent to the amount of natural gas used by U.S. residencies over the course of one year. The released methane gas is combusted to generate CO_2. Although CO_2 is considered a greenhouse gas is it less harmful than methane which has about 20 times the green house impact on CO_2 per pound. Currently, the CO_2 is either released to the atmosphere or captured and stored underground in empty oil wells. The latter is energy intensive and alternative uses for expelled CO_2 do exist. For example, CO_2 can be manipulated to produce carbon monoxide from which the Fischer-Tropsch process can be used to react carbon monoxide and hydrogen to produce larger hydrocarbon for alternative fuel.

The purpose of this project was to use Density Functional Theory calculations to study the interactions between CO_2 and a metal to help identify a catalyst for CO_2 activation and eventually CO_2 conversion to larger hydrocarbons. The first step requires CO_2 adsorption and activation followed by either hydrogenation or breakage of the C-O bond to form carbon monoxide and adsorbed oxygen. Calculations were completed using Vienna Ab-initio Simulation Package (VASP) utilizing projector augmented wave pseudopotentials. The ground state energies of CO_2 adsorbed on a copper surface and different adsorption sites were calculated and were ultimately used to assess interactions between CO_2 and the surface. Through assessing the interactions, the likelihood of a catalyzed reaction between CO_2 and a copper surface can be determined.

One specific copper surface was generated for the calculations by manipulating a face centered cubic (FCC) unit cell. A surface cut was also imposed to create a 3x3x3 copper FCC(100) surface using the atomic simulation environment. Additionally, a vacuum space of 15Å in the z-direction above and below the bulk was induced. The coordinate system was tessellated infinitely by VASP in the x, y, and z directions to generate multiple repeated surfaces in the z direction parallel to the x and y plane. The molecular adsorption was allowed on one side of the metal and the vacuum space prevented interactions between the adsorbate and the bottom part of the slab. The FCC(100) copper bulk was initially minimized without an adsorbate in order to obtain a calculated lattice constant which was compared to literature. Initially the top two layers of the surface were relaxed without an adsorbate and then kept frozen for all adsorbate interaction calculations. Three systems of the same FCC(100) copper bulk were created to calculate the CO_2 surface interactions on three unique adsorption sites available on a face centered cubic unit cell. This was done in order to determine if the conversion was preferentially more favorable on one absorption site over another.

MATLAB Based Analysis of EDS Spectral Imaging Acquired by AC-STEM

Eric Romero, Ping Lu, Sandia National Laboratories, PO Box 5800, Albuquerque, NM 87185

Email: edromer@sandia.gov

Abstract

Here we present a MATLAB based program designed for fast and efficient data interpretation of energy-dispersive X-ray spectroscopy (EDS) spectral imaging datasets. The EDS dataset, collected by a scanning electron nanoprobe using an aberration-corrected scanning transmission electron microscope (AC-STEM), contains useful chemical information for the characterization of all types of materials. However, the means of quick and effective data interpretation has been a challenge due to the large amount of data collected as well as the lack of statistically viable signal-to-noise (S/N) associated with individual data points. The program developed uses the overall EDS image along with the total EDS spectrum to perform multiple data analyses. Quick extraction of various elemental-maps may be obtained by selecting energy regions on the EDS spectrum. Also, the spectrum from a selected region may be viewed. This program has multiple built-in functions such as lattice-averaging and regional averaging for improving the S/N ratio of atomic-scale chemical maps, RGB surface display, chemical intensity profiling, and the deconvolution of overlapping EDS spectrum peaks allowing for pure elemental imaging. The applications presented here include: improving S/N ratio of atomic-scale chemical mapping by lattice averaging, resolving chemical species with similar characteristic X-rays via pixel-by-pixel deconvolution as well as ongoing research on development of a MATLAB based strain mapping technique in an effort to understand the relationship between the strain and chemical composition.

Sandia National Laboratories is a multi-program laboratory managed and operated by Sandia Corporation, a wholly owned subsidiary of Lockheed Martin Corporation, for the US Department of Energy's National Nuclear Security Administration under contract DE-AC04-94AL85000.

AIChE Poster Abstract 2014

Hannah Bulovsky

The Effect of Stabilizers on Zero Valent Iron Nanoparticle Stability, Reactivity, and Toxicity

Zero valent iron nanoparticles are effective at remediating groundwater contaminants due to their ability to catalyze the transformation of chemicals and the enhanced surface area available for nanoparticle-contaminant interactions to occur. However, this high reactivity also contributes to the particles rapidly agglomerating and becoming less effective for remedial applications. A stabilizing compound can be added to the particle solution during synthesis to overcome the attractive forces between the particles. In this study, we examined the effect of two widely used stabilizing agents, gum arabic (GA) and sodium carboxymethyl cellulose (CMC), on particle stability, reactivity, and toxicity. The stability of the particles was assessed by measuring particle hydrodynamic diameter using dynamic light scattering. The average hydrodynamic diameter of the unstabilized was over 1000 nm, significantly larger than the CMC and GA stabilized particles, which were around 100 nm. Their ability to degrade trichloroethylene was assessed using gas chromatography following a 24 hour incubation period as a measure of particle reactivity. The GA and CMC stabilized particles reduced the trichloroethylene concentration by 9.7% and 6.1% respectively, while the unstabilized particles did not reduce the trichloroethylene concentration at all. An embryonic zebrafish assay was used to investigate the toxicity resulting from particle exposure over 5 days. The CMC controls caused 51% mortality, and 40% solutions of unstabilized and CMC particle solutions both caused over 50% mortality. In comparison, the GA controls caused no mortality, and 30% particle solutions caused less than 30% mortality. However, GA controls and particle solutions caused the greatest delay in hatching. The small size of the particles synthesized with a stabilizing compound indicated that GA and CMC were stabilizing the particles. The small size of the stabilized particles provided more reactive sites per particle, leading to the reduction of trichloroethylene in the incubated samples. The CMC stabilized particles and the unstabilized particles caused similar mortality in embryonic zebrafish, which may be due to the oxidation of the unstabilized particles, and the high concentration of CMC used to stabilize the particles. While the GA stabilized particles caused the highest hatching delay, the GA controls and particle solutions caused the lowest mortality and therefore were the least toxic.

398961 Cerium Oxide Adsorption Efficiency of Antimicrobial Drugs in Water Using High Performance Liquid Chromatography

Abstract:

Antimicrobials drugs like fluoroquinolones and sulpha drugs are used across the world for both human and veterinary applications to treat and prevent pathogen growth thus protecting public safety. Among this, broad spectrum antibiotics such as fluoroquinolones and sulpha drugs are employed in the farming industry to reduce bacterial infection and promote healthy growth of poultry and cattle. Despite their ability to control infections these agents are often excreted intact enabling them to leach through soil onto the water table enabling the risk for bacterial resistance and release of potentially harmful by products. Moreover, these substances are associated with some rare but serious secondary effects that can affect some human body systems: cardiovascular, musculoskeletal, endocrine, renal and central nervous systems. This study analyzes the use of cerium oxide nanoparticles adsorption efficiency of sulfadiazine, levofloxacin, and ciprofloxacin. High Performance Liquid Chromatography was used as the technique for the analysis and validation of the method, evaluating linearity, reproducibility and accuracy for the abovementioned antimicrobials. Results demonstrate effective validation of the method with excellent values correlation factors close to 1 and accuracy. Most of the reproducibility results are under the acceptance criteria of relative standard deviation (%RSD <5). The percent of removal of the cerium oxide nanoparticles filters were 98.28% for sulfadiazine, 90.12% for levofloxacin, and 94.32% for ciprofloxacin. Drug removal plots showing the adsorption efficiency of each antimicrobials drugs will be presented.

Gold Nanoparticle Interactions with Model Cell Membranes

Christina Bailey, Elaheh Kamaloo, Kellie Waterman, Jeniece Macedonio, Kathleen Wang, Terri A. Camesano

Department of Chemical Engineering

Worcester Polytechnic Institute

Nanoparticles are investigated for drug delivery, gene therapy, and imaging techniques; however, the fundamental question of how they will interact with cell membranes is still in question. Ultimately, NPs will come in contact with our water and soil systems, making it crucial to understand how NPs of different sizes will interact with natural environments for proper and safe use. The goal of our studies is to quantify the interactions between gold nanoparticles and biological membranes. A Quartz Crystal Microbalance with Dissipation monitoring was used to measure frequency changes on a lipid bilayer's surface. Nanoparticle sizes of 2, 5, 10, and 40 nm were chosen to give a range of small and large particles to investigate. This project focuses on gold nanoparticles due to the low chemical toxicity of bulk gold, allowing us to concentrate on a membrane disruption toxicity mechanism. Gold nanoparticles were suspended in water and Poly(methacrylic acid) (PMAA). Nanoparticles in water simulate a control lab condition while coating nanoparticles in PMAA mimics a soil-like environment. Using mathematical modelling, mass change data on the bilayer surface gives us insight to the nanoparticle – bilayer interactions. Our results show that when nanoparticles are suspended in water mass is lost from the bilayer due to energetic relationships occurring at the surface. When gold nanoparticles were coated with PMAA, a net mass increase resulted; 2 nm and 40 nm particles interacted the most with the bilayer. Thus, a size dependent relationship was observed during interaction with NPs in a PMAA environment. With this information we can make informative suggestions to ensure the safe use and disposal of gold nanoparticles in the environment.

Removal of Phenanthrene From Aqueous Solution using Fenton's Reagent as an Oxidant and Bio-Char as an Adsorbent

Phillip Stringer[1], Kendra Green[1], Shamim A. Begum[1]*, Jonathan C. Mbah[1] and Nosa O. Egiebor[2]

[1]Department of Chemical Engineering, Tuskegee University, Tuskegee, Alabama 36088
[2]Department of Chemical Engineering, University of Mississippi, Oxford, MS 38677-1848

*Author for communications. Email: begums@mytu.tuskegee.edu

Phenanthrene, a Polycyclic Aromatic Hydrocarbon (PAH), is an environmental pollutant comprised of multiple benzene rings which can either be natural or man-made. PAHs can be generated largely from the combustion of fossil fuels and other organic materials. As a result, PAHs are ubiquitous and a common constituent of many natural materials such as crude and refined oil. PAHs can be found in surface water, ground water and wastewater. PAHs are known to have mutagenic, teratogenic and potentially carcinogenic effects on health. Therefore, USEPA has listed the PAHs as a priority pollutant that should be monitored in the environment. The hydrophobic nature of PAHs makes their removal very difficult from the water as they persist for a long period of time. In this study, the removal efficiency of phenantherene was evaluated by using Fenton's Reagent as an oxidant and bio-char as a sorbent in batch tests. Various organic solvents were tested to identify it's suitability to prepare the aqueous solution of phenantherene for batch tests. The concentration of Fenton's reagent (i.e Hydrogen peroxide and Ferrous iron) was varied to obtain the maximum removal of phenanthrene from aqueous solution. The effectiveness of phenanthrene removal with bio-char was also observed at various pH values of 3, 7 and 9. Results show that acetonitrile is the most suitable solvent to prepare the aqueous solution of phenantherene, which can be detected by using UV-VIS Spectrophotometer. The removal of Phenanthrene was about 100% in 1 min with Fe(II) of 1 mg/L and H_2O_2 of 30 mg/L at an initial phenanthrene concentration of 100 μg/L. Using bio-char of 500 mg/L, the removal percentages of phenanthrene were 33%, 83% and 66% at pH values of 3, 7 and 9, respectively at an initial phenanthrene concentration of 200μg/L.

396719 Use of Sericulture Subproducts As Adsorbent and Absorbent of Water Pollutants

Abstract:

Use Of Sericulture Subproducts As Adsorbent And Absorbent Of Water Pollutants

Abstract: In this work the use of silkworm (Bombyx mori) pupa and bisu, which are subproducts in the industry of sericulture, were investigate to remove water pollutans. This study was divided in two parts: the first one, the silkworm pupa was tested as adsorbent of acid red 14 dye, and the second one, the bisu was tested as absorbent of crude oil, motor
oil 20w 50 and kerosene. The adsorbent removed the dye in water solution where the final concentration was
M,
otherwise, the results in the assays with bisu had a sorption capacity between 30 and 60 percent for the different hydrofobic water pollutants. In conclusion this materials shows a high performance as a low cost for bioremediation in contaminated water.
Keywords: Silkworm pupa, bisu, crude oil, acid red 14, kerosene, motor oil.
1. INTRODUCTION
The first records in sericulture come from China and date from about 2600BC. Historically the greatest sales of silk were from China to Japan where it was used to make clothing until now (Hill [1]), in Colombia the production of silk is in three regions: Valle del Cauca, Risaralda and Cauca, where the companies in sericulture consist in farms with 20 employees (Red Andina de la Seda [2]). In order to realize a great goal of silkworm industry in this days, we must meet new challenges (Kai [3]), for example, bioremediation is a technology that uses microorganism, fungi, plants or animals to treat contaminants through natural mechanism (EPA [4]), Colombia doesn't use this technology in his industries, so is necessary work in this

field as a way to develop sustainable process and solve environmental issues, because the poor treatment of wastewater in the industry, is a serious environmental problem due to their potential human toxicity, one of the cases are oil spill, this can happen in many places when oil is being transported by oil tankers or through pipelines (Chang et. al, [5]). Also, textile industries discharge to the surrounding environment visual pollutants without any further treatment, they also cause extensive toxicity to the aquatic and other forms of life (Khan [6]).

Silkworms go through complete metamorphosis in 4 stages: egg, larva, pupa and butterfly (Schaffer [7]), we are interested in pupal stage, is an inactive stage in life cycle of the silkworm (Singh and Saratchandra [8]), where the cuticle is made of chitin a natural cationic polysaccharide (Zhang et al, [9]), The amino acid nature of the pupa provided a reasonable capability for dye removal (Noorozi [10]) Otherwise, the bisu is most of the pierced cocoons or very dirty cocoons of the silkworm are discarded as industrial wastes, and can be obtained as a low cost material (Moriwaki et al, [11]).

There is much research that involve silkworm like develop vaccines for animals (Otsuki et al, [12]), reducing blood glucose level (Cong et al, [13]), control of lipids in Wistar rats (Mentang et al, [14]) and studies of his mechanical properties (Zhang et al, [15]), the main objective of this work is use the silkworm subproducts for removal water pollutants.

2. METHODOLOGY

Determination of absorbent capacity: the bisu was dried in an oven at 107 ° C for 120 minutes, was arranged in arrays of layers (10 layers, 20 layers, 30 layers) which is then pressed at 1 ton per square inch, once have compacted cut into pieces of 5x5mm and 2x2cm for absorption tests. Three immiscible contaminants in water were selected: kerosene, crude oil and motor oil (20w-50), the tests are performed at different times (2, 5 and 10 min), measured the weight of pollutant removed in order to evaluate absorbent capacity of each array by weight difference, defined as the removal rate ratio of the mass of pollutant removed and the initial mass.

Determination of adsorbent capacity: the cocoon is immersed in boiling water to remove the pupa, which is then dried at 110 ° C for 35 minutes; pupa then ground to a fine powder. Eight solutions of organic dye with four different concentrations are prepared, and the absorbance of each was measured by the spectrophotometer. Two tests were conducted, the first with the same retention time (2 hours) by adding 0.02 g of pupa and said next time varying (15, 30, 45 and 60 minutes) by adding 0.01 g. Subsequently, the solutions are filtered to remove the pupa and again used the spectrophotometer to measure the absorbance of the filtered solution.

3. RESULTS

Is observed for the absorbent material is stable in water and less dense than the latter, remaining in the surface and thereby facilitating the absorption of contaminants.

The removal percentage varies with the number of layers and the contact time, the latter being proportional to maximum absorbency after 10 minutes. Also the arrangement of 30 layers had the best performance in the absorption of the three pollutants with removal percentage: 70.1%, 50.2%, 37.6% for crude oil (Fig. 1), kerosene, and motor oil respectively in a time of 10 minutes. Additionally, a blank test was determined that the absorption of water by the material is less than 1%.

The dye solutions, by adding the same amount of adsorbent at different times can be observed gradually decreases of the color intensity. Also the concentration of the dye tends to drop at the same point regardless of the initial concentration . In treated solutions pupa traces which could not be separated by filtration was observed. The results are a sample of the absorbing power of the pupa, where a precedent for future investigations brand.

4. CONCLUSIONS

The properties observed both bisu and pupa show that an application can give an application to these raw materials that are treated as waste currently. The bisu works efficiently as a sorbent for oil, kerosene, and motor oil, polluting high-impact environmental levels. Besides the pupa serves as adsorbent of organic dyes dissolved in water, therefore, could become exploited in the treatment of effluents from the silk industry and others where these pollutants are found.

REFERENCIAS

[1] D.S. HILL , "The economic importance of insects" Champman & Hall. 1997.

[2] Red Andina de la seda. Proyecto de Instituto Italo-Latinoamericano. Sericultura en Colombia. Redandinadelaseda.org 2007
[3] Li Kai. "Chemical, Biological and environmental engineering". World Scientific.2010
[4] EPA. "Use of Bioremediation at Superfund Sites". September 2001
[5] M. Chang, C. Crane, K. Hamalainen, L. Jones. "Oil spill disaster" Scholastic. 2010
[6] A. A. Khan, Q. Husain. "Decolorization and removal of textile and non-textile dyes from pollutes wastewater and dyeing effluent by using
potato soluble and immobilized polyphenol oxidase". ELSEVIER 2007
[7] D. Schaffer. "Silkworm". Life Cycles. 1999
[8] B. Saratchandra, T. Singh. "Principles and techniques of Silkworm Seed Production". Discovery Publishing House. 2004
[9] M. Zhang, A. Haga, H. Sekiguchi, S. Hirano. "Structure of insect chitin isolated beetle larva cuticle and silkworm pupa exuvia. ELSEVIER,
December 2000
[10] B. Noroozi, G. A. Sorial, H. Bahrami, M. Arami. "Equilibrium and kinetic adsorption study of a cationic dye by a natural adsorbent".
ELSIELVER, 2007.
[11] H. Moriwaki, S. Kitajima, M. Kurashima, A. Hagiwara, K. Haraguchi, K. Shirai, R. Kanekatsu, K. Kiguchi. "Utilization of silkworm cocoon
waste as a sorbent for the removal of oil from water". ELSEVIER. 2008.
[12] T. Otsuki, J. Dong, T. Kato, E. Park. "Expression, purification and antigenicity of Neospora caninum-antigens using silkworm larvae
targeting for subunit vaccines". ELSERVIER. 2013.
[13] L. Cong, G. Cao, X. Renyu, P. Zhonghua, Z. Xiaojian, W. Zhou, C. Gong. "Reducing blood glucose level in TIDM mice by orally
administering the silk glands of transgenic hIGF-I silkworms". ELSERVIER. 2011.
[14] F. Mentang, M. Maita, H. Ushio, T. Ohshima. "Efficacy of silkworm (Bombyx mori L.) chrysalis oil as a lipid source in adult
Wistar rats". ELSERVIER. 2011.
[15] K. Zhang, F. W. Si, H. L. Duan, J. Wang. "Microstructures and mechanical properties of silks of silkworm and honeybee". ELSERVIER.

397240 Use and Evaluation of Chitosan As a Water Recovery Agent in an Inorganic Ionic Solution

Abstract:

The recovery of water from contaminated bodies is a major concern because of the upcoming scarcity of this resource. Chitosan is an organic and biodegradable polymer that occurs in crustaceous shells. For this purpose, we prepared chitosan beads by drop addition in an alkaline bath, evaluating its time of assembly as a function of pH, showing its best resolution at pH=12. Then, using a single-solute, 10 ppm, solution of As, the chitosan beads were submerged in it, and after that, the remaining concentration of As was measured using ICP-MS. The remaining concentration of the samples were higher than before the elution, thus confirming the capacity of Chitosan to retrieve water from a single-solute inorganic ionic solution.

Removal of Organic Dyes from Wastewater Using Porous Semiconductor Hexaniobate Nanotubes

Roya Edalatpour [1], Maryam Zarei Chaleshtori [1], Victor Correa [1], Natalie López [2], Manuel Ramos [1], Nancy Rondeau[1], Russell R. Chianelli [1]

[1] Materials Research and Technology Institute, The University of Texas at El Paso, El Paso, Texas, 79968

[2] Universidad de Puerto Rico-Mayaguez, 259 Blvd. Alfonso Valdez Cabian, Mayagüez, Puerto Rico, 00680

Keywords: Nanotubes, Niobium, Photocatalysis, Wastewater Decontamination

A major challenge for many textile industries has become the cost of eliminating organic waste contaminants, mainly dyes in water. Therefore, this results in the disposal of these compounds and contamination of a large quantity of water. Current methods for the renewal of contaminated water add considerable cost and delay to the production system. These processes include the use of filtration and activated carbon and the photo-fenton reaction.

Photocatalysis has proven to be a very promising technology for the elimination of pollutants in wastewater. UV light chemically activates photocatalytic materials, usually composed of wide band gap semiconductor. When those materials are exposed to UV light, they go into an oxidant state and thus eliminate water borne contaminants. After conducting a literature review, it was found that the most commonly employed photocatalyst in industrial processes is TiO_2 in mixed phase (70% anatase and 30% rutile). However, a fundamental problem that is faced when using this kind of material is that particle size seems to be relatively small, provoking agglomeration and consequently damaging of filters easily by clogging.

In consequence, cost-efficiency is compromised because of the inability to remove TiO_2 from filters. Due to the aforementioned reason, the search for better structure/function materials that act as photocatalysts has been an active field of study. An ideal photocatalytic material would be one that posses a high surface area and catalytic activity and strong morphological features (*i.e.*, nanotubes, nanowires and laminar materials). In order to achieve these properties with high precision, modern methods of synthesis are extremely important.

Previous reports have shown that porous metal oxides made up of titanium and niobium metal oxide such as $HTiNbO_5$ with wide band gap (3.3 eV) can be separated easily from solution, and therefore prevent clogging problems such as those that are faced when using TiO_2 nanoparticles alone. The porous-like metal oxides made from Niobium such as $H_4Nb_6O_{17}$ can form a stacked nanosheet final structure, due to positive ions such as H^+ or K^+ which lay interstitially between each niobium oxide layer. The nanosheets can spread out and separate to form a colloidal solution, by adding a large molecule that possesses a high electron density, such as tetra (*n*-butyl) ammonium hydroxide (TBA^+OH^-). Furthermore, such sheets can be converted into a nanotubular structure using exfoliation methods, in a process where TBA ions are removed, as described previously.

In this research, we present the chemical synthesis of hexaniobate nanotubes using two methods, (1) starting material $K_4Nb_6O_{17}$ and (2) parent material of $H_4Nb_6O_{17}$ via ion exchange. The as-synthesized materials were exfoliated by adjusting the pH to 9–10 using tetra-n-butylammonioum hydroxide (TBA^+OH^-), leading to a formation of hexaniobate nanotubes. In order to understand morphology, a full characterization was conducted using SEM, HRTEM, BET and powder-XRD. The photocatalytic activity was evaluated using photolysis method using Methyl Orange (MO) and Bromocresol Green (BG) as model contaminants. The results indicate a nanotube porous oxide with large porous and surface area; the photocatalytic activity is about 95% efficient when compared with commercial TiO_2.

AIChE Annual Student Conference
November 14-17, 2014

"Carbon-Based Renewable Hydrogel Nanocomposites for Water Purification"

Erik J. Hurley[1], Samantha A. Meenach[1]

[1]Department of Chemical Engineering, University of Rhode Island, Kingston, RI 02881

Since the dawn of the industrial revolution indiscriminate dumping of wastewater from textile mills, metallurgical factories, and other manufacturing facilities have contaminated our water ways with toxic dyes and heavy metal ions. The removal of contaminants from wastewater has been successfully completed using many carbon-based materials such as C_{18} silica and activated carbon but these materials often suffer from low adsorption capacities and lack of reusability. Over the past two decades new carbon-based nanomaterials, carbon nanotubes (CNT) and graphene oxide (GO), have presented a possible solution to this problem. To fully utilize the large surface area of these nanomaterials for water purification applications, CNT and GO nanoparticulates were suspended in the matrix of a poly (ethylene glycol) (PEG) hydrogel. The objective of this project was to evaluate the adsorption potential of PEG hydrogels loaded with multi-walled carbon nanotubes (MWCNTs), GO, or a combination of GO and MWCNTs. Three model contaminants were used to test this; methylene blue (cationic), 4 nitrophenol (anionic), and Eosin Y (organic compound/dye). The hydrogels were composed of poly (ethylene glycol) (400) dimethacrylate (PEG400DMA) as a crosslinker and poly (ethylene glycol) (200) methylether metharcylate as the backbone of the gel. The polymer solution was mixed with 1-1.5wt% MWCNTs 1-2wt% GO powder prior to addition of free radical initiator and catalyst to allow for hydrogel formation. The hydrogels used were made with 1-2mole% PEG400DMA as this composition resulted in a maximum swelling ratio while allowing for mechanical stability. The swelling ratio was determined by weighing uniform gel samples at different temperatures, between 25 and 80° C, and dividing that weight by the dried weight of each gel. The chosen gel compositions were then evaluated for absorption of each of the model dyes; where the gel sample was placed in an aqueous solution containing the dye and UV-Vis absorbance readings were taken over time to determine a change in concentration. The results from the adsorption experiments showed that the GO-loaded hydrogels were the best at adsorbing methylene blue where 0.05mg of MB was adsorbed per gel sample from a starting concentration of 0.014mg/mL and a starting volume of 5mL over a 24 hour period. The GO-loaded hydrogels had an average initial weight of 0.1917g. The GO/MWCNT composite gels showed slightly less adsorption towards methylene blue compared to GO loaded gels. For Eosin Y there was nearly equal adsorption seen between the loaded gels and the blank control gels. As for the studies looking at the adsorption of 4-nitrophenol little change was seen in the concentration across all of the gels. The interaction of the reactive groups in the GO loaded hydrogels showed the most promise in the adsorption of cationic dyes like methylene blue. Overall, these materials show great promise in their use for water purification applications.

398804 Functionalized Membranes As a Flexible Platform for Water Detoxification

Abstract:

Membrane based water remediation techniques have many inherent advantages over traditional methods including high selectivity, continuous operation, and low energy consumption. These advantages warrant study into advanced applications of membranes for water remediation. Heavy metals such as selenium, cadmium, lead, and chromium are disposed of by electronics manufacturing and coal-fired power plants and pose a growing environmental danger as they easily reach toxic concentrations. For this reason, a highly selective and sustainable treatment for heavy metal removal has been examined by using functionalized membranes. Synthesis of iron nanoparticles within the membrane structure has proven to be an effective platform for the detoxification of selenium contaminated water. Further, by a similar functionalization it is hypothesized that the degradation of specific toxic organic compounds which pose a threat to the environment could also be introduced as an application of iron nanoparticle functionalized membranes. This study examines the possibility of the application of functionalized membranes for both heavy metal removal and the degradation of toxic organic acids in water.

DESIGN OF PHOTOREACTOR FOR THE TREATMENT OF SYNTHETIC WASTEWATER

Priyanka Padmanabhan[1]*, Ankit Kumar[2]

[1]A 301, Siesta Dwelling, Thaltej, Ahmedabad, 380059 , India
11bch054@nirmauni.ac.in

[2]602, Antique Tower, Morbi, 363641, India
11bch047@nirmauni.ac.in

Introduction

Across the world, one major topic among the educated society is the non-conventional energy sources, and more specifically the optimization of this aspect to practical usage. Solar energy has been at the vortex of this topic as its availability is abundant and relatively easy to access in comparison with the other sources of energy.

"Being continually replenished by nature, solar energy is an energy source that cannot be depleted. Assuming 100% absorption of solar radiation, it would only require 88 minutes of solar radiation to cover for humanity's total energy consumption for a year". Our sun provides an immense amount of potential energy to our world, but it needs to be harnessed and transformed to be used [1].

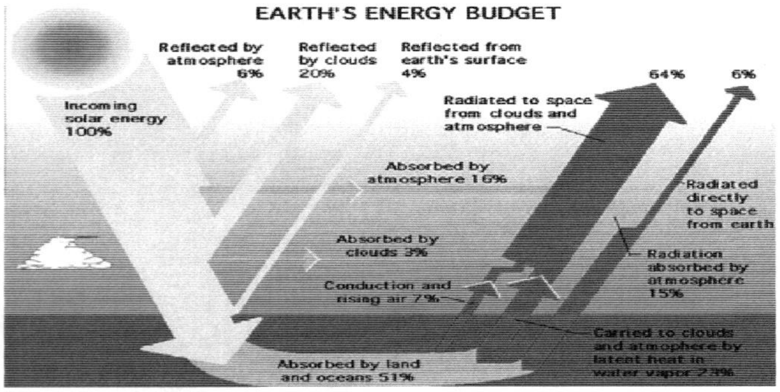

Figure 1: The Earth's Energy Budget [2]

Across the world, there continues to be huge volumes of wastewater pumped directly into rivers, streams and the ocean itself. The impact of this is severe, aside from the damage to the marine environment and to fisheries it can cause, it does little to preserve water at a time when many are predicting that a global shortage is just around the corner.

Objective

The objective is to develop an effective reactor for the treatment of synthetic wastewater generated by various industries. The aim is to design a reactor that degrades the contaminants using the natural solar radiations only. The possible designs have been proposed later in the document keeping in mind the feasibility and economics of the system.

Features required of the effective reactor

The following are the features required in the photoreactor to be designed:-

1. Maximum contact between light, catalyst, waste water and air

The reactor design should be such that the contact between the photons, wastewater, catalyst and oxygen should be maximum to increase the efficiency of the process. This requires the development of a thin layer of the process stream in the reactor that will come in contact with the photons for the reaction.

2. Maximum turbulence in the system

The process stream should be under considerable velocity in the reactor for a good rate of reaction. This can be achieved by incorporating obstructions in the path of the fluid.

*Presenting author: Priyanka Padmanabhan[1]

3. Providing a residence time of about 10-15 minutes with/without recycling

The reaction mechanism is such that it requires about 10-15 minutes to complete the reaction. This amount of time has to be given to the stream to come into contact with the photons. This can be done by providing a recycle tear in the system.

4. Increase in temperature is not the objective

The objective of the reactor is not to increase the temperature but to obtain the maximum optical efficiency. So, the reactor design will be based on the non-concentrating solar collectors.

5. Effective regeneration and recycling of the process fluids should be attained

The catalyst would be of nano size and thus, costly. The regeneration of the catalyst is thus an important factor in the process. Moreover, effective recycling should be done for the completion of the reaction.

Our Proposed Designs

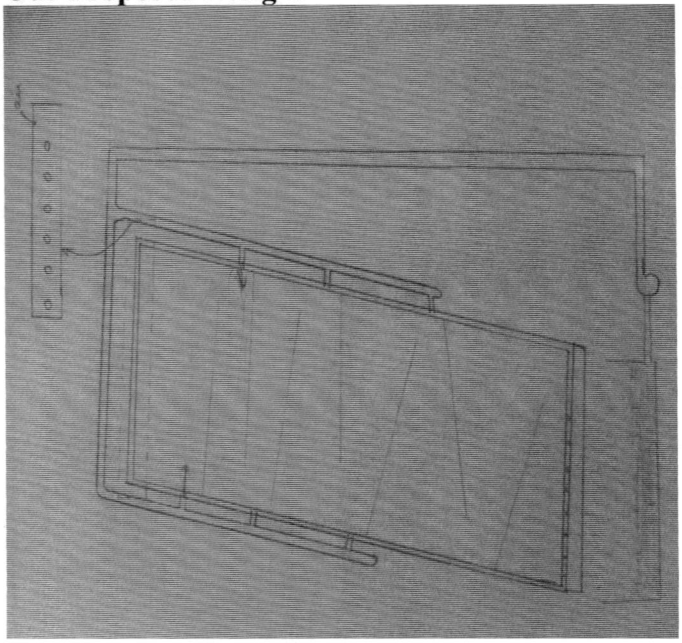

Figure 2: Proposed Design 1

Water and catalyst mixture is pumped into tubes placed along the side of an inclined plate. These tubes have holes in them at different heights that will introduce the fluid into the plate. The plate has slab-like obstructions placed at an increasing height difference with a tilt. Each slab will have a turbulator placed in it and the whole plate is placed inside a casing. The fluid that enters the plate will encounter the obstruction, thus, increasing the turbulence. This fluid flows along the length of the plate where new fluid is introduced along the length. This makes sure of intermixing of the fluid and the concentration gradient is always present as a favorable condition for the reaction. The treated fluid is then collected and recycled till all of the feed is treated. The casing of the plate is left open on the top to allow air to enter.

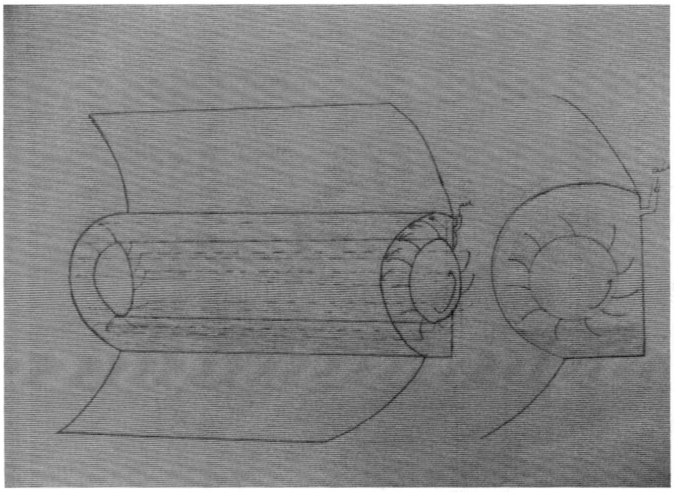

Figure 3: Proposed Design 2

A hollow cylinder is attached to a parabolic mirror. Here, the curved surface area of the cylinder is greater than the aperture of the mirror. This makes sure that all the light that hits the mirror falls on the surface of the cylinder. Inside the hollow of the cylinder, a rotating cylinder with spatulas attached to it is placed. A pool of liquid is maintained inside. The rotating cylinder inside would collect the water and throw it onto the inner surface of the fixed cylinder. This creates a thin film of water and creates the reaction environment. The rotation is left to be done until all the fluid in the pool is reacted.

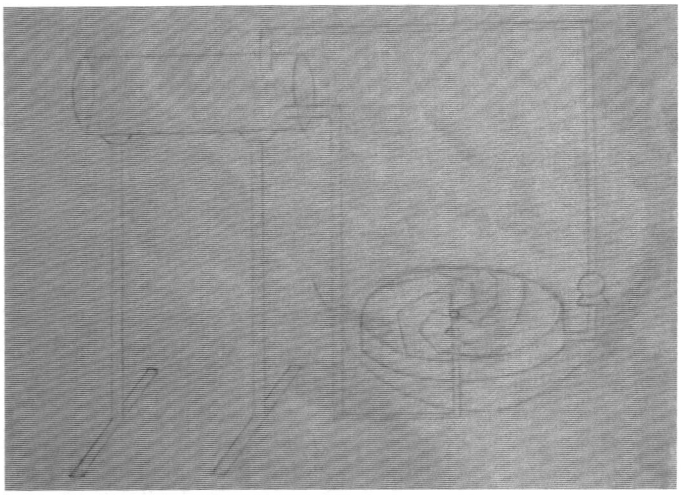

Figure 4: Proposed Design 3

The wastewater and catalyst mixture is stored in an overhead tank. The stream due to the head will flow into the disk vane through the centre. A rotor rotates this vane that will project the incoming fluid radially. This divided the stream into thin ones. These divided streams will hit the obstructions in the reactor thus, increasing the turbulence and the residence time. The material of construction would be of a transparent material and a parabolic mirror would be placed under the vane to project light. The reacted stream then falls into the casing and recycled back into the process until the feed is fully treated.

References

[1]http://www.renewablegreenenergypower.com/solar-energy-pros-and-cons/, 12.08.2014, 22:03

[2]http://www.nasa.gov/images/content/57911 main_Earth_Energy_Budget.jpg/, 10.08.2014, 22:23

398979 Digital Farming for Small Scale Agricultural Applications

Abstract:

This project aims to investigate the use of technology based solutions for farmers and gardeners aimed at improving agricultural production with sustanable practices.

The automation and use of sensors in agriculture is a relatively new field and is mainly used in large scale applications. Small farmers are not currently able to use much technology in their farming practices due to the costs associated with sensors, microcontrollers and other associated infrastructure.

Several established companies provide solutions for gardeners to automate watering of their crops. Communication over the internet and through SMS technology is also widely available and is utilized in both large scale greenhouse farming and some gardening applications.

A suite of technology solutions are needed for the typically small to medium farm. These include but are not limited to: water usage/storage, soil moisture, air humidity, temperature, pH and alkalinity.

The ultimate aim is to improve the management of agricultural resources in small landholdings and particularly rural areas. Due to climate change, there are projected decreases in rainfall and lengthening of drought periods in some tropical countries like Jamaica. Additionally, projected global population increases will require significantly higher agricultural output than is possible today. It is therefore imperative for small farmers to be more efficient in the management of their crops and resources such as water. Technology will be a great asset in this regard. Cost is a major inhibitor in the aim to increase the use of technology in farming however. This project therefore aims to demonstrate the possibility of utilizing technology to realize more sustanable farming and gardening.

Through the use of low cost microcontrollers, communication infrastructure and sensors as well as renewable energy, rural farmers will be able to monitor and control their assets in a cost effective and environmentally friendly way.

395317 Magnetophoresis of Iron Oxide Nanoparticles for Oil Recovery from Emulsions

Abstract:

Oil-in-water emulsions pose a serious pollution problem which entail a difficult and expensive clean up and separation. In the natural environment, born out of oil spills in a turbulent environment, the resulting oil-in-water emulsions have a tendency to persist, endangering wildlife and ecosystems. In industrial plants, oil-in-water emulsions from an oil pipeline leakage can clog pipework and downstream distillation columns. Traditional methods to remove oil spill from water include containment and skimming, *in situ* burning, and chemical and biological dispersants. However, these approaches target mainly the oil layer on the water surface instead of emulsified oil spheres, and are often labor-intensive and environmentally harmful. A novel and efficient method to recycle oil from oil-in-water emulsions utilizing oleylamine-capped iron oxide nanoparticles has been developed. Fe3O4 nanoparticles were synthesized by the reductive thermal decomposition of iron (III) acetylacetonate in benzyl ether andoleylamine, washed with ethanol and re-suspended in toluene for use and storage. Suspensions of Fe3O4 nanoparticles in toluene were shown to diffuse through the oil phase in oil-in-water emulsions of different oil contents, after which an external magnet was used to assess the efficiency of oil removal. Overall, the use of iron oxide nanoparticles is a promising method to remove oil from oil-in-water emulsions by making oil droplets magnetically active.

396123 Studies on Briquetting of Chromite Ore with Waste Plastic Binder

Abstract:

This work focuses on the development of new binders with the help of waste plastics for briquetting of ore fines and to measure its cold compressive strength (CCS). Chromite is an important mineral, during the mining operation lot of fines are produced which cannot be easily used by existing technology because most of the technology are cost intensive. Briquetting is considered a cheap and easy method but it needs suitable binder. Plastic is usually synthetic and non-biodegradable material. Now a days a large quantity of plastic is being generated as a waste, utilization of these wastes are very difficult to carry out. One of the best use of these plastic waste is to develop as a binders for ore fines briquetting. In the present study common synthetic waste material i.e., different types of plastic, viz., Polyethylene (P.E.), Poly Vinyl Chloride (PVC) and Low Density Polyethylene (LDPE) is used as binder with molasses and lime. Briquettes were prepared and cold compressive strength (CCS) of these briquettes was estimated. From the current study it was observed that the among the new binders waste plastic i.e., PE seems to be very promising for the briquetting of chromites ore.

396839 Configuration and Development of TSI Powersight Solid State Laser-Based LDV System

Abstract:

The study of dense gas dispersion at atmospheric conditions is critical to understand the safety concerns of industrial operations where large amounts of toxic or flammable materials are stored and generated. A large airborne release of these materials into the atmosphere could have catastrophic effects. The Chemical Hazards Research Center (CHRC) at the University of Arkansas performs important research on dense gas dispersion. The CHRC houses the world's largest low speed wind tunnel which is used to simulate dispersion at atmospheric conditions. Substantial amounts of data, including velocities, are recorded for modeling real life gas dispersion events. Before conducting experiments, an accurate characterization of air flow within the wind tunnel, especially at the lower boundary layer, must be established. The goal of this project was to install TSI's Powersight Solid State Laser-based LDV System and use the system to characterize the velocity profile of the wind tunnel. Hotwire anemometry velocity measurements were also taken and used to determine the accuracy of the LDV system's velocity data. The major phases of the project include the physical installation of the LDV system, alignment of the laser probes within the tunnel, initial LDV velocity measurements, hotwire anemometry measurements, and finally any adjustments needed to properly calibrate the LDV system. Once the LDV's measurements are established as accurate, data will be collected to characterize the wind tunnel.

Algal Remediation of Landfill Leachate

S. G. Myers*, S. G. White, D. E. Hackleman. School of Chemical, Biological, and Environmental Engineering, Oregon State University. shanna.g.myers@gmail.com

This poster will explore the potential of leachate treatment using algae to remove nutrients and metals from pretreated landfill leachate. Microalgae have been shown to inexpensively and efficiently treat waste streams containing high loads of organics, as well as to adsorb metals. Landfill leachate is considered a high strength waste, typically containing high levels of ammonium and trace amounts of metals. With pretreatment to remove herbicides and toxic organics, microalgal treatment of leachate has the potential to greatly reduce leachate treatment costs. *Scenedesmus dimorphus* that had been cultured in a modified BG-11 growth media was grown in various dilutions of pretreated and untreated leachate. Pretreatment consisted of distillation to reduce ammonia concentration and filtration through dead brown algae. Nitrogen concentration and algae density were monitored using a Hach assay kit and hemocytometer cell counts, respectively. Algal populations were found to increase in low dilutions of untreated leachate with media. Some nitrogen removal was witnessed in all dilutions of pretreated and untreated leachate, but assay results were not precise. Even in pretreated leachate, dilutions containing high percentages of leachate showed adverse effects on algae. With adequate pretreatment and dilution, microalgae could thrive in landfill leachate and serve a double purpose of treating high strength waste streams as well as producing marketable biomass for non-food uses. This study was supported in part by a generous donation from the Camille and Henry Dreyfus Foundation.

Pharmaceutical Disposal: Assessing the Impacts on Your Community

An active pharmaceutical ingredient (API) is the portion of a drug product that is used to treat and prevent disease. While a fraction of such ingredients are consumed and transformed by the patient's body, significant amounts of APIs either remain untaken or pass through the body unchanged. As a result, both the APIs that pass through the body unchanged and those that are untaken are disposed of through flushing, treated by the wastewater system, and subsequently directed to rivers, resulting in contamination of natural waterways. Concerns have become more prevalent since the discovery of mutated bass in the Susquehanna River Watershed. In addition to contaminating waterways, inappropriate use of narcotics and other untaken prescriptions leads to illegal behavior and health risks, such as their theft and abuse by explorative teenagers and drug addicts. These crimes lead to increased costs for law enforcement, courts, and emergency medical services.

This project concentrates on mathematically modeling the flows of pharmaceutical waste in a mass balance, which was constructed using information obtained from literature as well as information gathered from local law enforcement, healthcare professionals, and wastewater specialists. By utilizing the flow information, we are able to use the model to compute potential costs based on the impact to the human body, rivers, landfills, agricultural land, and the atmosphere. Using these cost assessments, our clients in the community can look to drive change through legislation and education. These changes could lead to better disposal methods, improved surface water quality, and reduced drug-related crime.

Two case studies are used to demonstrate the model: one on addictive opioid drugs and one on drugs that impact the environment. The opioid case study shows that by increasing the take-back program budget or taking preventative measures to decrease physician overprescribing habits, the local region will save millions of dollars. For example, raising the amount spent on take-back programs by $40,000 results in roughly $2,000,000 in savings, half of which stems from regaining lost productivity. It also shows that take-back programs are currently under 1% of the total costs incurred by excess opioids, which emphasizes how underutilized they are. The environmental case study shows that the natural waterways are impacted by pharmaceuticals, but most contributions are through unprocessed drugs excreted by humans and through agricultural runoff, not from unused medications. After considering how waste water treatment upgrades would improve the quality of river water, further investigation in regards to the value of the river has begun

Growth of Anaerobic Cultures under Diluted Conditions to Monitor Changes in Activity and Microbial Community Structure

Jenny Green, Dr. Mohammad Azizian, Dr. Lewis Semprini
Oregon State University; School of Chemical, Biological, and Environmental Engineering
Corvallis, OR 97331

ABSTRACT

Chlorinated aliphatic hydrocarbons (CAHs) such as trichloroethene (TCE) were commonly used organic solvents, and their improper storage and disposal resulted in widespread groundwater contamination. Anaerobic bioremediation is a promising solution to this problem and is investigated at Oregon State University. Sequential dehalogenation of TCE into *cis*-dichloroethene (cDCE) and vinyl chloride (VC), and finally ethene is receiving special attention because ethene is a non-toxic end product of this process. *Dehalococcoides Mccartyi* is recognized for its ability to obtain energy for growth from this process, which is called halorespiration, and it is the only known organism with this ability. The Evanite (EV) culture that contains *D. Mccartyi* is grown in 2-Liter (EV-2L) and 5-Liter (EV-5L) reactors (chemostats) in the OSU Bioremediation laboratory, and is the focus of the following study. The objectives of this study were to 1) compare how two different cultures transform TCE in batch reactors containing diluted culture samples from the chemostat, 2) observe overall effect of dilution on microbial activity, 3) observe evidence of microbial growth, and 4) compare how different electron donors affect kinetics and growth. The end goal of this study is to construct a small-scale version of the reactors already under operation.

Experiments were performed with culture samples obtained under anaerobic conditions from the chemostats, and the samples were then diluted in ratios of 1:5, 1:10, and 1:100 with standard growth media. Over the course of months, gas samples were taken regularly from each bottle to monitor the mass of CAHs, ethene, and H_2. These data were utilized to determine when hydrogen (H_2) should be added and provide evidence that growth was occurring. The data was also used to estimate transformation rates using a rate model programmed in Excel.

Objective 1) was satisfied with CAH, ethene, and H_2 mass monitoring. The biggest risk when diluting these cultures was the possibility of "shocking" the bacteria with a new environment, which is made apparent in slow or zero dehalogenation. This did not occur, as both cultures survived even the highest dilution. Dilution did lower dehalogenation rates—theoretically, relative to the dilution ratio as a result of lower biomass. Objective 3) was also satisfied with mass monitoring. It was anticipated that the step from VC to ethene would be the slowest, but would provide the greatest opportunity to demonstrate growth. More TCE was added over the course of the experiments with the objective of observing changes in transformation rates. In many cases, ethene production rates increased after the second addition of TCE, indicating microbial growth.

Objective 4) was satisfied by constant H_2 mass monitoring, as the presence of H_2 was required as an electron donor for growth. H_2 was added and was viable until pressure built up and disrupted atmospheric conditions. Formate as a direct H_2 source and propionate as an H_2 source via fermentation were used in place of H_2 gas. While both sources worked, the extremely slow release of H_2 by propionate limited dehalogenation compared to formate.

Objective 2) is ongoing because microbial population is being monitored at Stanford University using new DNA-based molecular methods on samples taken from the large-scale dilutions. They will examine changes in population size and community structure over the course of the experiments.

Ionicity of Silylamine Reversible Ionic Liquids

Researcher: Juan D. Jimenez, Sungyup Jung, and Elizabeth J. Biddinger
Research Advisor: Elizabeth J. Biddinger
Department of Chemical Engineering, The City College of New York,
City University of New York, 140th ST. and Convent Ave., New York, NY 10031

Abstract

Ionic Liquids (ILs) are molten salts with a melting point below 100°C. Ionic liquids are best characterized by their high conductivity, high viscosity and low vapor pressure. Reversible ionic liquids (RevILs) are unique in the fact they can be converted from a molecular state to an ionic state, which display the defining characteristics of a traditional ionic liquid, and can be reverted back into a molecular by adding heat. The novelty of RevILs is the fact the molecular and ionic states have different polarities, allowing for facile separations after the desired electrochemical process has been carried out in the reversible ionic state.

The RevIL that will be discussed in detail is (3-aminopropyl)triethoxysilane which upon addition of CO_2 forms an ammonium-carbamate ion pair. The properties that will be discussed will be conductivity, molar conductivity, and viscosity. These properties were chosen since they provide the sufficient information to determine the ionicity of an electrolyte, where ionicity is defined as the extent of ion dissociation of an ion pair with respect to 0.010M KCl. The key components required to determine ionicity are viscosity, which influences ion mobility, and conductivity, which is a measure of a substance's ability to carry a charge or ion flow. In order to normalize the conductivity with respect to concentration, molar conductivity will be used to determine ionicity, which is conductivity of the solution per mol of ions in solution. The Walden Plot then plots molar conductivity vs. (1/viscosity) on a log scale which compares the ionicity of the solution with 0.010M KCl, which is regarded as an ideal ionic species. However, when the RevIL is in its pure ionic state the conductivity is completely inhibited by the viscosity and ion pairing. In an attempt to elucidate the ionicity of the RevIL several different mixtures with varying solvents based on their dielectric constants will be investigated and their effects on both conductivity and viscosity.

The solvents that will be investigated will be methanol, acetonitrile, DMSO, and THF, which cover a wide range of dielectric constants as well as differences in protic vs. aprotic solvents. All of the solvent mixtures displayed a maximum in conductivity between 0-10mol% RevIL, which is consistent with what is found in the literature for tradition ionic liquids in organic solvent. It was found that after the addition of methanol to the RevIL (~1-10mol% RevIL mixture range) the conductivity of the solution increased by about 3 orders of magnitude, bringing the relatively non-conductive pure RevIL state into an acceptable region under which electrochemical processes can be carried out. In the case of the other 3 solvents, each exhibited a similar trend in conductivity as a function of mol% RevIL, except 1 one order of magnitude lower conductivity, and similar trends in viscosity as a function of mol% RevIL.

398565 Biodegradable and pH-Responsive Nanoparticles Designed for Site-Specific Delivery in Agriculture

Abstract:

While pH-responsive materials have been extensively studied in the realm of biological and medicinal fields, less attention has been given to the use of pH-responsive materials within agricultural science. Responsive nanoparticles have significant potential as vehicles for the delivery of pesticides and nutrients to plants, due to the higher pH of the delivery site (phloem) compared to its surrounding environment. While the most well-known and studied pH-responsive polymers (PAA, PMAA, PDMAEMA, etc.) contain non-degradable polymethylene backbones, biodegradability is particularly important for delivery to plants. The construction of a nanoparticle suitable for delivery to plants thus becomes more complicated—the particle must be biodegradable, responsive to basic pH, and capable of encapsulating hydrophobic, and ideally, hydrophilic compounds as well.

Polysuccinimide (PSI) was chosen as a starting material due to the biodegradable and hydrophilic nature of its derivatives, namely poly(aspartic acid) (PASP) and poly(hydroxyethyl aspartimide) (PHEA). Nanoparticles of assorted size and functionalization were prepared via the nanoprecipitation technique, incorporating either hexylamine (HA) or 2-(2-aminoethoxy)ethanol (2AEE) nucleophiles. Nanoparticle size distributions were approximated via dynamic light scattering (DLS), and size trends were explored by modifying functionalization and organic phase dilution. It was seen that dilute organic phase solutions led to small average hydrodynamic radii following nanoprecipitation, while degree of functionalization held little influence. Because nanoparticle size depended on dilution rather than functionalization, we were able to tune the pH responsiveness of the nanoparticles while retaining the required diameter of 30 nm or less. To demonstrate their potential as controlled-release delivery vehicles, functionalized nanoparticles were loaded with a model hydrophobic molecule and subsequently exposed to buffer solutions of varied pH. It was found that above pH 7, PSI-nanoparticles hydrolyzed and released encapsulated materials. The release rate significantly increased with increasing pH, and decreased with increasing degree of functionalization. Finally, plant toxicity studies suggested that the polymer materials exhibit little to no toxic effects at biologically relevant concentrations.

398720 Thermodynamic Study of Anthracene + Phenanthrene Solid Mixtures

Abstract:

The combustion of carbon-based fossil fuels currently provides the United States with nearly 70% of its useable energy. Production and use of these fuels often involves concerns regarding polycyclic aromatic hydrocarbons (PAHs). PAHs usually occur in mixtures both naturally (e.g., in petroleum) and as byproducts of combustion or gasification (e.g., in coal tar) that are sometimes released to the environment. The US EPA considers many PAHs to be hazardous to human health due to their potential to cause cancer. Studying behavior that might affect the environmental fate of PAHs is important in understanding the risks associated with exposure to these contaminants. This study focused on the mixture properties of model PAH mixtures. Differential scanning calorimetry and x-ray diffraction were used to assess the fusion and solidification behavior of mixtures of anthracene and phenanthrene at varying mole fractions. In addition, the vapor pressure of these mixtures was studied via the Knudsen effusion technique and compared with the observed melting and solidification behavior to infer mixture phase behavior. Mixtures with greater than 20% anthracene showed distinct two-phase behavior, but neither phase was pure anthracene or pure phenanthrene. Mixtures with less than 20% anthracene showed behavior somewhat like a single mixture phase with constant enthalpy of fusion and melting temperature. The temperature at which solid anthracene and phenanthrene mixtures first began to melt (the thaw temperature) remained constant and very close to the melting temperature of pure phenanthrene for all mixture compositions. These measurements suggest the presence of a constant fusion enthalpy, phenanthrene-rich phase that behaves similarly to pure phenanthrene and can incorporate up to 20% anthracene into its crystal structure. This information was used to compute the enthalpy of fusion for what is thought to be an anthracene-rich phase that co-exists in the presence of the phenanthrene-rich phase for all mixture compositions with greater than 20% anthracene. It was this phase that determined the final melting temperature (the liquidus temperature) of the mixture. The enthalpic contribution of this anthracene-rich phase was always significant and greater than that of the phenanthrene rich phase, even when only a small mole fraction of that phase was present. The vapor pressure and x-ray diffraction analyses further confirmed that neither pure anthracene nor pure phenanthrene phases exist in these binary mixtures. However, the phenanthrene-rich phase showed vapor pressure behavior similar to that of pure phenanthrene, while that of the anthracene-rich compared with pure anthracene. This information is useful in developing a more complete understanding of the thermodynamics of PAH systems, which can lead to improved predictions of how mixtures of PAHs interact with their surrounding environment.

398387 Asphalt Recovery from Used Roofing Shingles

Abstract:

Over eleven million tons of asphalt roofing shingles are discarded in the United States annually, and there is interest in separating asphalt from the other components of the shingles. Separations based upon dissolving asphalt in solvents and upon mechanical agitation in non-solvents at elevated temperatures were investigated. Agitation in boiling water and boiling propylene glycol was not effective in separating asphalt from other components, and produced suspensions of particles that were difficult to separate from the liquid. Asphalt solubility in five common organic solvents was tested, and toluene was selected by qualitative observation of the solutions and consideration of potential process hazards of tetrahydrofuran. Experiments were performed to quantify mass transfer in the solution process. Samples were taken as 1-6 g samples of shingles were stirred with 75-150 mL of toluene for 10-40 minutes. A colorimeter was used to determine the asphalt concentration. Stir rate, shingle particle size and type were also varied during testing. Analysis is ongoing, and the results will be applied in a preliminary process design for a large-scale process.

398457 Investigating the Effects of Biodiesel Composition on Emissions from a Compression Ignition Engine

Abstract:

For biodiesels to be considered as replacements for conventional fossil fuels, they must be shown to produce cleaner emission profiles than existing fuels. When evaluating the environmental footprint of alternative fuels, the production of harmful pollutants, such as NO_x, vapor-phase hydrocarbons, and particulate matter (PM) are considered. However, a biodiesel's exhaust composition depends strongly on its fatty acid methyl ester (FAME) content. Depending on the feed stocks used, FAME content can vary significantly among biodiesels, resulting in different physical properties that affect combustion and pollutant production.

To examine the effects of FAME content on exhaust composition, six fuel blends were prepared by mixing coconut and canola biodiesels in various proportions. The blended fuels were then combusted using a single cylinder, Yanmar compression ignition engine, while emission data was recorded. The chemical make-up of coconut and canola biodiesels are quite different; while coconut biodiesel contains mostly short, saturated FAMEs, canola biodiesel contains longer, unsaturated FAMEs, allowing the effects of FAME content and associated physical properties on fuel performance and emissions to be observed. Ultra-low-sulfur diesel (ULSD) was tested as a control to normalize the injection timing of each fuel blend to that of ULSD, and to compare performance and emission results with those of a conventional fuel.

The emission results were very nonlinear. At both 50% and 100% engine loads, NO_x levels peaked at around 80% - 90% canola biodiesel. Vapor-phase hydrocarbon emissions were lowest using a blend of 50:50 coconut to canola biodiesel at 50% load, while no change between blends was observed at 100% load. Particulate matter levels were unusually low across all fuel blends for both loads. The non-linear behavior of emission production with respect to biodiesel blending suggests the possibility of using coconut biodiesel as an additive to improve the emission profiles of better-performing biodiesels.

The municipal landfill produces 10^6 m^3 of methane per year and 10^8 m^3 liquid leachate per year. The remediation of these by-products from the decomposition of municipal solid waste is the focus of research because of the significant environmental impact, ground water contamination, and contributions toward climate change. This study investigated the effect of a two-stage distillation process to remove high levels of ammonia and volatile organic compounds from the leachate. It was determined that 85% of the ammonia and 80% of the volatile organic compounds were removed. Additionally, the double-distillation process reduced the total soluble solids in the leachate by 65%.

398653 Measuring Surface Tension of Atmospheric Secondary Organic Aerosols Using Atomic Force Microscopy

Abstract:

Atmospheric aerosols are particles on the nanoscale that are suspended in air. Aerosols play a major role in cloud development and thus are relevant for climate modeling. Specifically, certain atmospheric aerosols are capable of forming cloud condensation nuclei, or cloud droplets. Models that predict particles' capacities to become cloud condensation nuclei require knowledge of the surface tension of the particles. However, obtaining direct measurements of particle surface tension using traditional methods is challenging, because there is insufficient volume of liquid in aerosol particles to use in typical instruments that measure surface tension. Recently, a method of measuring the surface tension of droplets using atomic force microscopy has been developed. In this study, that method was utilized to directly measure the surface tension of liquid aerosols generated in the lab. The liquid aerosols were generated by exposing vapor-phase volatile organic compounds to ozone in an isolated and particle-free environment. The aerosols were collected via impaction, where the resulting film was probed with a specifically-designed tip in the atomic force microscope (AFM) to calculate its surface tension. As a result of this work, the surface tension of oxidized alpha-pinene aerosols was determined.

Abstract

Chicorral is a small, indigenous community in rural highlands of Guatemala. As with many rural communities in the country, Chicorral suffers from a severe lack of infrastructure: over half of the approximately 200 residents do not have access to an improved water source. These residents must climb down a steep, 100-meter ravine to reach the community's main reliable water source. Women and children bear the responsibility of taking this trip up to 5 times a day, carrying up to 10 liters in each trip. The community requested support in having water pumped up to the communal center. In 2006, EWB-USA JHU made the first assessment trip and proposed to design and build a photovoltaic-powered system that would pump water from the ravine to a communal tap. Continuing social assessments revealed, however, that expanding the system to include household water distribution was important for three reasons: (1) household distribution would be more equitable, and residents would thus be more willing to pay a water usage tariff; (2) by having household distribution for at least 85% of its residents the community could be recognized as an independent village (aldea) rather than a semi-independent community (paraje), which is an important goal for residents; and (3) a design that includes household distribution could be integrated into regional water plans and allow the project to receive funding from the Municipality of San Bartolo, to which it belongs. Therefore, in 2012, the team agreed to extend its original design. The project now entails pumping water to a distribution tank 170m above the ravine, from which water will be distributed by gravity into each of the 22 participating households.

Our team has designed the water distribution system to meet World Health Organization guidelines[1] for improved drinking water sources. This includes protecting the water source from outside contamination and providing a minimum of 50 liters of water/person/day for household consumption and sanitation use. With this design basis, our team expects that the households participating in this project will benefit in regard to the following two metrics:

1. Per capita water consumption (as needed for hygiene as well as drinking);
2. Time spent collecting water.

To further community involvement in the project, EWB-USA JHU helped residents create a legal body, the Water Committee of Chicorral, to manage and provide oversight of the overall project sustainability. The committee is comprised of one member from each of the 22 involved households and has an Executive Board, or Junta Directiva. It has acquired the rights to the water source and all other property where system infrastructure will be placed, and has overall ownership of the project. The Water Committee will oversee the project's implementation and be responsible for long term maintenance. Recognizing the importance of community involvement in the project, we have conducted informative meetings with the Water Committee on each of our trips.

Furthermore, the team and community have developed working relationships with the Municipality of San Bartolo, which has agreed to lend its financial and political power to the project, and Agua del Pueblo, a Guatemalan NGO with over 40 years of experience in rural water projects in the country. In 2013, the team and these three partners signed a formal Memorandum of Understanding (MOU) outlining each party's responsibilities throughout the system's life.

During implementation, the Water Committee will be principally responsible for providing lodging and the required unskilled labor for construction. After implementation, the EWB-USA JHU team and Agua del Pueblo

Commented [CU1]: What was the purpose of these meetings? To brief the Water Committee on project progress? What kind of input did you seek from the Water Committee at these meetings? Concurrence/approval on the direction of the project? Something else?

[1]World Health Organization. *Guidelines for Drinking-quality Water.* Rep. 4th ed. Malta: Gutenberg, 2011.

will train the residents to use and maintain the system. Agua del Pueblo specializes in community organization and training; with their assistance, the team will develop an education program focusing on water quality and use of the system. The maintenance program will focus on procedures for routine cleaning and troubleshooting problems and will train a community maintenance team in charge of conducting basic maintenance on all parts of the system. This will include work such as routine cleaning of parts and repairing small leaks. For serious emergencies the community maintenance team will be instructed to call Agua del Pueblo for advice or, if necessary, to send out the proper response team from Agua del Pueblo's headquarters, located approximately 2 hours away from Chicorral.

In addition to supporting the community maintenance team, the Water Committee will also be responsible for collecting a water tariff which will be used for system maintenance. Over the long term, the Municipality of San Bartolo has agreed to support the system by paying for the replacement of pumps if they fail, saving the community significant maintenance costs. The water tariff will cover the rest of O&M costs, including trips by our in-country partners at Agua del Pueblo and Alternativas Energeticas Renovables for technical checkups or to resolve non-routine issues.

The schedule for implementation of the water distribution system has been divided into 3 phases. Phase 1 of the system consists of the construction of the spring box and the pump tank, the installation of one pump, the construction of the storage tank, the layout of the piping from the pump tank to the storage tank, and the installation of the photovoltaic array and all its electrical components necessary to support the first pump. Phase 2 consists of the distribution network extending from the storage tank; however, this part of the system will be constructed and financed by a third party (the Municipality of San Bartolo). Phase 3 consists of the completion of the system by installing the second pump and associated photovoltaic and electrical components to provide enough capacity to last the community for 20 years. This number is based on a usage of 50L of water per day per person, with a population growth of 3%.

In April 2014, the team and its in-country partners began construction on the system and have now completed the spring box water capture tank and pump tank. Construction is now shifting to laying the piping and construction of the distribution tank, which is expected to be completed by November 2014.

In addition to the three partners mentioned above (Water Committee of Chicorral, Municipality of San Bartolo, and AdP), EWB-USA JHU receives professional and financial help from partners such as Dr. Debora Ley, Alternativas Energeticas Renovables (AER), and the Baltimore Rotary Club. Dr. Ley first introduced EWB-USA JHU to the project. She is an expert in rural renewable energy systems, and has assisted the team throughout the entirety of the project. AER is a Guatemalan firm specializing in rural photovoltaic projects and will be supplying all electrical components for the system. The company's owner has also provided technical assistance to the system's electrical design and has received training on US codes and standards for solar systems. The Baltimore Rotary Club has assisted the project financially since 2013 and is now working with the Towson and Quetzaltenango Rotary Clubs to apply for additional support through the Rotary International Global Grant Program.

To date, the team has raised approximately $51,000. As previously described, funding has largely come from the individual donors, the Davis Project for Peace, the Johns Hopkins University Alumni Association, and the ERM Group Foundation. An additional $13,000 in emergency funds held by the EWB-USA JHU chapter is available to the project, if deemed absolutely necessary, bringing the total available funds up to as much as approximately $64,000. This past year, the team has spent $50,000 of the $51,000 to pay for one trip to the community and all construction costs related to the concrete structures and piping for Phase 1. The team still needs to raise money for the key electrical components so that the system can be functional for the community.

Abstract

The largely unregulated administration of antibiotic-supplemented food to animals raised in concentrated animal feeding operations (CAFOs) results in consistent sub-therapeutic exposure, yielding elevated levels of microbiome derived antibiotic resistant bacteria (ARB). ARB are then potentially widely distributed via the sale of manure as fertilizer, creating a pathway of exposure that can endanger the health and wellbeing of consumers. The purpose of this research is to analyze ARB derived from a variety of fertilizers whose manure was sourced from conventionally farmed chicken and steer. Also analyzed were a compost sample and a sample derived from Mt. Baldy. Isolates were cultured from fertilizers and replica plated onto three nutrient broth agar types: plain, penicillin supplemented and erythromycin supplemented. Antibiotic concentrations were varied from 2ug/ml to 200 ug/ml. Resistance, defined as growth comparable to a control, was observed to be consistently high among varying concentrations of penicillin, for all sample types. Susceptibility, defined as growth that was visually less prominent when compared to a control, was found to be almost nonexistent in the compost sample studied but played an important role in other samples. Resistance and susceptibility levels were observed to fluctuate among the steer samples, with chicken falling between steer readings. The compost sample was found to exhibit highest resistance, while steer and chicken manure derived fertilizers were found to be comparable. Similar but more drastic patterns were observed for Erythromycin, implying that there is a clear difference in ARB levels among fertilizer source types and compost. This work identifies fertilizer as a source of ARB and furthers our understanding of exposure routes to consumers and landscape workers.

The rapid expansion of horizontal drilling and hydraulic fracturing for natural gas mining has created concerns about the potential for environmental impacts. Flowback wastewater (FBW) from these mining sites contains a specific concern with radioactive content from naturally occurring radium isotopes present in the natural gas mines. This inadvertent extraction of radium from gas mines results in the need to create an appropriate disposal method. Current methods for disposal of this wastewater consists of dumping into abandoned coal mines which results in leakage into the environment. Determining the affinity of radium to either a soil based or aqueous media is important to understanding its fate in the environment. In cases of unintentional spills, being able to identify radium's location in the environment allows for effective methods of clean up to be produced.

Polysorbate-80/Lecithin Mixtures

As possible alternative Oil Spill Dispersants

John Neilsen[1], Joseph Rocchio[1], and Geoffrey Bothun[1]

[1]Department of Chemical Engineering

University of Rhode Island, Kingston, RI 02881

The Deepwater Horizon Oil Spill of 2010 led to the release of nearly 5 million barrels of crude oil into the Gulf of Mexico. To treat the spill, approximately 2 million gallons of a chemical dispersant (or surfactant) package were used to disperse the oil into the ocean. Dispersant application is the most widely used method for treating oil spills because it yields oil droplets dispersed in water that can be biodegraded by naturally-occurring marine bacteria. However, there is growing evidence that conventional dispersant packages such as Corexit 9500A, which was used in the Deepwater spill, may be toxic to marine life and pose safety risks to coastal communities. The concerns relating to conventional dispersant packages highlight the need to develop environmentally friendly and effective dispersants to treat future oil spills. This study characterizes the dispersion efficiency of mixtures of polysorbate-80 and lecithin dissolved in ethanol or dispersed as aqueous micellar solutions. These two surfactants were chosen due to their applications in the food and pharmaceutical industry, their proven capabilities in forming stable emulsions, and the fact that both surfactants are environmentally friendly and have no known hazardous effects. Dispersion efficiency has been examined using baffled flask tests with synthetic seawater and Pennsylvania crude oil (used as a model crude oil) as a function of the dispersant or surfactant to oil ratios and the dispersant concentration. Our results show that an optimized polysorbate-80 + lecithin dispersant solution is capable of matching the initial dispersion effectiveness of Corexit 95000A. The findings obtained in this study will help to better understand the mechanisms of polysorbate-80/lecithin emulsification, and to determine the optimal concentration of the environmentally friendly dispersant system for crude oil.

394470 CO2 Valorization Via Solar Thermochemical CexZryHfzO2-Cycle

Abstract:

The concentration of CO_2 in environment has reached upto 392 ppm by volume as of 2011, which was 280 ppm in pre-industrial times and if this trend continues, by 2050 CO_2 concentration is expected to raise upto 550 ppm. Because of the adverse effects of CO_2 discharge on the global environment, the development of strategies for the utilization of captured CO_2 has become increasingly important. As a promising option, the liberated CO_2 can be re-energized into CO via metal oxide based solar thermochemical cycles. The CO produced via solar thermochemical CO_2-splitting can be combined with H_2 derived from solar thermochemical water-splitting process to produce solar syngas which can be further processed to liquid fuels such as Methanol, Diesel, and Kerosene via the Fischer-Tropsch process.

In this investigation, solar thermochemical $Ce_xZr_yHf_zO_{2-\delta}$ cycle was examined towards conversion of CO_2 into CO. At first, various combinations of $Ce_xZr_yHf_zO_{2-\delta}$ redox materials were synthesized via co-precipitation method and subsequent calcination at different temperatures in air. The calcined powders were characterized by powder X-ray diffraction, BET surface area analysis, scanning (SEM) and transmission electron microscopy (TEM), etc. Synthesized $Ce_xZr_yHf_zO_{2-\delta}$ redox materials were further examined towards thermochemical CO_2-splitting reaction with the help of a high-temperature thermogravimetric analyzer (TGA). Multiple thermal reduction and oxidation cycles were performed at various operating conditions by using TGA. The O_2 and CO produced during multiple cycles was quantified by gas chromatography. The findings indicate that $Ce_xZr_yHf_zO_{2-\delta}$ is capable of producing constant and high yield of CO in comparison with the other ceria based redox materials in multiple thermochemical cycles. The results related to synthesis, physiochemical characterization, and thermogravimetric experiments will be presented in detail.

AIChE 2014 Annual Student Conference Abstract
November 17th, 2014

"Evaluating nanocomposite microparticles for the delivery of therapeutics for pulmonary arterial hypertension"

Julie L. Cuddigan[1], Samantha A. Meenach[1,2]

[1]Department of Chemical Engineering, [2]Department of Biomedical & Pharmaceutical Sciences, University of Rhode Island, Kingston RI 02881

Pulmonary arterial hypertension (PAH) is a chronic, incurable, and life-threatening disease that is characterized by abnormally high blood pressure in the arteries leading from the heart to the lungs. Patients often suffer from chest pain, shortness of breath, fainting, dizziness, and in many cases heart failure and ultimately death. The therapies currently approved to treat PAH are most commonly available in oral or intravenous form, but often result in undesirable systemic side effects. An effective treatment needs to be designed to deliver targeted therapeutics to the lung periphery to ensure maximum localized bioavailability and to decrease the side effects associated with systemic delivery. The goal of our research is to develop and evaluate dry powder aerosols comprised of nanocomposite microparticles (nCmP) containing the newly-investigated PAH drug tacrolimus to be delivered directly to the lung and surrounding arteries. These particles are designed to overcome mucociliary clearance in the lung and to avoid alveolar macrophages to allow for longer retention time in the lungs. The production of these aerosols began with synthesizing the biodegradable polymer, acetalated dextran (Ac-Dex), with an appropriate cyclic acetal conversion (65-75%). Then using a single emulsion evaporation method, Ac-Dex was used to form nanoparticles, which were characterized for their size, charge, and drug encapsulation. The final step of synthesizing the nCmP was spray drying a solution of nanoparticles suspended with mannitol (an excipient) in water. The nanoparticles were successfully optimized to be of appropriate size to avoid macrophage uptake (150 - 230 nm). The drug-loaded particles had an encapsulation efficiency of 32.7%, resulting in a drug loading of 32.75 μg tacrolimus/mg acetalated dextran. The nanoparticles were loaded with the fluorescent dye FITC to verify their successful encapsulation into the microparticles. We were able to demonstrate the successful synthesis and the morphology of the dry nCmP using confocal fluorescence and scanning electron microscopy. The development of the nCmP could lead to the improved treatment of PAH for many patients afflicted by this life-threatening disease.

Citrate-capped Gold Nanoparticles: Synthesis, Characterization, and Effect of BSA Addition on Macrophage Uptake

Thomas Hercula
Department of Chemical Engineering and Materials Science
Wayne State University
Detroit, Michigan

Gold nanoparticles (GNPs) are quickly becoming a major point of interest in the treatment of various medical conditions. The GNPs have the ability to deliver drugs targeted to specific areas of the body in order to improve efficacy to toxicity ratio. Their small size and generally low cytoxicity mean GNPs have the potential to be a great asset in the medical field. My presentation will focus on the synthesis of GNPs using the citrate reduction method, a method known for its simplicity and efficiency. Characterization of citrate-capped GNPs will also be discussed, including numerous ways to determine the size of these GNPs. Lastly, my presentation will describe the effect of bovine serum albumin (BSA) on the GNPs agglomeration control and their uptake by macrophages.

GNPs are currently being analyzed, among other ways, as a potential drug carrier that can target cancerous cells in certain areas of the body. There are several benefits of GNP use as a drug carrier. By being able to target a specific area (unlike other treatments, such as chemotherapy), the efficacy to toxicity ratio greatly increases. Also, GNPs can be easily manipulated by size and shape in order to maximize the surface area to which drugs can be attached. GNPs, due to their small size, can maneuver easily within the body. The release of the drugs attached to GNPs can be controlled by infrared light which prevents improper release. Lastly, GNPs are chemically inert, and the side effects that may result from their use are greatly reduced compared to other treatments currently available. There are numerous ways to synthesize GNPs for use in drug delivery; my research focuses on the citrate reduction method resulting in citrate-capped GNPs.

One of the most common ways to create GNPs is the citrate reduction method, which combines chloroauric acid, a gold salt, with sodium citrate, the reducing agent. This method is preferred because of its versatility; by changing the ratio of gold salt to reducing agent, it is possible to create GNPs ranging in size from 8.5 nm to 75 nm and larger. The citrate reduction method is also a very quick and simple process. GNPs can be created in as little as fifteen minutes using this method, adding simplicity to its versatility. Once created, these citrate-capped GNPs can be characterized using several different methods, including (but not limited to): UV-visible spectroscopy, transmission electron microscopy (TEM), zetasizer, and x-ray diffraction (XRD). Citrate-capped GNPs can also be qualitatively characterized by a visible color change in the GNP solution. My presentation will show the results of several of these characterizations and the comparisons between them.

Although GNPs, and specifically citrate-capped GNPs, have shown a lot of promise in the medical field, there are still challenges facing their widespread use. One of the biggest drawbacks that has been discovered involves the tendency of citrate-capped GNPs to agglomerate when introduced to cell media. The issue may be related to the large Hamaker constant of GNPs, which indicates a high interaction strength between them. DMEM, the cell medium used in my research, has a large amount of salt and protein and therefore has a high ionic strength. In solutions of high ionic strength, the Van der Waal forces between GNPs outweigh the steric repulsion of the attached ligands (e.g: citrate), and the GNPs agglomerate in these solutions. This is obviously a major issue because the agglomerated particles not only lose their desired mobility, but their effective surface area as well. It is imperative to find a solution to this agglomeration problem; my research contends that bovine serum albumin (BSA) may be the answer.

BSA was initially used as a potential solution to GNP agglomeration because of the increased steric repulsion it provides. Since citrate alone is not enough to prevent agglomeration, BSA is added to a GNP solution; it is believed that, when combined with the citrate ligand, BSA provides enough repulsion to prevent (or seriously hinder) GNPs from agglomerating. BSA is known to bind easily to GNPs; although the mechanism has yet to be determined, studies have shown that BSA binds easily to negatively charged surfaces, which the citrate ligands provide. This information provides researchers with another positive of using BSA. Also, BSA is safe for use on humans, providing little to no side effects. Previous research has shown very promising results; however, there is only a limited number of reports documenting its success. My research furthers these studies by analyzing the effectiveness of BSA in preventing agglomeration as well as the effects of delayed BSA addition on agglomeration. My results will show that BSA addition to citrate-capped GNPs does indeed reduce the GNP agglomeration. It will also show that delaying the addition of BSA to GNP solutions dramatically increases the potential for GNPs to agglomerate. Results will be presented using imaging evidence of any changes as well as UV-vis spectroscopy and zetasizer results. Higher cell uptake of GNPs has been observed when BSA is used to prevent agglomeration.

396729 Targeted Drug Delivery with Peptoid- Based Nanospheres

Abstract:

Although medicinal technology has advanced greatly over the last several decades, many drugs have negative side effects. Undesirable side effects could be significantly reduced if non-systemic drug delivery systems were used because diseased cells would be harmed at a higher rate than healthy cells. One possible improved delivery system is peptoid nanospheres. These nanospheres can be linked to molecules engineered to target diseased cells. This research project is focused on designing peptoids to form nanospheres in solution. Four specific peptoids were initially synthesized because they were previously shown to form nanospheres of the appropriate size when dried on a solid substrate. After synthesis and purification, the peptoids were dissolved in a 4 to 1 solution of methanol and water for the remaining test. Circular dichroism was used to determine that all the peptoids were helical. Then scanning electron microscopy (SEM) showed that one of the four peptoids formed spheres when dried on a silicon chip. After (SEM), dynamic light scattering was used to determine that two of the four peptoids were possibly forming spheres. These two peptoids were frozen in solution and viewed using transmission electron microscopy (TEM). One of the two spheres showed sphere formation when viewed using TEM. Further analysis will be conducted to determine the size of the spheres formed. Once the testing is complete on these peptoids, more peptoids will be designed to form spheres of the appropriate size. When designing these peptoids, biocompatibility and cell specificity will be considered. The new peptoids will be synthesized and tested in the same manner as the original peptoids.

Role of Zinc Oxide Nanoparticles Against UVB-induced DNA Damage and Apoptosis in Human Keratinocytes

Lilia Rusu[1], Sumit Arora[2], Srinivas Palanki[1], Ajay P. Singh[2]

[1]Department of Chemical and Biomolecular Engineering, University of South Alabama

[2]Mitchell Cancer Institute, University of South Alabama

ABSTRACT

Zinc oxide (ZnO) nanoparticles (NPs) have been approved for use in sunscreens, mainly because of their intrinsic ability to filter ultraviolet (UV)A and UVB spectra and thereby enhance the protection capability of sunscreens. In this research, we conducted *in vitro* studies on a skin epidermal immortalized, non-tumorigenic keratinocytes (HaCaT) cell line to evaluate whether ZnO NPs also have chemopreventive activity against UVB radiation-induced DNA damage and apoptosis. Cell viability experiments suggested that the concentration of 1 mg/L ZnO NPs was well tolerated by the HaCaT cell line. Dot-blot assay analysis indicated that ZnO NPs did not provide any protection against UVB exposure-induced DNA damage, but instead promoted the formation of cyclobutane-pyrimidine dimers (CPDs) in HaCaT cells. Pretreatment of ZnO NPs and subsequent exposure to UVB radiation resulted in G2 phase arrest of cell cycle in HaCaT cells. These findings suggest that despite their optical (UV filtering) properties, ZnO NPs may actually cause direct DNA damage, and/or promote DNA damaging activity of UV radiation. These findings raise concerns about the prevalent use of ZnO NPs in sunscreens and underscore the need for the development of more effective and safer modalities for prevention of skin carcinogenesis.

397237 Synthesis and Characterization of Anticancer Nanoparticles

Abstract:

Our lab focuses on research related to anti-cancer nanomedicine. Our end goal is for every cancer patient to have access to personalized medicine, and an increased chance of survival.

We are developing and optimizing procedures for encapsulating anti-cancer drugs in polymer- or lipid-based nanostructures. By carefully selecting our materials and optimizing our procedures, we are endeavoring to create a single nanostructure which can be used for both diagnostic and theraputic or "theranostic" purposes. This can be achieved with careful attention to nanostructure size, zeta potential, and drug encapsulation efficiency.

Working on the nanoscale makes it extremely difficult to analyze our structures for the above characteristics. To characterize our particles effectively, we use two different techniques: dynamic light scattering (DLS) and spectrophotometry. The former is used to gather data on size and zeta potential; the latter aids in determining how much drug was successfully taken up by the particles. Transmission electron microscopy (TEM) allows us to image the particles, enabling more accurate sizing, and giving visual proof of important processes which take place on a scale too small to otherwise detect.

Colton Martinez
Chemical Engineering
Houston, Texas
Senior

Water in Oil Microemulsion for Nanocomposite Particles and Its Application in Oral Drug Delivery

Colton Martinez, Xin Fan, and Dr. Allan E. David
Department of Chemical Engineering
Auburn University, AL 36849

Nanocomposite particles, synthesized from silica and the polysaccharide alginate are to be used in enhancing oral drug delivery. The mucoadhesive properties of the polysaccharide can prolong the drug residence time in the gastrointestinal tract as well as provide a means of drug release. Silica is biocompatible and can improve the stability of the nanocomposites. Also silica can easily be modified for further specific targeting delivery. These composite nanoparticles are synthesized using the sol-gel reaction occurring in a water in oil microemulsion. The microemulsion utilizes the thermodynamic stability created between the water and oil phases and the surface-active agent, the surfactant. The micelles that are formed serve as platforms for the encapsulation of the polysaccharide as the silica precursor, tetraethyl-orthosilicate (TEOS), undergoes condensation reactions to produce a silica network. Trials with differing weight ratios of alginate have produced monodisperse particles in nanosize diameter when viewed on a transmission electron microscope (TEM). With the current alginate nanocomposite, drug uptake and release can begin to be modeled using a model drug. This model drug can be added to the microemulsion process and integrated into the nanocomposite utilizing the same sol-gel reaction. Future work involves determining how the reagents used in the microemulsion factor into the shape, size, and composition of the nanoparticle. Additionally, the current alginate nanocomposites are to be tested for their uptake of the model drug, with the hope that an actual drug can be loaded into the particle in the near future.

398631 Evaluation of the Cancer-Preventive Effect of Resveratrol-Loaded Nanoparticles on the Formation of Tumor Spheroids

Abstract:

Lung cancer is the most common type of cancer worldwide, accounting for 13% of all deaths annually. Survivors of this cancer nearly always live with the possibility of cancer recurrence. The objective of this research was to develop a drug-loaded nanoparticle that decreases the likelihood of lung cancer recurrence by preventing the reformation of lung tumors. Nanoparticle encapsulation has been shown to increase drug bioavailability, enhance tumor targeting for localized delivery, and allow for more sustained release of drugs compared to free drug administration. The nanoparticles used for this project were resveratrol-loaded acetalated dextran nanoparticles (RSV:Ac-Dex NPs). The NPs were formed using a single emulsion process, encapsulating the hydrophobic drug RSV inside the tunable degradable polymer Ac-Dex. These particles were characterized in terms of morphology, size and zeta potential, drug loading, in vitro drug release and degradation. Particles loaded with RSV showed successful encapsulation of the drug, with particle size ranging from 150 - 250 nm in diameter. Particles also demonstrated sustained release of drug at both pH 7.4 and pH 6.5 to represent release in normal physiological conditions and local tumor environment conditions, respectively. The efficacy of RSV in A549 lung adenocarcinoma cell line was evaluated in both two-dimensional (2D) and three-dimensional (3D) cell culture conditions. Dose-response curves from the drug applications was compared with the cellular response to application of the RSV:Ac-Dex NPs. The diameters of 3D lung cancer multicellular spheroids (MCS) were measured when RSV was applied either before or after MCS formation in order to compare the effects of RSV as a preventive or treatment drug. Cellular response in A549 cells exhibited significant cell death with concentrations of RSV ranging from 50 - 250 μM, and yielded significantly smaller or nonexistent MCS when RSV was applied before formation. The delivery of the actual NPs for in vitro cellular response tests will be performed for comparison to free drug application. Overall RSV:Ac-Dex NPs are expected to be a useful and efficient chemotherapy to prevent recurrence of lung tumors for cancer patients in remission.

Polyanhydrides are a class of biodegradable polymers that have been utilized as drug and vaccine carriers. We report a new automated high throughput method for synthesizing protein-encapsulated polyanhydride nanoparticles. This method is based on a multiplexed homogenizer that can handle parallel streams of monomer and/or polymer solutions for synthesis of polymers and nanoparticles. Copolymers based on sebacic acid (SA), 1,6-bis(*p*-carboxyphenoxy)hexane (CPH), and 1,8-bis(*p*-carboxyphenoxy)-3,6-dioxactane (CPTEG) polymer chemistries were synthesized using the high throughput method. With the same method, anti-solvent nanoprecipitation was used to synthesize the nanoparticles with a solid/oil/oil emulsion technique. A surfactant was incorporated during nanoparticle synthesis, and it's concentration was varied in order to study its effect on nanoparticle formation, protein release rate, and protein encapsulation efficiency. Nanoparticles synthesized from 20:80 CPH:SA copolymers were smaller and more monodisperse in the presence of increased surfactant concentrations. It was observed that the surfactant did not have a significant impact on protein release, but greatly enhanced protein encapsulation efficiency. The 20:80 CPTEG:CPH nanoparticles were significantly smaller than equivalent nanoparticles made using a batch process. Increased surfactant concentrations resulted in 20:80 CPTEG:CPH nanoparticles that were slightly larger and more aggregated. These results demonstrate the functionality of the high throughput system as an alternative to traditional batch methods and the system has the potential to rapidly synthesize polymers and nanoparticles for preliminary screening.

398682 Synthesis of Monodisperse Silica Nanoparticles for Drug Delivery to Solid Tumors

Abstract:

Silica nanoparticles have been identified as ideal carriers for drug delivery due to their low toxicity and their high range of functionalization. Due to the systemic side effects of chemotherapy, appropriation of these therapeutics onto the surface of silica nanoparticles will lead to a safer means of treating cancer. While utilizing these nanoparticles for tumor treatment, it is important to account for their circulation and residence in the blood, accumulation in the tumor, off-target toxicity, therapeutic activity, and drug release from the nanoparticle to the tumor. Research has shown a correlation between particle size and all of these characteristics. There is potential for maximizing drug efficacy by controlling the size of the nanoparticle. Current research is focusing on adapting the Stöber method and optimizing reaction conditions to synthesize optimal sizes of silica nanoparticles in high concentrations. Through this research, advances will be made in solid tumor treatment by further elucidating the benefits and efficacy of delivering therapeutics to the site of the tumor via silica nanoparticles of controlled size.

Erika Staskevicius

Abstract: Internalization and Trafficking of Protein Nanoparticles in Dendritic Cells

Protein nanoparticles have recently emerged as an efficient form of intracellular protein delivery for a range of applications including vaccination (Wang 2013). Not only are they relatively easy to produce, but they are also biocompatible, biodegradable, and have a high loading capacity (Estrada 2014). Given their high internalization and rapid dissolution in cells, protein nanoparticles lead to a quick therapeutic response and good bioavailability (Bamrungsap 2012). Acidic degradation of antigens is detrimental to successful immune responses, and preventing it is a major challenge in designing nanoparticle vaccines. The primary mechanism for antigen degradation is endosome acidification. Determining if endosome acidification is occurring and if it is size dependent can lead to better nanoparticle vaccine delivery. Fluorescent ovalbumin (OVA) nanoparticles are used to view the internal routing of particles. OVA-FITC (green) is acid-sensitive, and consequently its fluorescence decreases with decreasing pH. OVA-AlexaFluor647 (red) is acid-insensitive. Therefore, at decreasing pH levels, the ratio of green to red fluorescence from fluorescent protein nanoparticles decreases. Past studies have found that the ratio of the fluorescent intensities of OVA-FITC to OVA-AlexaFluor647 correlates with the phagosomal pH inside cells (Tran 2009). This ratio can be used to discover the intracellular fate of nanoparticles. By incubating fluorescent OVA protein nanoparticles in antigen-presenting JAWS II dendritic cells (an immortalized dendritic cell line), it has been observed that longer incubation time leads to lower fluorescence ratios, which correspond to lower pH values. Over time, the nanoparticles appear to be taking an acidic processing route within the dendritic cells. From these results, it is assumed that nanoparticles are undergoing acidic degradation after being internalized. This study highlights that endosome acidification affects all nanoparticle sizes tested. Correlating kinetics of protein nanoparticle degradation to endosomal acidification can yield additional insight into the antigen protective properties of protein nanoparticles.

References:

Wang L, Hess A, Chang TZ, Wang YC, Champion JA, Compans RW, Wang BZ. "Nanoclusters self-assembled from conformation-stabilized influenza M2e as broadly cross-protective influenza vaccines." *Nanomedicine.*

Herrera Estrada, L, Chu, B and Champion, J. "Protein Nanoparticles for Intracellular Delivery of Therapeutic Enzymes." *Journal of Pharmaceutical Sciences 103*, 1863-1871.

Bamrungsap, S, Zhao, Z, Chen, T, Wang, L, Li, C, Fu, T and Tan, W. "Nanotechnology in Therapeutics: A Focus on Nanoparticles as a Drug Delivery System." *Nanomedicine 8,* 1253-1271.

Tran, KK and Shen, H. "The role of phagosomal pH on the size-dependent efficiency of cross-presentation by dendritic cells." *Biomaterials 30*, 1356-1362.

MAXWELL, LEORA LIU,YUNG-WAY and PASCAL, JENNIFER Mathematical Modeling of Electroosmotic Flow in Tumor Cells for Tumor Treating Fields (TTF) Therapy
Tennessee Technological University, Cookeville, TN.
Tumor treating fields (TTF) is a new type of cancer therapy that was recently approved by the Food and Drug Administration (FDA) in March 2011 for use in glioblastoma multiforme (GBM) patients that dramatically increases quality of life, while still being just as effective as traditional chemotherapy. TTF therapy uses alternating (AC) applied electrical fields to interfere with tumor cell division by disrupting mitosis via the movement of charged molecules within the cells, altering their internal structure. The processes of electrophoresis, electroosmosis and dielectrophoresis play a key role in the effectiveness of this therapy. Thus, here we will present a mathematical model that describes electroosmotic flow within a single, dividing tumor cell during anaphase and cytokinesis by using idealized geometrical structures to describe the cell. Time dependent velocity profiles as a function of electrical field, cell phase and geometrical parameters indicate situations in which flow reversals can occur within the cells. These results can aid in ultimately predicting and optimizing tumor cell response to TTF therapy.

Format: POSTER
Presenter: Leora Maxwell

Modeling Breast Cancer Extravasation Using Nanoarrays In-Vitro

Kristina Maxwell[1], Jenn Fischer[2], Mark Carver[2], Tom Dziubla[2], Christine Trinkle[3], Chris Richards[4], Kim Anderson[2]
[1]University of Arkansas, Department of Biomedical Engineering
[2]Univeristy of Kentucky, Department of Chemical and Materials Engineering
[3]University of Kentucky, Department of Mechanical Engineering
[4]University of Kentucky, Department of Chemistry
Research completed as part of the University of Kentucky REU program on Bioactive Interfaces and Devices

Abstract

Metastasis of a primary tumor is the leading cause of death for breast cancer patients. Extravasation is a part of hematogenous metastasis of tumor cells when the cells invade new organs in the body via the circulatory system. We have used human umbilical vein endothelial cells (HUVECs), to mimic the lining of a blood vessel within the body, and two types of breast cancer cells lines of different metastatic potentials to compare the invasive and migratory properties of the two cells lines. The two cell lines include MCF-7 (non-invasive) and MDA-MB-231 (invasive) breast cancer cells. For the migratory assay, 7,000 cells were seeded directly onto the nanoarrays for each cell line, with the addition of a collagen-based attachment factor; lacking a cellular barrier to study the migratory properties of each cell line. The nanoarray consisted of 7 different sizes of wells ranging from 85-200nm. The assay was normalized by dividing the amount of pseudopodia found in each well size by the area covered in cells on top of each well size. A similar assay was used to observe the invasive properties of each cell line. A monolayer of HUVECs was grown on the nanoarray with 10,000 cells and the addition of the attachment factor. The HUVECs were grown to a monolayer in 24 hours and then 7,000 of each breast cancer cell line was individually seeded on top of the HUVEC monolayer on the nanoarrays. Each assay was imaged 24 hours after seeding. Our results concluded that the MDA-MB-231 breast cancer cell line showed more pseudopodia in the migratory assay, which is likely due to their higher invasive nature when compared to the MCF-7 breast cancer cell line. The amount of pseudopodia observed within the 7 different sizes of wells also directly correlated with the size of the well. These studies can be used in future work for evaluating the effectiveness of anti-cancer drugs as well as protein analysis. Nanoarrays and nanotechnology are beginning to advance the cancer research field due to their significant applications in biomolecular analysis.

398917 Development of a Novel Embryonic Zebrafish Xenograft Model for Early-Stage Glioblastoma Tumor Cell Invasion

Abstract:

Background: Glioblastoma multiforme is the most common primary brain tumor in adults ages 45-70, with over 22,000 Americans diagnosed in 2010 alone. Despite decades of research efforts, glioblastoma is still one of the most difficult cancers to treat because the tumor cells have the ability to invade into the surrounding tissue with finger-like projections, making it difficult to completely remove all of the tumor tissue without resulting in more damage. But by slowing down the invasion of the tumor, a patient's survival rate can be increased. Still, the 5 year survival rate remains low -- less than 5%, and without treatment, median survival time is only 4.5 months after diagnosis. To study this, we have developed a new embryonic zebrafish xenograft model to better understand key factors in tumor cell invasion and decision-making. These include the usage of specific attachment sites, links between tumor invasiveness and distance from blood vessels, and the velocity and directionality of the tumor cells. Zebrafish are ideal model organisms for this process due to a variety of characteristics including, but not limited to their transparent nature, presence of blood circulation and vertebrate anatomy, as well as a dense microenvironment comparable to that of a human brain's. Specifically, transgenic fli fish were selected. These fli fish express Green Fluorescent Protein (GFP) in the endothelial layer of their blood vessels and were used to help monitor the interaction of tumor cells with blood vessels within the brain.

Methods: Our current imaging protocol involves an efficient seven day process. Transgenic fli fish 0 days post fertilization (dpf) were acquired from the Sinnhuber Research Laboratory. Starting 1 dpf, the fli fish were continuously treated with phenylthiourea to further inhibit melanogenesis, enhancing image clarity and reducing background. At 3 dpf, the fish were sedated and microinjected with fluorescently dyed cell cultured glioblastoma cells (U251 & U87). From 4-7 dpf, 18 hours of live, overnight, multidimentional images (both GFP and red fluorescent CM-Dil) of the fli fish were captured using a High Content Imager. Finally, after imaging, the GFP and CM-Dil images were overlayed and analyzed. Specifically, this new model will be useful for the quantification tumor cell velocity and colocalization of blood vessels and glioblastoma cells over time, and the investigation of whether certain morphological characteristics determine directionality and affect tumor cell decision-making.

395972 Combining Radiation and Hyperthermia Methods for the Vitro Cancer Treatment of Multicellular Tumor Spheroids

Abstract:

Radiation is a common form of cancer treatment. However, with radiation therapy comes extreme side effects for patients like hair loss, fatigue, infertility, blistering at the treatment site, nausea, and vomiting. Hyperthermia, a supplemental cancer therapeutic which exposes cells to an elevated temperature between 42-45°C, has been shown in conventional monolayer studies to enhance the effects of radiation. These 2-dimensional cell culture models are insufficient representations of *in vivo* tumors. 3-dimensional (3D) multicellular tumor spheroids are more appropriate *in vitro* models; they retain the architectural and morphological characteristics of their tumor counterparts, and are helping to bridge the gap between cell-based *in vitro* studies and animal studies. This investigation aims to evaluate the efficacy of radiation therapy alone and in combination with hyperthermia using a multicellular tumor spheroid model. It is hypothesized that the two will combine to reveal a synergistic effect, meaning that the combination of treatments will result in a "more-than-additive" decrease in viability compared to the modalities alone. A synergistic effect of combined treatment would indisputably prove that parallel to a high dose of radiation, a lower dosage could be used in conjunction with hyperthermia to result in the same cell death. Thus, the severity of side effects of radiation treatment could ultimately be reduced in the clinical setting, well-represented by the 3D multicellular spheroid *in vitro* model.

Efficacy of Select Viability Assays in Multicellular Tumor Spheroid Culture

Jonathan F. Coburn, Sarah E. Patterson, Anastasia K. Hauser, Kimberly W. Anderson, J. Zach Hilt

Department of Chemical and Materials Engineering, University of Kentucky, Lexington, KY

Research completed as part of the University of Kentucky REU program on Bioactive Interfaces and Devices

REU Engineered Bioactive Interfaces and Devices

University of Kentucky College of Engineering Department of Chemical and Materials Engineering

Abstract

Multicellular tumor spheroids (MCTS) mimic the physiological state of tumors *in vivo* more accurately than do cell monolayers. Essential tumor characteristics such as hypoxia, drug resistance, and the extracellular matrix may collectively be included in cancer therapy screening through the use of this complex *in vitro* model. Attendant with the increased complexity of the MCTS model is increased difficulty of application, particularly in assaying MCTS viability. The objective of this study was to investigate the efficacy of the acid phosphatase, resazurin, and live/dead assays in A549 MCTS culture. Experiments were conducted to examine consistency and accuracy as well as usefulness for future hyperthermia and chemotherapy studies. The results presented herein contribute to the development and validation of assays necessary to the application of MCTS in cancer research.

Background: Multicellular Tumor Spheroids

- Tightly packed spherical clusters of cells
- Used in cancer research to screen cancer therapies and to elucidate associated mechanisms
- More accurate physiological model than cell monolayers[1-2]

Physiological characteristics unique to MCTS

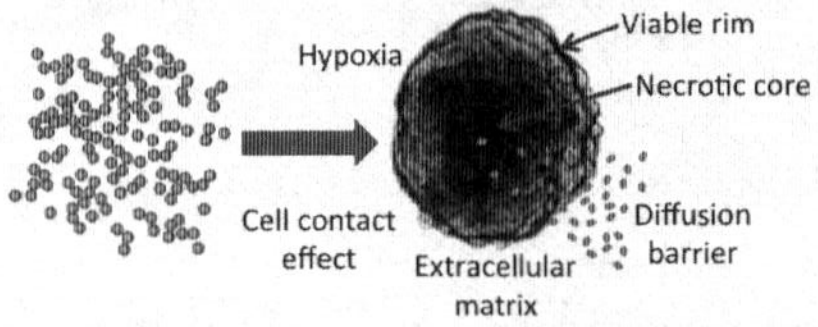

- Variety of formation techniques, spheroid sizes, and cell types in use
- Disadvantages:
  - Increased time and cost requirements
  - More difficult to obtain results of treatment

Common MCTS assays with associated disadvantages[3]

Assay	Problems
Change in volume over time	Lengthy, does not correlate with viability
Colony formation assay	Lengthy, does not correlate with viability
Chemical assay	Poor penetration of spheroid
Dissociation before chemical assay	Loss or damage of cells

Objectives

To evaluate the efficacy of select viability assays in A549 MCTS culture, by examining the assays' responses to the following:

- Varying number of spheroids
- Varying spheroid seeding densities
- Varying concentrations of cisplatin treatment
- Dissociation of spheroids after cisplatin treatment
- Combined cisplatin and hyperthermia treatment

Viability Assays

Acid phosphatase (APH):
- *p*-nitrophenyl phosphate is metabolized by live cells to *p*-nitrophenol, an absorbant compound (peak 405 nm)

Resazurin:
- Resazurin is metabolized by live cells to resorufin, a fluorescent compound (excitation/emission 530/590 nm)

Live/dead:
- Calcein AM is trapped within live cells and metabolized to a fluorescent compound (excitation/emission 495/515 nm)
- Ethidium homodimer-1 enters dead cells through ruptured membranes; DNA binding induces fluorescence (excitation/emission 528/617 nm)

Assay mechanisms of action

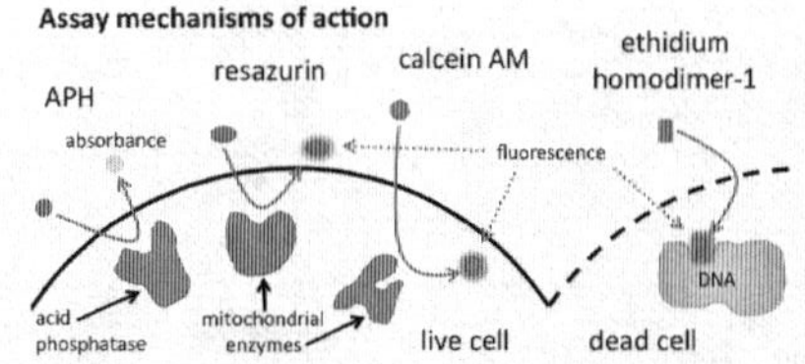

Known characteristics

Assay	Pros	Cons
APH	High-throughput, inexpensive	Concentration-sensitive
Resazurin	High-throughput, inexpensive	Concentration- and light-sensitive
Live/dead	Provides images for visual confirmation of results	Low-throughput, light-sensitive, expensive

Multicellular Tumor Spheroid Formation

- A549 cells seeded at 5,000 cells/spheroid or as specified
- 20 μg/mL collagen to encourage aggregation
- Nunc Sphera 96-well plates, coated with a hydrophobic polymer
- Spheroids of about 400 μm diameter formed in 48 h

Spheroid Number Validation

Wells containing one to five spheroids were assayed (n≥3).

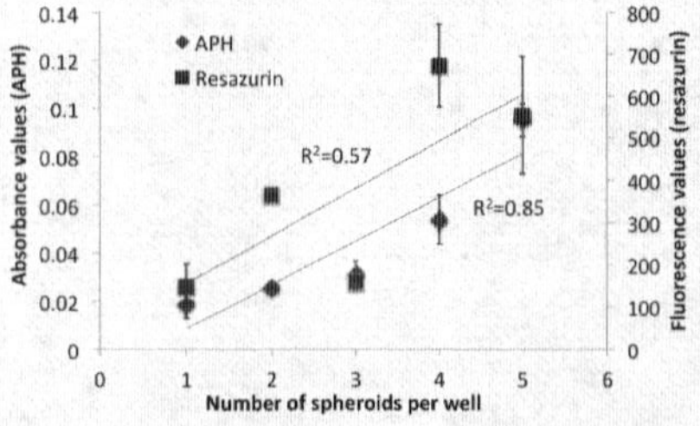

Although neither assay responded linearly, the APH assay displays a more consistent trend than the resazurin assay.

Seeding Density Validation

Spheroids seeded at various seeding densities were assayed (n≥4).

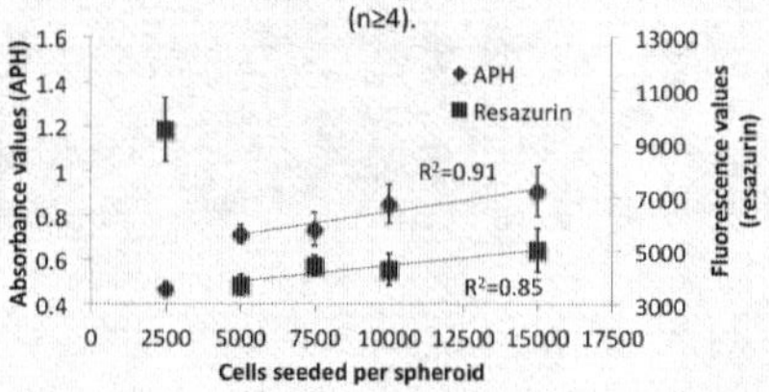

A difference in linearity would suggest a difference in penetration, but no difference in linearity was observed.

Cells did not cluster when seeded at 2500 cells per well; resazurin responded strongly to unclustered cells whereas APH responded weakly.

Cisplatin and Dissociation

- Cisplatin (CDDP) is a chemotherapeutic agent that induces apoptosis through DNA binding
- 72 h incubation with 0, 20, and 2000 μM CDDP
- Dissociation of some spheroids with 0.1% trypsin after CDDP incubation period
- Each of the seven assay groups was normalized to the 0 μM CDDP control (n≥5)

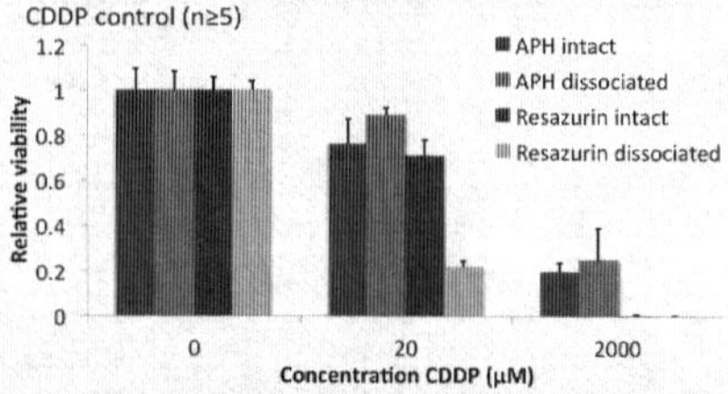

Results for resazurin are inconsistent between dissociated and intact at 20 μM CDDP, whereas APH is consistent at each concentration.

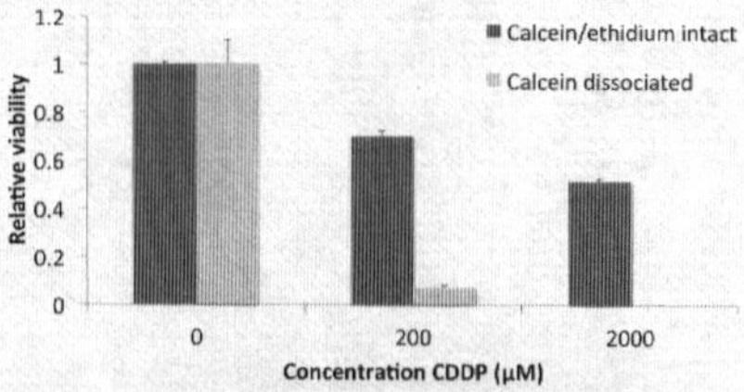

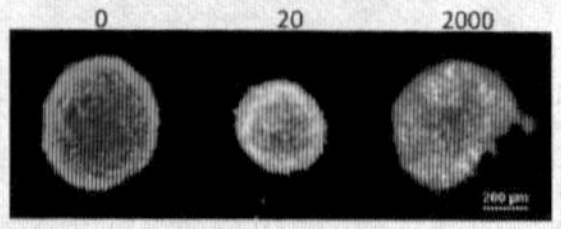

This analysis does not match the dissociated calcein trend.

Cisplatin and Hyperthermia

- In a cancer context, hyperthermia is elevation of temperature to 41–45 °C to achieve a tumor-specific cytotoxic effect
- Hyperthermia interacts synergistically with many chemotherapeutic agents such as CDDP[4]
- 72 h incubation with CDDP between 0-2000 μm (n≥5)
- Hyperthermia induced immediately after CDDP addition for 30 min in an incubator at 42–43 °C
- APH assay and supporting live/dead images

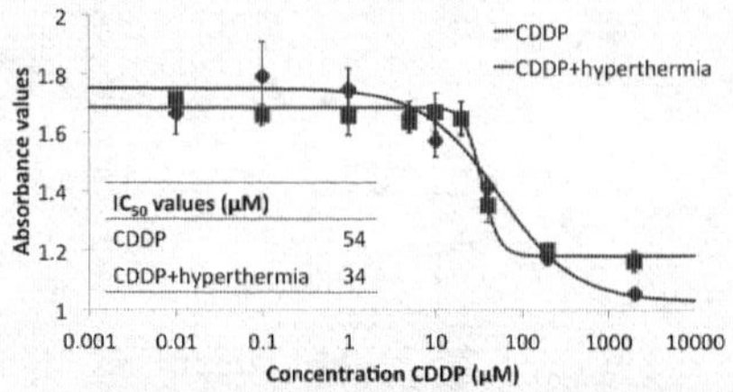

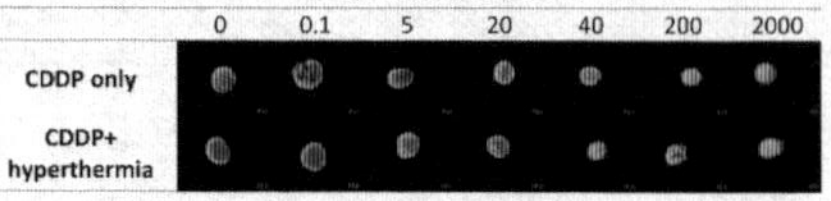

Fits the four parameter logistics curve:

$$\text{relative viability} = \min + \frac{\max - \min}{1 + \left(\frac{\text{absorbance}}{IC_{50}}\right)^{-\text{hillslope}}}$$

Little effect of hyperthermia except a slight decrease in IC_{50}. Live/dead images support general trend.

Conclusions

- Response to spheroid number, unclustered cells, and dissociation suggest APH is superior to resazurin assay
- APH assay has the potential to successfully assay studies involving hyperthermia and CDDP
- Live/dead assay, though unsuitable for high-throughput analysis, provides a useful imaging basis for qualitative confirmation of results

Future Work

- Repetition of the above studies to confirm results
- Optimization of the APH assay for A549 MCTS
- Application of the APH assay to hyperthermia studies

References

1. Ivascu, A. and M. Kubbies. *Rapid generation of single-tumor spheroids for high-throughput cell function and toxicity analysis.* Journal of Biomolecular Screening, 2006. 11(8): 922-932.
2. Minchinton, A. I. and I. F. Tannock. *Drug penetration in solid tumours.* Nature Reviews Cancer, 2006. 6(8): 583-592.
3. Friedrich, J., R. Ebner & L. A. Kunz-Schughart. *Experimental anti-tumor therapy in 3-D: Spheroids - old hat or new challenge?* International Journal of Radiation Biology, 2007. 83(11-12): 849-871.
4. Issels, R. D. *Hyperthermia adds to chemotherapy.* European Journal of Cancer, 2008. 44(17): 2546-2554.

Acknowledgments

Research supported by the National Science Foundation REU Program # EEC-0851716

Computational Modeling of the *In Vivo* Effects of Cannabinoids on Glioma Tumors

Daniel Bier, Colleen Fridley, Megan Kelly and Pavlos Pachidis
Chemical and Biomolecular Engineering, Johns Hopkins University, Baltimore, MD

Cancer patients have been using cannabis for decades to help alleviate side effects of chemotherapy. Recently, the FDA has approved several cannabinoid-based drugs, including Nabilone and Dronabinol, which research shows can prevent the nausea and appetite loss associated with most cancer chemotherapy treatments.[1] However, numerous studies suggest that cannabis can also influence the tumors' apoptosis, angiogenesis, and cellular maintenance and growth processes.[2] By combining the empirical results of these studies with known biological pathways, we developed a model that predicts the degree of reduction in tumor size when cannabis is used as a treatment in conjunction with chemotherapy and also as a standalone therapy.

Many experiments related to cannabis and tumor growth focus on brain tumors, namely gliomas. From a biological perspective, it seems plausible that cannabis could influence tumor growth because brain tumors, like healthy brain tissue, express cannabinoid receptors, a class of G-protein coupled receptors. Normally, the brain mainly contains CB1 receptors, with very little expression of CB2 receptors. Gliomas are exactly the opposite, with high expression of CB2 receptors.[2] Our model is based off of mechanisms likely involving the CB1 and CB2 receptors.

We modeled three mechanisms by which cannabis can reduce the size and growth of gliomas through interactions with the CB1 and CB2 receptors. These are: *increased apoptosis* of tumor cells through increased production of the pro-apoptotic molecule sphingolipid ceramide, *reduction of angiogenesis* through decreased production of molecules in the vascular endothelial growth factor (VEGF) pathway, and *increased chemotherapy (temozolomide) uptake* by activation of a transient receptor potential vanilloid 2 (TRPV2) receptor.[2] Our model shows that for a 55 kg Grade IV brain cancer patient with 5 L of blood, cannabis can decrease the growth rate of tumors in some cases. Theoretically, elimination of angiogenesis and a high rate of apoptosis could kill a tumor without the need for supplementary chemotherapy. However, our data suggests that to produce sufficient tumor cell death, cannabinoids would need to be present in unrealistic concentrations at the site of the tumor. Therefore, our recommendation is that patients should smoke cannabis in conjunction with a chemotherapy regimen. Future studies should focus on combined effects of cannabis treatment with a chemotherapy regimen.

Literature Cited:

1. "Cannabis and Cannabinoids." National Cancer Institute. N.p., n.d. Web. 1 Oct. 2014.
2. Velasco, Guillermo, Cristina Sanchez, and Manuel Guzman. "Towards the Use of Cannabinoids as Antitumor Agents." Nature (2012): n. pag. Web.

IOWA STATE UNIVERSITY
Department of Chemical and Biological Engineering[1], Department of Materials Science and Engineering[2]

2014 AIChE Annual Conference
November 17, 2014

Rachel Lieser[1], Hannah Bygd[2], Paige Dettman[1], Dr. Kaitlin Bratlie[1,2]

Targeted Drug Delivery and Macrophage Reprogramming for Tumor Suppression

Introduction and Background

Current cancer treatments rely on systemic drug delivery where chemotherapeutic drugs kill both healthy and cancerous cells. This leads to negative side effects such as alopecia, nausea, fatigue, and anemia. Targeted and controlled delivery of chemotherapeutic drug could mitigate side effects and improve cytotoxicity to cancer cells. Additional tumor suppression can be achieved through reprogramming of macrophages. Macrophages can be polarized to M1 or M2. M2 macrophages promote tumor growth through the release of angiogenic molecules, while M1 macrophages produce molecules that are toxic to tumor cells[1]. Tumor-associated macrophages (TAMs) are thought to be M2[2]. Certain material properties are known to reprogram macrophages from M2 to M1[3]. Our project looks to use controlled drug delivery in conjunction with macrophage reprogramming in order to improve cancer treatment efficacy. Polymeric systems are being investigated for their potential to reprogram TAMs and to deliver anti-cancer therapeutics to tumors. Here, we focus on alginate nanoparticles coated with modified poly-L-arginine (PLR).

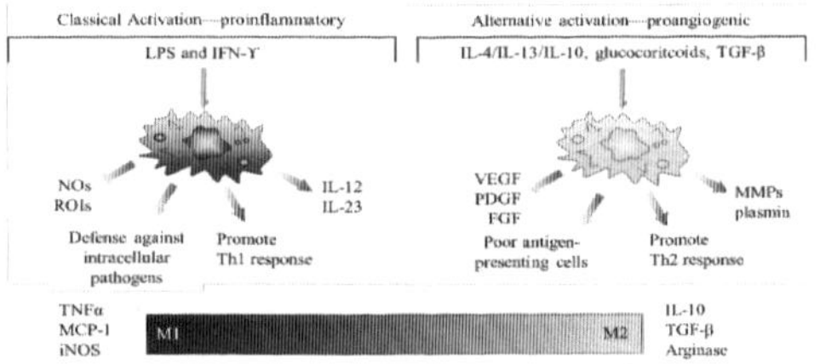

Material Modification

To synthesize the alginate nanoparticles 18mM Calcium Chloride is added to 0.12% low viscosity Sodium Alginate dropwise. This mixture is allowed to stir for one hour before 0.1% Poly-L-arginine (PLR) is added dropwise and stirred for two hours to create unloaded nanoparticles. The nanoparticles are refrigerated overnight before being centrifuged for 45 minutes. They are then freeze-dried. To modify the particles, they are suspended in a 0.05% modified PLR solution and incubated for five minutes at 37° C. The particles are then centrifuged, washed twice, and resuspended in phosphate buffered saline (PBS) to create a 1% solution.

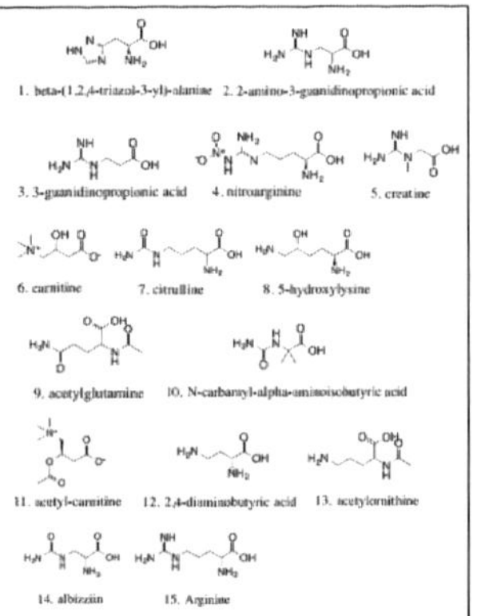

Drug Release

A library of 15 modified alginate particles were loaded with doxorubicin and suspended in 7.4 pH PBS. The particles were put in dialysis tubing and submerged in 100 mL of PBS. Samples were taken from the external solution over the following week. Absorbance was used to quantify the concentration of doxorubicin outside of the dialysis tubing. The release kinetics of representative modifications are shown below along with the percentage of doxorubicin loaded into all modified particles.

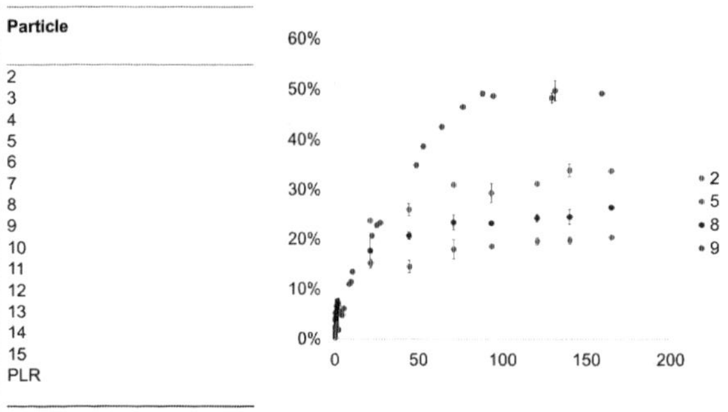

Cell Viability

Cytotoxicity of all materials used are considered by performing MTT assays to understand the viability of the cells upon exposure to the different functionalized polymers. The viability of Raji cells when exposed to PLR-coated alginate nanoparticles is represented below.

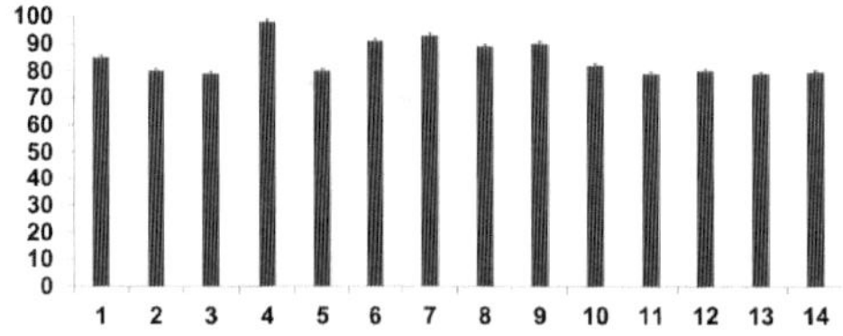

The viability of activated and non-activated macrophages when exposed to alginate nanoparticles is shown below.

Cytotoxicity

The half maximal inhibitory concentration (IC_{50}) was determined to measure the effectiveness of doxorubicin at inhibiting Raji cell function. Differing concentrations of doxorubicin were incubated with Raji Cells overnight. An MTT assay was performed after 48 hours to quantify the cell viability after exposure to doxorubicin.

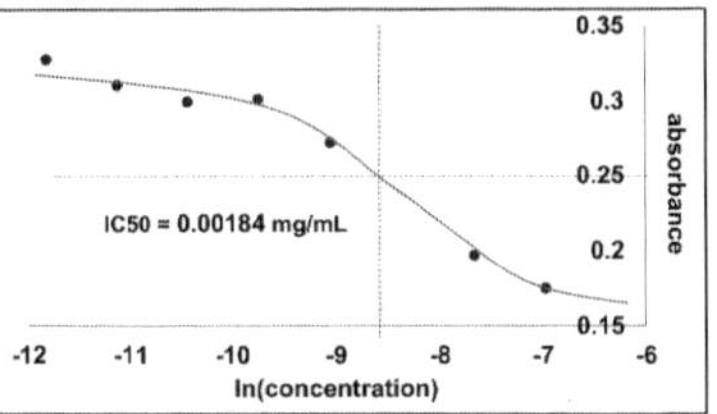

Particle Internalization

Modified alginate nanoparticles are stained with Fluorescein isothiocyanate (FITC) and incubated with cells to promote particle internalization. The cell solution is stained with Red DNA stain and read on a Tali® Image-Based Cytometer to determine the number of particles associated with each cell. Cellular uptake of the Raji cells is represented below.

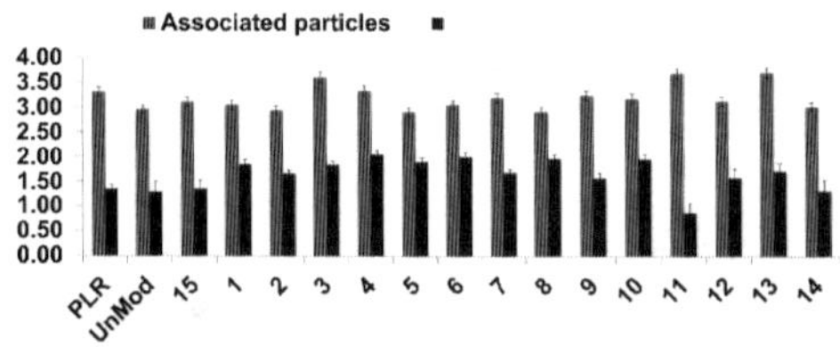

The ability of activated and non-activated macrophages to internalize nanoparticles was also analyzed and the results are highlighted below.

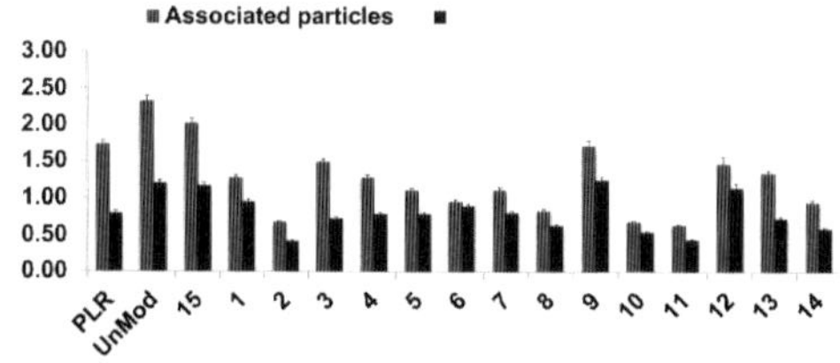

Conclusion

Notable Results

- Macrophages and Raji cells are viable when exposed to PLR-coated alginate nanoparticles
- Alginate nanoparticles bearing different functional groups can alter the release kinetics of chemotherapeutic drug
- Differing functional groups have the ability to alter the number of particles internalized by cells

Future Research

- Use MTT assays to test the cytotoxicity of nanoparticles loaded with doxorubicin on Raji cells
- Complete TNF ELISAs and arginase/iNOS assays to determine if PLR-coated alginate particles have the ability to reprogram macrophages from M2 to M1
- Determine the mechanism of particle internalization for Raji cells and macrophages by blocking specific uptake mechanism (phagocytosis, endocytosis, macropinocytosis)
- Perform drug release experiments in an external environment of pH 5

Acknowledgements:
National Science Foundation . Grant Number: CBET-1227867
Roy J. Carver Charitable Trust . Grant Number: 13-4265
NSF ARI-R2 (the renovation of the research laboratories used for these studies) . Grant Number: CMMI-0963224

References:
1. Gordon, S. Alternative activation of macrophages. *Nat. Rev. Immunol.* **3**, 23–35 (2003).
2. Schmid, M. C. & Varner, J. A. Myeloid cells in the tumor microenvironment: modulation of tumor angiogenesis and tumor inflammation. *J. Oncol.* **2010**, 201026 (2010).
3. Brown, N. B. & Ratner, D. B. Macrophage polarization: An opportunity for improved outcomes in biomaterials and regenerative medicine. *Biomaterials*. 3792–3802 (2012).

Development of Gold-Lipid Nanocomposites as a Two-Stage Drug Delivery System

Introduction

Chemotherapeutic drugs such as doxorubicin are cytotoxic agents that are effective at killing cancer cells, but also cause non-target toxicity due to their lack of selectivity. Stealth liposomes have been used as an effective strategy for improving drug delivery and limiting systemic toxicity, but can still suffer from inconsistencies in formulation, poor drug exposure profiles and poor detection *in vivo*. Stealth liposomes can circulate in the body for periods as long as multiple days. The enhanced permeability and retention (EPR) effect is a characteristic of poor vascular development in cancerous tissues that leads to leaky vasculature, permitting enhanced extravasation and accumulation of nanoparticles (relative to free drug) due to their size. Gold nanoparticles have a host of advantageous properties for a drug delivery and imaging platform, including biocompatibility, chemical flexibility, tunable optical properties, and more, but are limited by very short circulation times in the body (approximately 4 hours or less). We hypothesized that a scalable solvent injection method could be used to create novel nanocomposite systems that consist of gold nanoparticles encapsulated within liposomes to improve drug delivery while also permitting non-invasive imaging.

Materials and Methods

Glutathione-capped gold nanoparticles were synthesized by a borohydride reduction of chloroauric acid and glutathione in a solvent composed of water and methanol. The relative amounts of each reagent were varied to investigate their significance as parameters useful for the control of the nanoparticles' properties. Purification was performed by centrifugation and rinsing. Liposome formulations were prepared by injecting a lipid-ethanol solution (9:5:1 lipid molar ratio of DSPC:Cholesterol:DSPE-PEG) into water with agitation, followed by heating under mild vacuum to remove excess ethanol. The ethanol-water volumetric ratio was varied between 10% and 30% in order to optimize the conditions for liposome formation and produce small (under 100nm diameter), uniform particles (polydispersity index less than 0.3). Nanocomposites were prepared in a similar fashion: Nanocomposites – liposomes with gold entrapped in the aqueous core – were generated by using a suspension of the gold nanoparticles in place of the water used to synthesize traditional liposomes. The resulting particles were characterized by dynamic light scattering to determine their size distribution and effective surface charge (zeta potential). UV-Vis spectroscopy was used to quantify the entrapped gold nanoparticles, with absorbance measured across the UV, visible, and infrared spectra. We examined the *in vitro* cytotoxicity of these nanocomposites using mitochondrial enzymatic (MTT) and total protein (SRB) assays, which were used to analyze the potential growth inhibition in human prostate carcinoma (PC-3) cells. The intracellular uptake into PC-3 cells was quantified by labeling the nanocomposites with DiR, a non-exchangeable fluorescent membrane probe that excites and emits in the near-infrared range. Tumor and tissue biodistribution analysis *in vivo* was performed in athymic NCR mice using a human xenograft model of human prostate cancer, with aggressive, hormone-refractory PC-3 cells. The nanocomposites were tracked by labeling with DiR and dosing *via* tail vein injection, followed by non-invasive imaging (IVIS Lumina XR) 48 hours post-treatment. The nanocomposites and gold nanoparticles were also both analyzed for optical properties using the IVIS Lumina XR system.

Results and Discussion

We have synthesized 2-nm, glutathione-capped gold nanoparticles via the borohydride method described above. These particles are uniform in size (with polydispersity below 0.5) and display fluorescence when excited by infrared light. Liposomes and nanocomposites with narrow size distributions centered around 100 nm were prepared using the ethanol injection method described above. Even with gold entrapped, the liposomes maintained a similar size distribution profile (polydispersity index around 0.25) and a zeta potential near 30 mV, indicating good colloidal stability. These size and surface properties are essential, not only for particle stability (prior to application and *in vivo*) but also for the nanocomposites to passively target tumors *via* the EPR effect. These data also suggest that interactions between ethanol and the lipids leads to apparent lipid aggregation with ethanol in water compositions at and above 20% (v/v), meaning that particles prepared under these conditions have undesirable surface properties and reduced stability. *In vitro* studies suggest that the nanocomposites are non-toxic up to millimolar concentrations (greater than the desired clinical dose) and intracellular uptake of the nanocomposites was confirmed by scanning confocal microscopy using the DiR label. Intracellular uptake was observed to be at levels similar to established liposome formulations, regardless of the presence of gold nanoparticles. Preliminary *in vivo* imaging studies suggest enhanced tumor deposition relative to tumor volume for the nanocomposite formulations *versus* traditional liposomes after 48 hours.

Conclusions: We have developed and characterized a novel golid-lipidic nanocomposite consisting of 2-nm glutathione-capped gold nanoclusters entrapped within liposomes. An ethanol injection method resulted in consistent manufacturing of nanocomposites with desirable surface (stability) and size properties for cancer drug delivery. Further, this method is simple and highly scalable, rendering it more industrially and clinically relevant than traditional liposome formulations. These particles were taken up readily by aggressive, hormone-refractory prostate cancer cells and the data suggest that they possess enhanced tumor deposition properties *in vivo* compared to traditional long-circulating sterically stabilized liposome (SSL) formulations. The conjugation of therapeutic and imaging moieties to gold nanoparticles has been established in the literature, and future work will focus on utilizing these methods to incorporate drugs and imaging agents into the nanocomposites for truly multifunctional particles with controlled release properties. Preliminary results show that the particles we have created are readily conjugated to poly-(ethylene glycol) and paclitaxel by Steglich esterification. This strategy has the potential to address the problem of drug resistance in tumors by simultaneously delivering multiple drugs to tumors (even poorly-soluble ones, such as paclitaxel). Through the use of established targeting strategies such as phage display or cell-penetrating peptides, it is possible to create long-circulating, targeted nanocomposite carriers with intracellular drug release mediated by the surface chemistry of the entrapped gold nanoparticles. For example, hydrazine bonds have been used to design gold nanoparticles that will release conjugated doxorubicin only at pH levels found in the intracellular environment and not in the bloodstream. These novel nanocomposites have the potential to address the challenges traditionally associated with liposomes and provide a highly flexible imaging and drug delivery system for a variety of cancers.

Acknowledgements: This project was funded, in part, by NIH grant 1R01-1EB016100-01 (RDA), as well as by the Auburn Undergraduate Research Fellowship Program (CSD and CMP).

398714 Targeting Key Genes of Cervical Carcinogenesis in Cancer Cell Lines Via siRNA

Abstract:

Cervical cancer is the second most common cancer among women worldwide with approximately 12,340 invasive cases diagnosed in 2013, in the U.S. alone. Unfortunately, even considering the anti-HPV vaccine related reduction to morbidity, cervical cancer will remain among the 10 most frequent cancers in women. Furthermore, the high mortality rate of cervical cancer patients (30% in the first five years) is attributed to the ineffectiveness of conventional therapeutic treatments. Thus, it is clear that a new and efficient therapeutic approach is highly warranted. It was recently proposed that simultaneously targeting of different carcinogenic pathways can be a promising therapeutic strategy that also prevents drug resistance. Moreover, it was shown that effective cancer treatment can be achieved by targeting genetic alterations of a given tumor harbored by individual patients. In a recent study, meta-analysis of genomic aberrations and global gene expression revealed key genes of cervical carcinogenesis, located in regions of chromosomal gains. It is hypothesized that blocking the expression of these key gene-drivers will result in the inhibition of tumor growth and metastatic activity. For this study, cervical cancer cell lines were transfected with the siRNA of key genes and *in vitro* assays were performed to evaluate different oncogenic pathways and expression of genes regulated by the key driver genes. Four genes (DUSP12, CKS1B, LAMP3, and RFC4) were chosen for initial testing in HeLa and ME180 cell lines that have chromosomal gains in the 1q and 3q region respectively. Cell viability results reveal that suppression of individual genes does not inhibit the growth of cancer cells dramatically. However, gene expression data does reveal co-regulation between specific genes which verifies the inferred networks of cervical cancer carcinogenesis. It is believed that combinations of siRNAs, targeted toward different key genes, will be able to achieve a larger effect on cancer growth and metastasis and at the very least, increase the susceptibility of cancer cells to other forms of treatment such as chemotherapy drugs. This conclusion is supported by cell viability and gene expression data, as well as other preliminary studies.

Computational Modeling of Cancer Stem Cell Sphere Formation

PURDUE UNIVERSITY

THE UNIVERSITY OF ALABAMA®

Erin Retzlaff-Roberts(1), David A. Dozier(2), Yonghyun (John) Kim(2), Ph.D.

(1) Department of Chemical Engineering, Purdue University, West Lafayette, IN;

(2) Department of Chemical and Biological Engineering, The University of Alabama, Tuscaloosa, AL

Abstract

A number of lab experiments and computational models were performed on cancer stem cell spheroids to gain a better understanding of their growth patterns. The lab experiments looked at cell diameter and cell count of brain tumor cell line U87. The models used mass transfer rate, diffusivity, and concentration to approximate the environment within a tumor spheroid. The knowledge gained from this work will help lay the foundation for future models, experiments, and procedures using tumor spheroids. If these spheroids can be modelled with higher precision, they can be studied and understood more quickly. This will allow them to be grown on a larger scale with a bioreactor which will allow research into cancer treatment and diagnosis to progress.

Background

Cancer stem cells (CSCs) have been grown in culture for many research experiments. It has been shown that cell lines grown in chemically defined media have more stemness characteristics and are more like CSCs *in vivo*. For example, glioblastoma brain tumor cells grown in defined media as three dimensional spheroids have higher population of CSCs compared to those grown using the conventional two dimensional monolayer cultures. While desirable for maintaining stemness, this method of growth creates a non-uniform environment for cells due to the formation of nutrient concentration gradient as spheroids grow in size. Once above a critical size, the inner cells lose access to nutrients and a necrotic core is formed. For optimal cell growth, it is desirable to passage or separate the cells into single cells to support continued cell growth and by forming new smaller spheres. Ideally, the passaging should happen before the necrotic core begins to form.

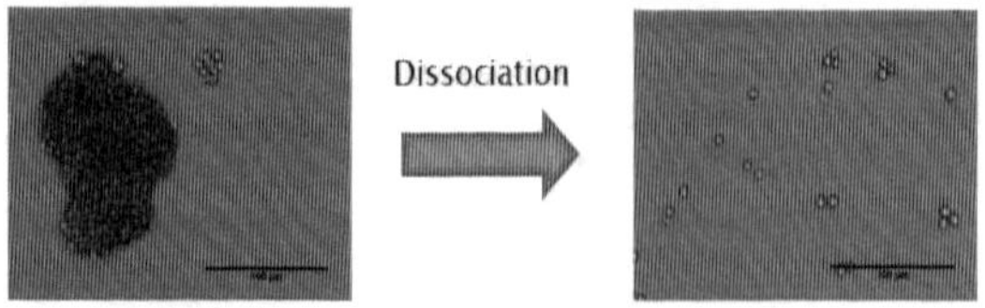

Figure 1: Image on the left shows U87 spheroid after five days with a diameter of about 100 μm culture immediately before passaging. Image on the right shows U87 on day zero about 4 hours after passaging as mostly single cells.

Hypothesis

Necrotic core forms due to hypoxic conditions in the center of spheroids. Modeling growth kinetics will find optimal time for passaging.

Experiment Design

U87 cells were grown over several weeks in flasks with neurobasal media supplemented with basic fibroblast and epidermal growth factors (NBE). The cells were passaged and counted every five to six days. Initially, 3.00×10^5 cells were placed in a T25 flask with 4mL of NBE. It was placed in the incubator at 37°C on day zero. On day three, the 2.5mL of NBE was added to the culture. On day five, the cells were passaged and counted. In this way a growth curve of the cells was obtained. To calculate sphere diameter over time the cells were grown in a similar condition and photograph on days zero, one, three, and five. ImageJ software was then used to analyze the pictures by calculating diameter in micrometers as the area times four divided by the perimeter.

All models were constructed using diffusion of dilute species physics, and time dependent calculations in COMSOL Multiphysics 4.4. This program uses a modified version of Fick's Law of diffusion. The program runs a modified version of this that uses a diffusion coefficient, peak initial concentration, mass transfer coefficient, a dimensionless constant, initial concentration, and an average of flux across a point shown in the table below. Currently the last three have place holder values. Mass transfer coefficient is currently based off as the mass transfer rate of oxygen into a bioreactor. Other constants are calculated for oxygen. The model shown below was made to have a diameter of 75 μm.

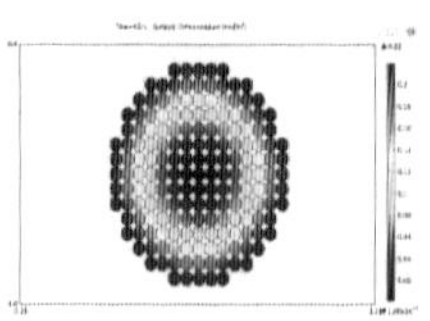

Figure2: Image above shows the model that was used to represent a cell spheroid at time 0.1 seconds. It is shown as a reference for the model.

Table 1: The table to the right shows the constants and variables used in the model and their values.

Constant or Variable	
Diffusivity coefficient	-9[m^2/s]
concentration	mol/m^3]
Sphere radius	3.5x10^{-6}m
	]
constant	1000
	exp(a(-
	k_f*c)

Results

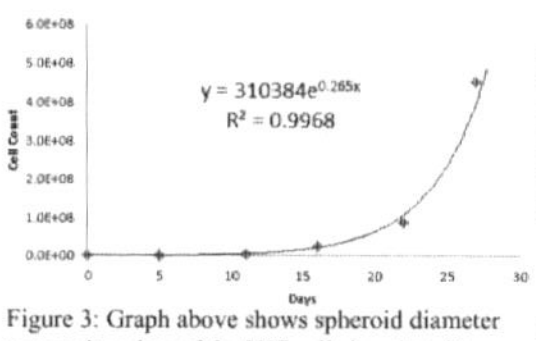

Figure 3: Graph above shows spheroid diameter versus time data of the U87 cells in spheroid culture as well as fitted trend line with error bars.

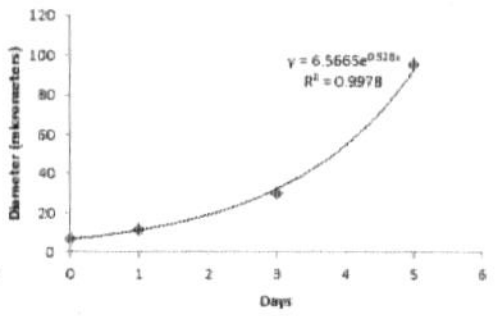

Figure 4: Graph above shows cell count versus time data of the U87 cells in spheroid culture as well as fitted trend line.

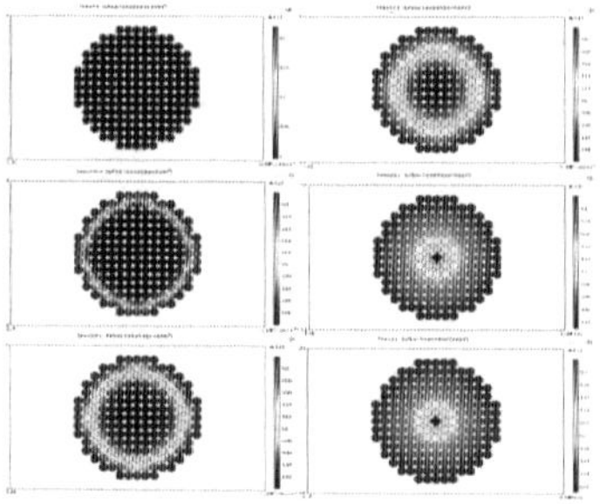

Figure 5: Images above show oxygen concentration increasing around the sphere to the point where only the center does not have oxygen. The final pictures show no change indicating a steady state of concentration throughout the spheroid.

Discussion

- Growth curve corresponds to typical yield of 1.27×10^6 cells in 5.16 days
- Sphere diameter corresponds with current lab procedure of passaging cells at 5 days when diameter is about 100 μm
- Greater variation in sphere size over time likely due to diffusion gradient
- Model shows sphere will reach critical diameter when insufficient oxygen in center for cells to live

Acknowledgements

- The University of Alabama Chemical and Biological Engineering Department
- The National Science Foundation (EEC-1358750)

398474 Biomarker Significance of Exosomes in the Initiation and Progression of Breast Cancer

Abstract:

Currently, breast cancer diagnosis involves invasive methods such as mammography and biopsy. Thus, there's a need for non-invasive, patient-friendly methods. Recent developments indicate much promise for the use of circulatory serum exosomes as potential biomarkers for diagnosis. Exosomes are found in most bodily fluids, and are more abundantly released from tumor cells compared to healthy cells. We examined if serum exosomes could be used as potential biomarker during the initiation and progress of breast cancer using ACI rats, which develop tumors upon low-dose continuous exposure to 17β-estradiol (E_2). Serum from E_2-treated and control animals after 3 weeks, 3 months, and 7 months was used to isolate exosomes. Exosomes were analyzed for size, protein yield, and exosomal surface and proliferation markers. Exosomes were in the size range of 40-200 nm. The mean exosomal protein yield from E_2-treated samples was higher than the control for each time period; however, the differences were insignificant due to small sample size. Serum exosomes were positive for hallmark exosomal proteins including Alix, CD44, CD63, CD81, and proliferation marker, EGFR. Levels of Alix and CD44 were elevated in E_2-treated serum exosomes versus control. E_2-treated serum exosomes enriched using CD63-captured magnetic beads exhibited higher protein yield and EGFR expression than controls; however, the differences were insignificant. Higher levels of serum exosomes positive for exosomal surface proteins and proliferation markers with E_2-treatment indicate their potential as biomarker during initiation and progression of breast cancer. (Supported from the NCI grant R25-CA134238, Duggan Endowment and Hemsley Funds).

396720 Dwr-SECM Testing of Pseudomonas Aeruginosa Biofilms

Abstract:

Pseudomonas aeruginosa biofilms are multicellular assemblies held together by an extracellular polymeric substance (EPS). *P. aeruginosa* begins to form biofilms once pyocyanin – a small toxic molecule excreted by *P. aeruginosa* – concentrations overcomes a certain threshold. Through quorum sensing, the bacteria will detect this chemical composition and subsequently change its defense mechanism by changing its phenotype. In a healthy human, *P. aeruginosa* lives within the mucus of the lungs without detriment to the human. In humans with cystic fibrosis (CF), these biofilms can make the person's mucus even more viscous, leading to serious health complications. We designed and tested a novel apparatus that combined double wall ring (DRW) interfacial rheology with scanning electrochemical microscopy (SECM) probe. By doing DWR-SECM rheology on the biofilm, we are able to directly correlate the presence of high concentrations of pyocyanin to the rheological increase in modulus resulting from the release of pyocyanin quorum sensing.

398592 Determination of Evaporative Loss in Revolving Algal Biofilm Systems

Abstract:

Microalgae has tremendous potential in the production of products ranging from biofuels to pharmaceuticals and food. Algal cultivation has been mainly performed in an open pond or bioreactors, but recently, an attachment based Revolving Algal Biofilm (RAB) cultivation system was developed that not only demonstrated a better biomass productivity, but had more harvestable area. The downside of this system however is the high evaporation rate. The objective of this research was to develop a model to accurately predict evaporation from the developed algal growth systems. A model was developed to evaluate the evaporative water loss rate for varying environmental parameters that include solar irradiance, ambient air temperature, wind speed and relative humidity. A week long simulation of the model was performed using the algal systems at a greenhouse facility at Boone, Iowa, USA. The range of the evaporative water loss from the system was approximately 1.89L/m^2day, 0.95 L/m^2day, 3.67 L/m^2day for the Raceway pond, RAB in Raceway system and Trough RAB respectively. The model was considered to a beneficial tool to explore the performance and utility requirements of a greenhouse raceway and cultivation systems under varying environmental conditions.

Introduction: Biofilms are a collection of immobilized bacterial cells inside of a complex extracellular matrix, which improves the survival rate of the pathogens by modifying their local environment [1]. While biofilm removal is required in most industries, there are very few commercially available devices and no government guidelines for evaluating the removal of mature biofilms, because they are difficult to image [2]. Studies of biofilms and their removal are particularly important for treating chronic infections. *P. aeruginosa* is a medically relevant bacterium, which is known to form biofilms on medical equipment and cause severe infections. Microfluidic devices are particularly suitable for the study of biofilms, because they operate at the scale of systems they model and are easily incorporated into preexisting microscope setups. In this study, microfluidic devices are used to study the effects of shear stress on *P. aeruginosa* biofilms using a standard upright microscope.

Materials and Methods: A master silicon wafer with the desired design was made using a previously defined process [3]. The design consists of a main channel, which is connected to seven food channels through sub-micrometer constrictions of varying sizes. Polydimethylsiloxane (PDMS) was poured onto the wafer, heat cured, and separated from the wafer to make a PDMS chip. Access holes were drilled into the main channel and food chambers of the PDMS. Oxygen plasma bonding was performed to attach the PDMS chip to a glass microscope coverslip.

Cyan Fluorescent Protein labeled *P. aeruginosa* was cultured overnight in lysogeny broth (LB) at 37 °C. The food channels were filled with LB and the main channel was filled with 100 mM phosphate buffered saline (PBS). The main channel of the filled device was then inoculated with *P. aeruginosa* and incubated overnight at 37 °C to allow biofilms to form at the constriction entrances. A Fluigent MFCS pressure-driven flow controller, connected to an in-house air line, regulated the pressure applied in the main channel of the chip. Pressure was applied to flow PBS around the biofilm. The applied pressure was increased until the signs of biofilm removal were observed. A CCD camera connected to a Zeiss upright fluorescence microscope was used for the imaging. Brightfield and fluorescence images were taken sequentially.

Results and Discussion: A biofilm grew in the main channel around the constriction in the device overnight after overnight incubation (Figure 1). Bacterial cells were attracted via chemotaxis to the constriction entrances.

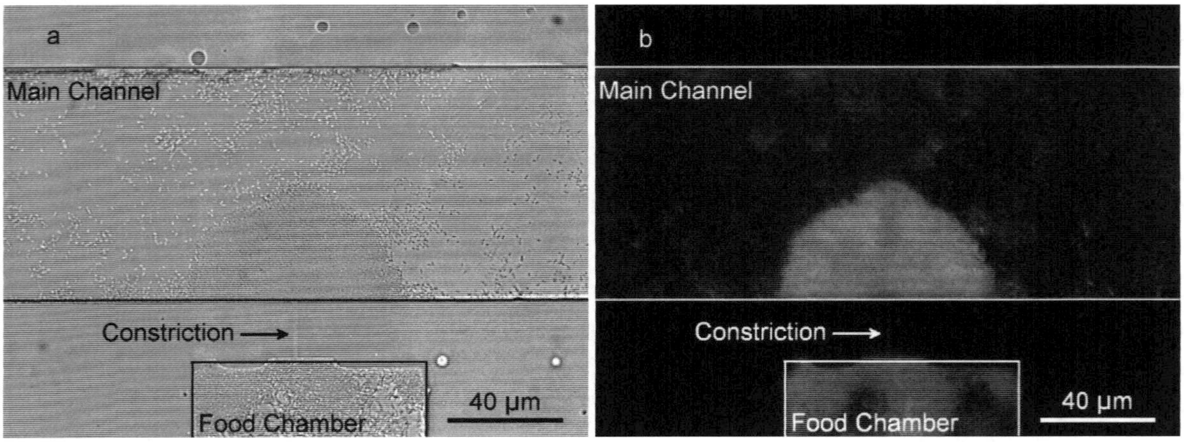

Figure 1: a) Brightfield and b) fluorescence images of the biofilm before flow is started.

PBS was then flowed through the main channel at 50, 75, and 100 mbar for ten minutes each. The experiment was stopped after applying 100 mbar because the biofilm began to be removed (Figure 2). At 100 mbar, a sufficient shear force was applied to the biofilm by the moving fluid to break apart the extracellular matrix. The removal is indicated by the decrease in fluorescence in the biofilm and a similar change in the amount of bacteria in the brightfield image where the fluorescence changed.

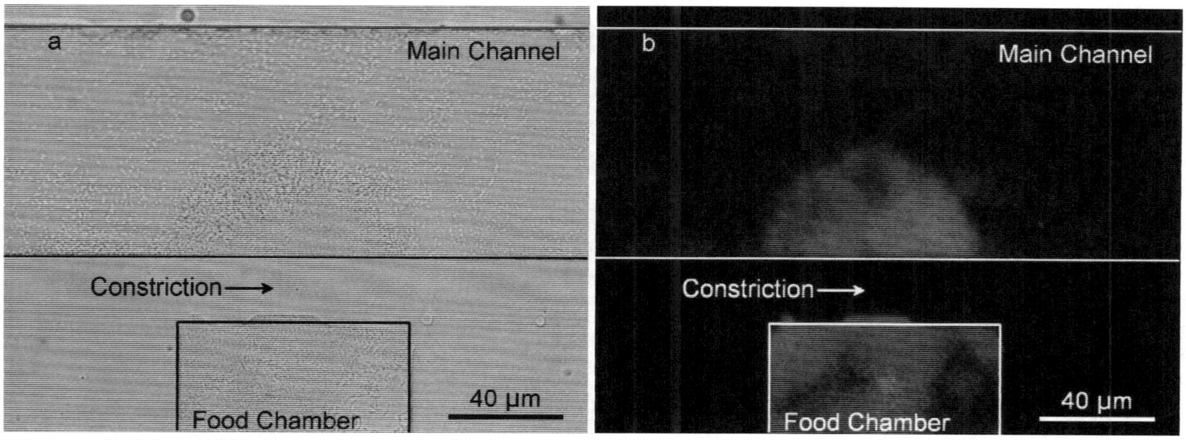

Figure 2: The PBS flows from the left to the right. a) Brightfield and b) fluorescence images of the biofilm after 10 min of flow at 100 mbar. A notable decrease in fluorescence is observed on the left side of the biofilm.

Conclusion: Biofilms were successfully grown and removed in microfluidic devices. The shear applied by the moving fluid detached the biofilm, which established a benchmark for *P. aeruginosa* biofilm removal. The pressure that the experiment was stopped at is the minimum threshold for biofilm removal. The device creates a biofilm that is turned horizontally so the flow is around the biofilm instead of over the biofilm. The horizontal plane allows the device to be integrated into a standard microscope setup. Further studies will combine antibiotics and sterilization products with fluid flow to evaluate their efficacy.

1. **Prüss, B. M., *et al.*,** "Environmental and genetic factors that contribute to *Escherichia coli* K-12 biofilm formation," *Archives of Microbiology,* **192** (9), pp. 715-728 (2010) .

2. **Vertes, A., et al.,** "Analytical Challenges of Microbial Biofilms on Medical Devices," *Analytical Chemistry,* **84** (9), pp. 3858–3866 (2012).

3. **Tandogan, N. *et al.*,** "Isolation of Microorganisms Using Sub-Micrometer Constrictions," *PLoS ONE*, **9** (6), pp. 1-7 (2014).

398713 Rule-Based Modeling of B-Cell Receptor Clustering

Abstract:

Transmembrane B-Cell Receptors (BCRs) can bind to extracellular antigens with high affinity and specificity, and initiate a signaling cascade leading to an immune response. While the cytoplasmic signaling components of B-Cells have been elucidated, the mechanism by which antigen binding transmits a signal across the plasma membrane is poorly understood. Several studies have shown that antigen induced clustering of BCRs on the plasma membrane is critical for B-Cell activation, suggesting that spatial reorganization is the mode of signal transmission. Experimental evidence has implicated the Cμ4 ectodomain of BCRs in their clustering, but no model has been reported which explains this phenomenon in detail. We have developed a rule-based model to describe how cooperativity between antigen binding BCRs and BCR dimerization at this domain leads to the formation of higher order oligomers. Using Monte Carlo simulations in the MCell software suite, we tested our model and statistically characterized the degree to which cooperativity leads to BCR clustering.

398676 Intracellular Heterogeneity: Seeing the Branches through Gaussian Mixture Modeling

Abstract:

Prior to comprehensive genetic profiling techniques, cancer-specific tumors were expected to be similar regardless of the patient; now, it is known that cancer tumors are in fact quite diverse and each tumor consists of diverse subpopulations. Common approaches to characterizing a cancer population include recovering the distribution of the average light intensity of channels per biopsy image; unfortunately, such approaches insufficiently characterize the biopsy. However, through genetic testing and in situ imaging of tissue sections via fluorescence-based immunohistochemistry, it is possible to analyze the spatial distribution of specific biomarkers within a single cancer biopsy. In this work, the intracellular heterogeneity (ICH) spatial distribution was analyzed via Gaussian mixture models and revealed the number of characteristics, or branches, required to suitably describe the distribution. Essentially, ICH spatial pattern recognition in cancer biopsies may aid clinicians' diagnosis and prognosis, and perhaps ease the development of personalized cancer treatments.

398140 Modeling the Compartmentalization of Tetrahydrobiopterin Production in the Yeast

Abstract:

Phenylketonuria (PKU) is a genetic disease that leads to an accumulation of the amino acid phenylalanine in the blood, resulting in stunted growth, seizures, and cognitive impairment. Some moderate forms of PKU respond to tetrahydrobiopterin (BH4) supplementation, which lowers the levels of phenylalanine in the blood, and are known as BH4-responsive PKU. Today, chemically synthesized BH4 (Kuvan®) is used to treat BH4-responsive PKU and is the only source of therapeutic BH4 in the US; however, costs can reach $137,000/year for an average patient. Our goal is to engineer the biological manufacturing of BH4 from glucose using Saccharomyces cerevisiae as the microbial host. By taking advantage of the different physicochemical properties of organelles, such as size, enzymatic activity, and pH, compartmentalization can lead to increased chemical yields. We propose to first localize the BH4 pathway to the mitochondria to investigate the effect of compartmentalization on enzyme activity and subsequent BH4 production levels. We have constructed a mathematical model in MATLAB loosely based off the Standard Compartment Model (SCM) to determine the placement of enzymes necessary for the optimal production of BH4.

Investigation of tailoring enzymes in pradimicin A biosynthesis

Eugenio Carlos Soto, Kandy Napan, Jixun Zhan
Department of Biological Engineering
Utah State University, 4105 Old Main Hill, Logan, UT 84322

ABSTRACT

Cloning and expression of four monooxygenases (PdmV, PdmE, PdmM, PdmM1) and one deacetylase (PdmP-3) to determine their roles in pradimicin biosynthesis.

BACKGROUND

Pradimicin A is a novel natural product synthesized by *Actinomadura Hibisca* P157-2. It functions as a broad spectrum fungicide and effective HIV entry inhibitor. It's unique lectin-like mechanism of action, shows potential to meet the challenge of multidrug resistance.

OBJECTIVES

- To determine the monooxygenase that hydroxylates C-14 in the biosynthesis of pradimicin A
- To test the removal of the methoxyl group at C-7 by an deacetylase in the biosynthesis of pradimicin A.

DESIGNS

In order to achieve our objectives we have three different but complementary designs.

A. **Expression of recombinant proteins, and in vitro reactions to prove the function of those enzymes.**
B. Co-expression of a set of pradimicin biosynthetic genes in a heterologous host.
C. Gene disruption in the pradimicin A producing strain using single and double crossover approaches.

METHODS

- PCR amplification of genes *pdmM, pdmE, pdmM1, pdmV,* and *pdmP-3* from the pradimicin biosynthetic gene cluster.
- Ligation of the PCR product into the pJET1.2 cloning vector.
- DNA sequencing to confirm correct amplification.
- Ligation of each gene into the expression vector pET28a.
- Transformation of the engineered plasmids into *Escherichia coli* BL21(DE3) and RIL.
- Induction of the T7 promoter by addition of 0.2 mM IPTG.
- SDS-PAGE to detect protein expression.
- Purification of the His_6-Tagged enzymes by affinity chromatography.
- Biotransformation of KN85a using IPTG induced *E. coli* BL21(DE3) /pES5 as a whole cell biocatalyst.
- Centrifugation of biotransformation supernatant was dried, then the cell pellet was extracted with methanol.
- LCMS to check biotransformation.

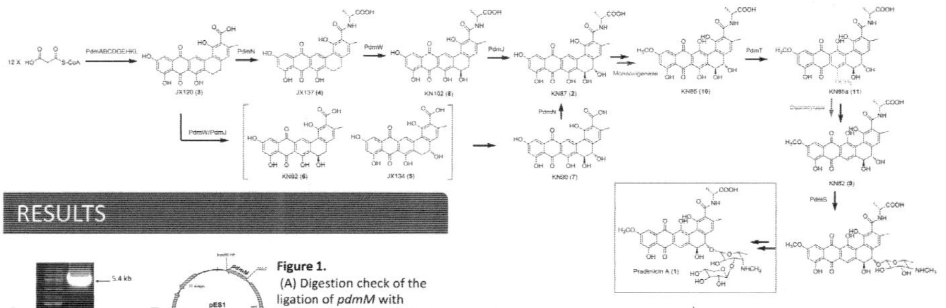

RESULTS

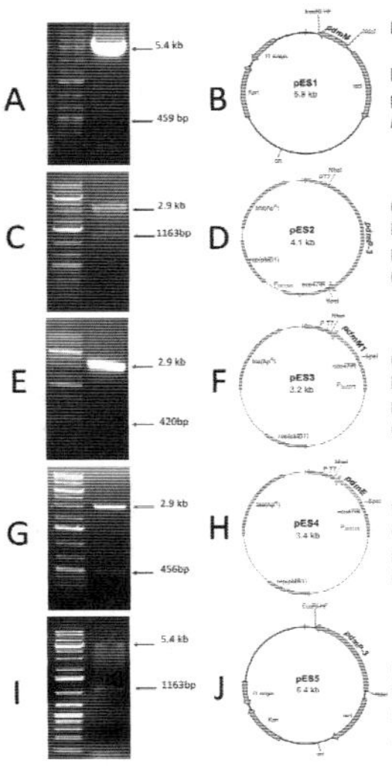

Figure 1.
(A) Digestion check of the ligation of *pdmM* with pET28a using *NdeI* and *EcoRI-HF*. (B) Map of pES1.
(C) Digestion check of the ligation of *pdmP-3* with pJET1.2 using *NheI* and *SpeI*. (D) Map of pES2.
(E) Digestion check of the ligation of *pdmM1* with pJET1.2 using *NheI* and *SpeI*. (F) Map of pES3.
(G) Digestion check of the digestion of *pdmE* with pJET1.2 using *NheI* and *SpeI*. (H) Map of pES4.
(I) Digestion check of the ligation of *pdmP-3* with pET28a using *NdeI* and *EcoRI-HF*. (J) Map of pES5.

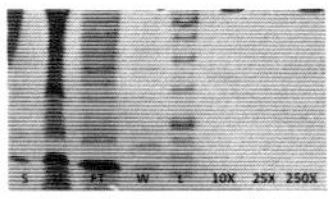

Figure 2. SDS-PAGE of PdmM.

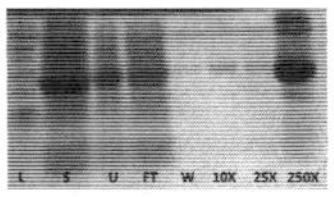

Figure 3. SDS-PAGE of PdmP-3.

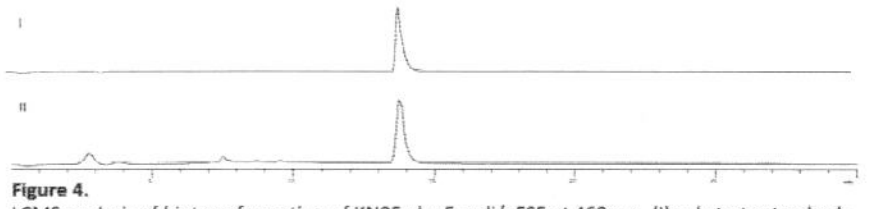

Figure 4.
LCMS analysis of biotransformation of KN85a by E. coli/pES5 at 460 nm. (I) substrate standard; (II) extract of the biotransformation broth.

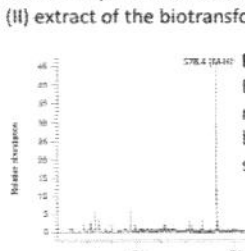

Figure 5.
ESI-MS spectra of the major peak in the biotransformation sample.

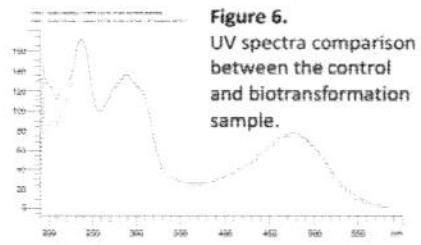

Figure 6.
UV spectra comparison between the control and biotransformation sample.

CONCLUSIONS

- Successful construction of pES1, pES2, pES3, pES4, and pES5.
- PdmM cannot be expressed in *E. coli* BL21(DE3) or RIL.
- PdmP-3 was overexpressed soluble in *E. coli* BL21(DE3) and purified using Ni-NTA resin.
- PdmP-3 could not biotransform KN85a in *E. coli* BL21(DE3).

FUTURE WORK

- Ligation of *pdmM1* and *pdmE* into pET28a.
- Expression of PdmM1 and PdmE in *E. coli* BL21(DE3).
- Biotransformation of KN87 by PdmM1 and PdmE in *E. coli* BL21(DE3).
- In vitro reactions with PdmP-3, PdmM1, and PdmE.

ACKNOWLEDGEMENTS

- Dr. Jixun Zhan and Dr. Ronald C. Sims for their great support.
- This work was supported by the National Institute of Allergy and Infectious Diseases (R15AI089347).

398327 Bionano Enzyme Conjugates for Efficient Bacterial Decontamination of Surfaces

Abstract:

Outbreak of severe infections due to unsuccessful decontamination of surfaces could affect both civilians and infrastructures thus leading to unfavorable socio-economic impacts. The development of the next generation of decontaminants needs to account for reduction of the logistical burdens associated with infection prevention when harsh chemical reagents are employed and should aim to use green-based technologies with minimum impact on the user and the environment.

In this work, we seek to create environmentally-friendly, self-sufficient, self-cleaning, enzymatic decontamination platforms that encapsulate bio-nano catalytic conjugates to prevent infectious outbreaks. Our technology uses two enzymes, namely glucose oxidase and chloroperoxidase, attached to a nanosupport to generate a potent decontaminant, hypochlorus acid (HOCl). The mechanisms of HOCl generation as well as the conditions that ensure enzyme immobilization, activity and maximum stability at the nanosupports, are investigated to determine the optimum circumstances that lead to maximum catalytic behavior at the nanointerfaces. The optimization is reported in terms of conjugates ability to efficiently decontaminate bacteria such as *E. Coli.* Implementation of such a technique could lead to self-sustainable catalytic interfaces with advanced decontamination capability and large consumer applications.

398627 Use of Peptoids Microspheres to Improve Enzyme Linked Immunosorbent Assay (ELISA) Microarrays

Abstract:

The pursuit of an efficient and economical technique for early detection of cancer has led to the development of sophisticated biosensors for antigen analysis. As biosensors become more complex, coatings are necessary and important. The preferred coatings need to have the following properties: high surface area for binding, resistance to biofouling, and flexible synthesis. This project focus on determining whether using peptoid microsphere-coated glass slides in the Enzyme Linked Immunosorbent Assay (ELISA) Microarray is a better technique than using regular commercially available glass slides. Current microarray slides are coated with reactive groups, such as amines, epoxides, or aldehydes. It is believed that using three-dimensional coatings will allow for increased binding efficiency of the capture reagent, and therefore increased dynamic range and sensitivity. Helical peptoids that are partially soluble in water have been shown to form microspheres when dried on a solid surface. In this project, peptoids will be synthetized, purified and characterized. Each peptoid will be dissolved in a protic solvent to stabilize the secondary helical structure and promote microsphere formation. The peptoids solutions will be administered onto the ELISA microarrays glass slides and allowed to dry in order to form uniform microsphere coatings. The peptoid microsphere-coated glass slides will be tested and compared against commercially available glass slides for use in ELISA microarray.

Introduction:
Both selectable and counter-selectable markers are useful tools for genome engineering. Selectable markers are mainly used for plasmid maintenance, engineered conjugation and genome manipulation. On the other hand, counter-selectable markers can aid in scar-less gene deletion, plasmid curing, and engineering double-crossovers. Since selectable and counter-selectable markers both have practical applications that can be of great utility, it is desirable to optimize dual-selectable markers, which can serve as both selectable and counter-selectable markers. Like counter-selectable markers, dual selectable markers also suffer from high counter-selection escape; this is one of the primary complications of finding suitable dual selectable markers.
Methods:
In this project we will aim towards the development of a selection marker for automated genome engineering. We will be working with the enzyme thymidine kinase (tdk) and demonstrate whether it can be used for coselection Multiplex Automated Genome Engineering (coS-MAGE). We will be alternating between selections that utilize either azidothymidine (AZT) or 5-Fluorodeoxyuridine (FDU) and thymidine through the cycles and determining where the break out point lies.
Results and Future Directions:
An effective dual-selectable marker is one that can survive for at least 10 cycles without breaking down. Our marker managed to survive and maintain stability throughout 10 cycles at three different concentrations (0.2μM, 0.5μM, and 2.0μM). The next steps in our project are verification and segment integration. If successful, tdk will serve as a useful tool for genome engineering applications and prospective synthetic biology research.

Mutational analyses of Amyloid beta using yeast model for Alzheimer's disease

Kris Casey[1], Pavithra Chandramowlishwaran[2], Yury Chernoff[2]
[1] School of Chemical and Biomolecular Engineering, Georgia Tech
[2] School of Biology, Georgia Tech

Alzheimer's disease (AD) is an age-dependent neurodegenerative disease that is characterized by the formation of amyloid plaques and tau tangles in the brain. AD is hypothesized to be triggered by an abnormal cleavage of amyloid precursor protein (APP) that produces Amyloid beta (Abeta), a protein that subsequently aggregates and forms deleterious plaques in the brain. Animal models for Alzheimer's disease are cumbersome to use owing to long incubation times to produce the disease symptoms and complexity of disease. Amyloids are also found in organisms such as yeast where they can form infectious amyloids called prions, thereby providing us a better model for studying amyloid formation by Abeta. The most well-characterized yeast prion is the prion form of Sup35, a translation termination factor. The prion domain of Sup35 (Sup35N) can readily misfold and aggregate to form prion in the presence of a pre-existing amyloid nuclei. Chernoff Lab at Georgia Tech has established an amyloid nucleation assay that consists of a fusion of Sup35N to Abeta, which upon overexpression can cause prion formation. The principle behind this is that Abeta provides amyloid nuclei that can readily convert Sup35N into a prion and this can be monitored as a phenotypic readout. The objective of my research is to study the effects of Abeta mutations on amyloid aggregation. I employed a molecular biology method called site-directed mutagenesis to create point mutations in Abeta to test against the wild-type Abeta using the established nucleation assay. Some of the mutations are familial mutations while others are predicted by recent protein-folding models. I was able to successfully study the effects of the predicted mutations in yeast and also biochemically characterize the amyloid aggregates. The data obtained from Abeta mutations in this yeast model can help us better understand the structure and folding patterns of Abeta, which can help up create better therapeutics for AD.

The RAS gene is a common oncogene and mutations in the RAS gene have been found in 30% of all human tumors. Therefore, targeting and inhibiting RAS signaling activity has been a common goal in many cancer therapy approaches. Although ligands that bind to RAS have been identified, a small molecule ligand that binds to a selective state of RAS has not been identified. Here, we describe the mRNA display method in generating a 10^{12}-10^{13} diverse DNA library that encodes numerous covalent peptides. The peptides undergo selection that simultaneously select for peptides with affinity to the active form of RAS and optimizes the binding and placement of the peptide ligand to RAS. A low molecular weight, high affinity peptide ligand that targets a selective state of the RAS protein has been identified through in vitro selection.

Development of a Paper-based Test for *Salmonella* dublin Detection in Cow Milk

Wyatt Self and Elain Fu
Chemical, Biological, and Environmental Engineering
Oregon State University
Corvallis, OR 97331

In the United States, the dairy industry is worth approximately $140 billion dollars. Dairy products are a huge nutritional part of many people's diet throughout the world. Diseases such as *Salmonella* dublin can cause health and economic issues for dairies everywhere. *Salmonella* dublin is also a zoonotic disease, meaning that it presents health issues for humans and other mammals, as well as cattle. As with all diseases, prevention and management are key factors for controlling disease severity. Current testing methods are lab-based, so limit accessibility.to testing. The purpose of this research is to create a rapid, simple to use, low cost, field use device that will test cow milk for *Salmonella* dublin.

This presentation will describe progress towards the development of a paper-based microfluidic test, which uses a colostrum sample and can present results in less than 20 minutes. The device design uses a sandwich immunoassay format in a completely disposable device composed of porous materials. Fluids move through these materials using capillary flow, so there is no need for pumping instrumentation. A simple card design has been developed and will be presented. Preparation of this device is quite simple. First a schematic is designed to with all of the parameters that are desired. Next the card and components are cut using a laser to ensure precision, one of the components requires additional processing before it can be incorporated into the card. Once prepared the components are added to the card. User steps are simple; dilute a colostrum sample, add fluids to appropriate pads, fold the device closed, and count the number of colored spots that develop. Due to its high viscosity colostrum does not flow well in the paper strips, therefore it must be diluted. Experiments investigating the flow properties of different colostrum dilutions and the effectiveness of different immunoassay capture approaches will be presented. Finally, key next steps will be discussed.

398483 Ultrasound Mediated Transfection of Yeast for Bio-Sensing Applications

Abstract:

Sonoporation is a facile method of transferring exogenous DNA into cells. In sonoporation, ultrasound is used to create transient pores in cell membranes to allow for the passage of foreign particles. This technique has proven feasible in several cell models, but it does not appear in the literature to have been tested on yeast. Yeast are of particular interest in this study because they are a very robust organism that is ideal for bio-sensing. The ability to easily and cheaply transfect yeast using sonoporation would be a large step towards an effective bio-sensor. Transfection is the process by which DNA external to the cell is passed into the cell and taken up into the nucleus. This process is important because it allows for the development of different types of bio-sensors and detectors. The full mechanism by which sonoporation works is not yet fully elucidated, but it is thought to primarily act through acoustic cavitation. The addition of microbubbles to the solution lowers the cavitation threshold and allows greater transient pore formation in the cell membrane. In this research, the sonoporation of yeast was studied using two cell impermeable fluorescent dyes, FITC-Dextran and Propidium Iodide. Ultrasound was applied to yeast colonies in the presence of microbubbles and fluorescent dyes. Optical and Fluorescence microscopy allows for the characterization of the uptake of FITC-Dextran and Propidium Iodide. The goal of this study is to determine if sonoporation is an efficient and easy way to transfect yeast.

Author: Alexis Prybutok
Advisor: Jennifer Maynard, Ph.D.
Department: McKetta Department of Chemical Engineering
Institution: The University of Texas at Austin

Evaluationg *in vitro* neutralizing activity of anit-Pertussis antibodies

In recent years, incidences of pertussis, or whooping cough, increased worldwide, particularly in infants.[1] This partly resulted from the introduction of a new acellular vaccine, which loses effectiveness long term, enabling pathogen growth in adults and transfer to infants.[2] Although generally mild in adults, pertussis causes hospitalization and/or death in infants too young to be vaccinated. Our lab develops a therapeutic consisting of two anti-pertussis toxin antibodies that help fight the disease by direct injection into sick individuals.

Pertussis toxin protein (PTx), secreted by *Bordetella pertussis*, majorly contributes to whooping cough symptoms such as promotion of leukocytosis, secretion of insulin, sensitization of histamines, and evasion of immune response.[3,4] PTx contains five subunits labelled S1-S5, where S1 is enzymatically active and S2-S5 bind to receptors. Two previously identified antibodies exhibit a high affinity for PTx: 1B7 and 11E6. The 1B7 antibody binds the S1 and S4 subunits, preventing toxin from entering the cell while 11E6 binds the S2 and S3 subunits, inhibiting toxin binding onto extracellular receptors[2,3] Due to the different epitopes and protective mechanisms, we hypothesized that the two antibodies might exhibit synergy, meaning they neutralize pertussis more effectively in combination than individually.

To test this synergy, I performed cellular assays, called Chinese Hamster Ovary (CHO) cell assays, to study the neutralizing ratio of various antibody combinations. CHO cells cluster together in the presence of active PTx, which I scored from 0-3 under a microscope based on cluster size and cell morphology after a suitable incubation time. Thus, I titrated in our antibodies with PTx solution and determined the antibody:PTx ratio at which clustering no longer occured, meaning antibodies fully neutralized the toxin. Using this assay, I compared mouse, human, and chimeric versions of 1B7 and 11E6, where chimeric antibodies contained the human Fc region and the mouse binding region. Data showed the humanized versions retained the ability to neutralize PTx. I also compared the required neutralizing ratio of 1B7 and 11E6, both alone and in various mixtures of the two antibodies. I confirmed that 1B7 and 11E6 exhibit synergy *in vivo*, and equimolar mixtures required the least amount of antibody to neutralize PTx. I also showed that this synergistic mixture required significantly less total antibody for neutralization than P-IVIG, the pertussis-reactive antibody fraction from pooled blood serum. Since P-IVIG is currently under investigation as a therapeutic option for pertussis, we believe that our 1B7 and 11E6 cocktail also exhibits therapeutic potential.[2] Currently, this mixture is being tested in baboons and thus far shows the ability to stop progression of the disease. I am also developing a new, more quantitative assay dependent on cell death counts to further verify all previous results as well as make the procedure more replicable.

References

1. Cherry, J.; Brunell, P.; Golden, G.; Karzon, D.; Report of the Task Force on Pertussis and Pertussis Immmunization—1988. *Pediatrics*. **1988**, 81, 933-984.
2. Sutherland, J.; Maynard, J.; Characterization of a Key Neutralizing Epitope on Pertussis Toxin Recognized by Monoclonal Antibody 1B7. *Biochemistry*, **2009**, 48, 11982-11993.
3. Sato, H.; Sato, Y.; Protective activities in mice of monoclonal antibodies against pertussis toxin. *Infect. Immun.*. **October 1990**, 58, 3369-3374.
4. Sato, H.; Sato, Y.; Ito, A.; Ohishi, I.; Effect of monoclonal antibody to pertussis toxin on toxin activity. *Infect. Immun.* **April 1987**, 55, 909-915.

398451 Development of Novel Sentinel Species-Based Bacterial Biosensors for Estrogenic Compounds

Abstract:

Estrogenic compounds are an important class of chemicals that have the ability to modulate the function of the nuclear hormone receptor, estrogen receptor beta (ERβ). These compounds can be found naturally in the environment or chemically synthesized, and have been linked to health risks ranging from altered sexual development to breast cancer. Our previous work resulted in an *Escherichia coli* cellular biosensor that is comprised of an engineered protein scaffold with an inserted human ERβ ligand-binding domain (LBD) fused to a thymidylate synthase (TS) reporter enzyme. Through growth in thymineless media, the biosensor can distinguish between estrogen agonists and antagonists based on the resulting *E. coli* growth phenotype. This research seeks to extend the ERβ biosensor method to include sentinel animal species, such as the zebra finch and wood frog, which serve as early indicators of the presence of environmental toxins. Preliminary findings using ERβ biosensors from several mammals demonstrated species-specific responses to ligands, suggesting species-specific susceptibility to certain endocrine disruptors. Using protein sequence alignments and PCR cloning, novel zebra finch and wood frog-based ERβ biosensors were created and each biosensor was validated against five known estrogens using the bacterial assay method. Mammalian and fish biosensors were also tested and a general trend of unique responses to weak and strong estrogens based on animal class was observed. The animal biosensors will be moved into higher sensitivity constructs in an attempt to improve ligand response. Adaptation of the ERβ biosensor to sentinel species, in particular, will enable the prevention and treatment of diseases through early detection of estrogenic compounds in the environment and expand the bacterial biosensor library. The creation of additional animal ERβ bacterial biosensors will also provide a broader understanding of how drug response deviates across species.

We have developed modular DNAzyme sensors that are capable of detecting oligonucleotides, small molecules and plasmids. These sensors are designed with individual target and reporter modules, which has allowed for rapid production of sensors capable of detecting unique targets. The inactive DNAzyme complex finds its destination and attaches through toehold mediated strand displacement. Combined with the use of a fuel sequence, which is universal for all of our sensors, the inhibitor will be fully displaced from the inactive DNAzyme. This allows for the now active DNAzyme to bind with the substrate strand and cleave the substrate. (Fig1)

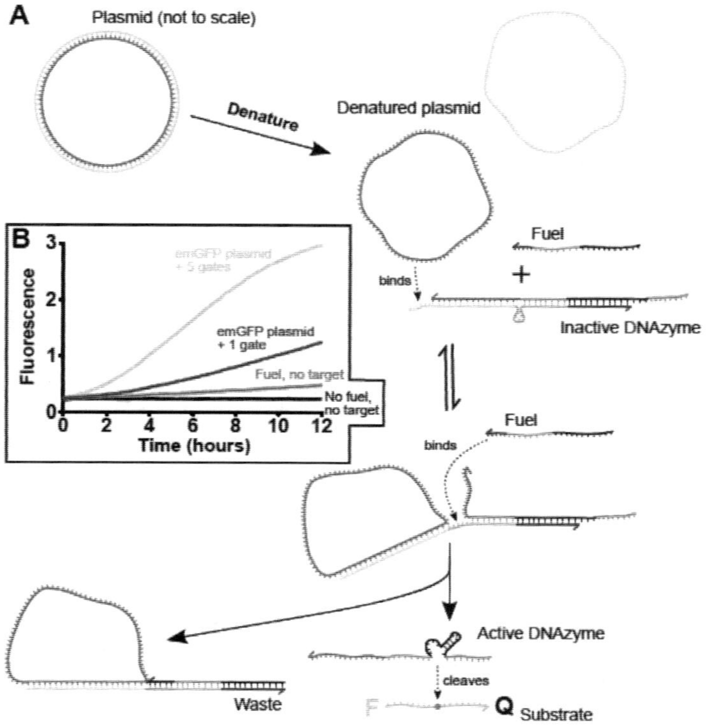

The most noteworthy achievement is the ability to detect plasmids. Using an isothermal alkali treatment of the plasmid containing solution, we denature the plasmid strands. Our modular design has allowed us to create five sensors that detect five different regions on plasmids. These five regions lie on two different plasmids, SNAP25 and emGFP. Having three identical regions on the two plasmids and one unique to each has allowed us to conduct controls for our sequences and to check for specificity in our sensors. The sensors are very target-specific and show a limit of detection nearing pM levels. We also have shown that replacing the target region with an aptamer structure sequence allows for us to detect ATP in solution, which shows the potential to detect other small molecules by using a design similar to that which we have already developed.

The ability to detect target regions in DNA opens doors for taking advantage of many viral and bacterial strains that are well known. There are possible applications in the medical and agricultural science fields. A potential example is the development of multiplexed assays that could detect pathogens in foods or to be used by medical professionals to find bacterial infection.

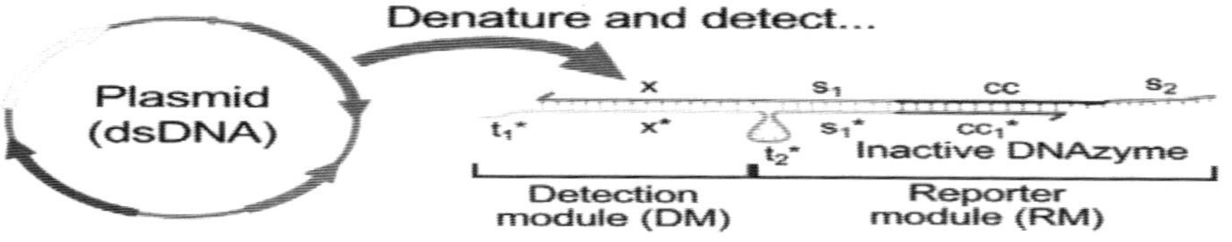

398370 Optimization and Effects of Sequential Mixing on Plasmid DNA-Polyethylenimine-Polyaspartic Acid Ternary Complexes As Synthetic Gene Delivery Vectors

Abstract:

Gene therapy is the process of inserting genetic material that codes for a therapeutic protein into disease affected cells with the intention of treating the disease. Originally modified viruses were used to deliver genetic material, but this proved to be unsafe due to severe immunoresponses from treated individuals and unspecific cell targeting of the viruses. Recently, complexes consisting of plasmid DNA, coated by a cationic material such as a polymer or lipid prepared by simple bulk mixing, have been studied and used as non-viral gene delivery vectors. Addition of a third material to binary complexes, creating a ternary complex, has been shown to increase efficiency and functionality of the vectors. In this study we optimize a ternary complex of pDNA, polyethylenimine, and polyaspartic acid prepared by conventional methods and then test the effects of novel preparations on the characteristics of the complex.

394736 Characterization and Optimization of α-Polyglutamic Acid and Branched Polyethyleneimine Polyplexes for Gene Transfection Via a Combinatorial Approach

Abstract:

Gene transfection holds tremendous promise for treatment of a wide range of acquired and genetic diseases via human gene therapy and for the production of recombinant proteins with post-translational modifications on an industrial scale. Implementation of these techniques is hindered, however, by the lack of safe and efficient methods for delivery of genetic material. Conventional cationic polymer/DNA complexes (polyplexes) appear unlikely to perform all of the functions necessary for efficient transfection. Ternary complexes, comprising multiple polymers, provide an opportunity to incorporate additional functions. To explore this approach, 40 unique polyplexes composed of DNA, polyethyleneimine (PEI), and α-polyglutamic acid (PGA) were prepared via a combinatorial approach by simultaneously varying the PEI:DNA and PGA:DNA ratios. The particle diameter and zeta potential of each polyplex were determined. Transfection efficiency and cell viability experiments were performed in growth media with and without serum. Only a 5% increase in the average hydrodynamic radius was observed upon addition of α-PGA to the PEI/DNA complexes. In serum free media, the optimal polyplex demonstrated 2.9 times the transfection efficiency of the optimal PEI/DNA complex and 24.3 times the transfection efficiency of Lipofectamine 2000. In media with serum, the optimal polyplex exhibited a transfection efficiency 6.9 times greater than that of the optimal PEI/DNA complex and 2.6 times greater than that of Lipofectamine 2000. Additionally, the α-PGA layer caused statistically significant increases in cell viability relative to PEI/DNA complexes and Lipofectamine 2000 in growth media with and without serum.

398918 Validation of a Systems Model for Human Bronchial Epithelial Cells with and without Cystic Fibrosis

Abstract:

Cystic fibrosis (CF) is an autosomal recessive disorder caused by defects in the cystic fibrosis transmembrane conductance regulator (CFTR) gene. This defect leads to absent or malfunctioning chloride channels on the airway epithelium. Sodium absorption across the epithelium is also typically increased through the ENaC channel. Together, these effects lead to increased airway surface liquid absorption, the accumulation of dehydrated mucus, and respiratory failure.

We seek to measure liquid and solute transport in a human bronchial epithelial (HBE) cell model in order to support the development of a systems model of airway epithelial physiology. Using a small radiolabeled molecule, diethylene triamine pentaacetic acid (DTPA), we are able to quantify airway surface liquid (ASL) volume changes in both CF and non-CF human bronchial epithelial cell cultures. Experiments used apical mannitol addition to modulate ASL absorption rate. Apical mannitol addition slows DTPA clearance by creating a transepithelial osmotic gradient determining direction of water absorption, and is used as a "zero convection case" which allows us to determine whether there are basic permeability differences between CF and non-CF epithelia. A hyposmotic case was also tested using apical addition of diluted (150 mOsm) Ringers solution, thereby creating an environment favoring liquid absorption. These results were compared to the findings for the same procedure using undiluted (300 mOsm) Ringers solution. An optical method is used to calculate these changes in ASL volume, which are compared to DTPA absorption findings. These studies are important tools to verify the accuracy of our systems model of cell volume regulation.

Validation of this model in response to osmotic challenges was performed using the data sets described previously. In response to hyposmotic Ringers addition, running the model under validation conditions produced the same trends that were shown experimentally: an increase in both DTPA and liquid absorption in both CF and non-CF HBE lines. In CF HBEs, the model predicts a subtle difference in absorption dynamics between the control and experimental groups, recognizing that CF cells hyperabsorb liquid even in an isosmotic environment. Model training in response to hyperosmotic mannitol showed deviations from the Ringers data. Validation of this parameter set against a separate set of filters obtained under identical conditions demonstrated a reduction in liquid and DTPA absorption in CF and non-CF HBEs – mirroring earlier outcomes of the model fitting data. So, while the model predictions and the data under mannitol challenge show similar trends, dynamic behavior, particularly at later time-points, is not matched between the model and the data, which may indicate the action of physiological mechanisms yet unincorporated in the model structure.

In vitro, apical mannitol addition is shown to reduce DTPA clearance and liquid absorption in CF and non-CF HBEs, while hyposmotic Ringers addition increases DTPA clearance and liquid absorption. Experimental data for these test cases serve as validation sets for cell behavior predictions via a systems model. These experiments and the development of a systems model are important tools for determining more effective and personalized treatments for CF patients in the future.

395971 The Efficacy of Poly(trolox ester) Antioxidant Therapy in Cancer Metastasis Inhibition

Abstract:

Circulating tumor cells that metastasize from a primary tumor are responsible for 90% of cancer-related deaths in the United States. These circulating tumor cells can adhere to the endothelial lining of the vascular system and extravasate to form deadly secondary tumors. Reactive oxygen species (ROS) are produced within the endothelial cells of the vasculature from shear stress of blood flow, but are balanced by natural occurring anti-oxidants. Loss of blood supply to endothelial cells from invasive primary tumor removal surgeries can cause an imbalance in the consumption and production of ROS causing oxidative stress and also upregulate cellular adhesion molecules. If circulating tumor cells are within the system while endothelial cells are experiencing oxidative stress, cancer cells may adhere to the endothelial lining more readily. Static studies were conducted on a monolayer of human umbilical vein endothelial cells (HUVECs) to investigate the role of oxidative stress and inflammation on cancer cell adhesion to endothelial cells. The HUVECs were exposed to factors related to the inflammatory response such as TNF-α and a hypoxanthine (HX)-xanthine oxidase (XO) reaction to simulate the conditions the endothelial cells experience during extensive surgeries. This project explains the correlation between the generation of ROS, the upregulation of cellular adhesion molecules, and the adhesion of breast cancer cells to a HUVEC monolayer. The studies were repeated using antioxidant nanoparticle treated HUVECs to determine the efficacy of Poly(trolox ester) antioxidant nanoparticles.

Opioid receptors have been known to dimerize on the cell surface, but it is unclear if they can dimerize in the endoplasmic reticulum before being trafficked to the cell membrane. The objective of this project is to find where these receptors dimerize and determine if these are homodimers or heterodimers. In order to do this, florescent protein was ligated into each opioid receptor subtype by a series of PCR, digestions, and transformations. Next, the plasmids were transfected into Human Embryonic Kidney 293 (HEK) cells. The membrane expression of the receptors was determined by Total Internal Reflection Florescence (TIRF) Microscopy. The next step in the project will be to conduct a single molecule study and a Forester Resonance Energy Transfer (FRET) test.

Identification of enhancer elements for Onecut1 and Onecut2 genes in retinal development

Abstract

Onecut1 (OC1) and Onecut2 (OC2) are two genes essential for the differentiation of retinal cells during early embryonic development. This project tries to identify potential enhancer sequences that may play a role in regulating the gene expression of OC1 and OC2 using chick and mouse models. To do so, several evolutionary conserved sequences across various species were identified as candidate enhancers. These sequences were then cloned into a reporter vector and electroporated into the retinal cells. The correlation of OC1 and OC2 protein expression to the reporter expression driven by potential enhancer sequences were monitored using immunostaining procedures. Several sequences were identified as likely involved in regulating the expression of both these genes using this criteria. Information obtained about these enhancers can be used to better understand early eye development and problems that may arise during development resulting in retinal diseases.

TITLE: APTT, PT, and spectrophotometric analysis of coagulation cascade activation and inhibition

Heidi Oldenkamp, Dr. Cristina Puy, and Dr. Owen J.T. McCarty
Department of Biomedical Engineering, Oregon Health and Science University, Portland, OR, USA

Background: Hemostasis is the cessation of blood loss from a damaged blood vessel. The damaged vessel wall is covered with fibrin and platelets to stop bleeding and to repair the vessel injury. Contrarily, thrombosis is the formation of a clot inside a blood vessel, obstructing the flow of blood. In the United States, 300,000 to 600,000 people are affected by deep vein thrombosis (DVT) every year, according to the CDC. Incidences are higher in the elderly (over 80 years of age). There are several parts of the coagulation cascade (hemostasis), the process that is activated when a blood vessel is injured and the endothelial cells are exposed to proteins in the blood that cause it to clot. The intrinsic pathway begins with the activation of blood coagulation factor XII (fXII), while the extrinsic pathway begins when factor VII (fVII) is activated by tissue factor. The two pathways meet at the activation of factor X (fX) and the process continues until the conversion of soluble fibrinogen to insoluble fibrin strands. Humans with severe factor XI (fXI) deficiency are protected from DVT, suggesting that fXI activation contributes to the pathogenesis of DVT.

Aims: To examine the effect of inhibitors that block the intrinsic pathway of the coagulation cascade in platelet-poor plasma and determine if these inhibitors could potentially be used as anticoagulant treatment for patients with DVT.

Experiments: Citrated platelet-poor plasma (with various concentrations of the four inhibitors) is activated by calcium. Two different tests (activated partial thromboplastin time [APTT] and prothrombin time [PT]) were run in the Trinity Biotech KC4 Delta semi-automatic coagulation analyzer. This machine uses micro-mechanical (magnetic) clot detection technology. The continuously rotating reaction cuvette contains a steel ball held in place by a magnet. The timer stops when the magnetic sensor can no longer hold the steel ball in place, indicating that a clot has formed in the plasma. The two tests use different reagents, activating different pathways of the coagulation cascade. A chromogenic substrate assay was also used, with testing in a spectrophotometer.

Results: The four inhibitors all act on specific parts of the intrinsic pathway of the coagulation cascade. Therefore, they all affect the clotting time differently. Inhibitor 1A6 increases the clotting time greatly because it affects the activation of two coagulation factors rather than just one. Inhibitor CTI blocks the active site of fXII, while 10C9 and 14E11 both affect parts of fXI.

Conclusions: The data shows that these inhibitors could reduce thrombus formation *in vivo,* and therefore could potentially have a use in anticoagulant therapy for DVT patients. The APTT and PT tests are useful for many applications in blood coagulation analysis.

Future Work: The next step for this project is rheological testing in Dr. Travis Walker's lab at Oregon State University. Examining the rheological properties of plasma coagulation with these inhibitors could potentially assist in the development of an anticoagulant therapy. This type of test is already used for point-of-care testing and when quick analysis of a patient is needed.

398356 Stochastic Electrotransport of Activity-Modulated Molecules for Rapid and Scalable 3D Phenotyping

Abstract:

For decades, sectioning-based two-dimensional molecular phenotyping techniques have been extensively used for investigation of tissue samples across biology and medicine. These techniques ensure that molecular targets in ultrathin tissue slices experience relatively similar reaction conditions (e.g. probe concentration and reaction time) to achieve complete and uniform labeling of given samples. However, three-dimensional information is lost unless a sophisticated and laborious 3D reconstruction is employed. CLARITY technology has succeeded in proving that an intact, full brain can be labeled with molecular probes, but penetration is poor because passive diffusion of molecular probes in the dense nanoporous mesh is slow (Chung et al., Nature 2013). Even near the surface, labeling is highly non-uniform due to large variations in probe concentration and in the probe-target interaction time across superficial layers. Moreover, labeling of the crucial inner structures remains incomplete even after months. These critical limitations of the current techniques have prevented us from obtaining system-wide quantitative molecular information from large scale intact tissues. Here we introduce a new technology (termed eTANGO) that addresses these challenges by integrating two novel concepts: stochastic electrotransport and dynamic affinity shift. The penetration problem is solved by stochastic electrotransport, which selectively and rapidly drives only charged molecules (e.g. antibodies and RNA probes) without disrupting the surrounding charged matrix (e.g. brain). Furthermore, probe-target binding affinities are dynamically modulated to synchronize reaction times brain-wide. Integration of these two concepts enables all the endogenous molecular targets in billions of cells to experience the same reaction condition (time and concentration). This yields complete and uniform staining of the intact brain within hours. We applied eTANGO to immunostain various cellular, molecular and structural markers a variety of molecular probes, including organic dyes, carbohydrate-binding proteins, and immunoglobulins. The unique strength of eTANGO -- complete homogeneous immunostaining -- opens the possibility of high-throughput brain-wide proteomic profiling at single cell resolution. Together with CLARITY, we anticipate eTANGO to facilitate integrated understanding of large-scale intact biological systems.

396352 Development of a Microfluidic Emulsion Generating Device for Increased Throughput

Abstract:

This paper deals with the creation of monodisperse emulsions using a microfluidic device. Emulsions can be used to simplify many tedious tasks such as PCR. In just 1 mL of sample volume, parallel reactions can take place in $>10^{10}$ droplets of fL – nL size, greatly reducing reagent consumption and time over standard plate reader assays. For many of these tasks, the interpretation of assay results relies on the assumption that all droplets receive the same amount of starting material. Therefore, droplet volume is an important factor. Previous microfluidic chips used, either slowly generated monodisperse emulsions, or rapidly generated more polydisperse emulsions. We have parameterized a multi-channel, emulsion- generating device to rapidly generate droplets, while improving the consistency of emulsion size. The parameters selected for the new device include the creation of a thinner chip, a minimization of the number of inlets needed, and the utilization of an O-ring. The new device generated a monodisperse emulsion with a diameter range of 20-60 μm. This device still needs further testing to determine its viability as a substitute for the single channel generator. This testing would include using the device to generate emulsions with different concentrations of DNA and running PCR on the droplets.

Interfacing Cancer Cells with Silicon Nanowires on Soft Substrates

Alex R. Heatherly, Elizabeth M. Haley, Yonghyun Kim and Hung-Ta Wang

Department of Chemical and Biological Engineering,
The University of Alabama, Tuscaloosa 35487, United States.

Silicon nanowires with a high aspect ratio provide direct access to cells' interior, which is a promising approach towards the long-sought goal of single cell proteomics. Because nanowires are grown in an epitaxial manner on a silicon substrate, cells would be intruded by nanowires while forming an extracellular matrix with the physical environment of a hard silicon substrate. Literature has shown that cells could sense the mechanical property of the substrate, and consequently influence their intracellular signaling. In this work, a novel nanowire platform on an elastic substrate of PDMS (polydimethylsiloxane) was fabricated. Cancer cells (Glioblastoma U87-MG cells) were cultured on top of the nanowire/PDMS substrate, and it was found that cells were normally proliferated and interfaced with nanowires by forming anchor points or being impaled. This work suggests that the novel nanowire/PDMS platform could be a better method of accessing intracellular information using nanowires.

Reliable Bonding of Microfluidic Devices to Gold for Surface Plasmon Resonance Imaging of Bacteria

Richard T. Crowley, Nil Tandogan, Pegah N. Abadian & Edgar D. Goluch

Introduction:

Prior research has shown that the combination of surface plasmon resonance imaging (SPRi) and microfluidics allows for high-resolution imaging of bacterial biofilm growth.[1] However, designing such a system is a daunting task. When fabricating PDMS microfluidic devices, pieces are often permanently bonded to glass by treating the glass and PDMS surfaces with oxygen plasma.[2,3] When the glass is replaced with gold, as is necessary for SPRi, this method will no longer work and a new bonding method is required.

Prior research has shown that the use of (3-mercaptopropyl)trimethoxysilane (MPTMS) as an adhesion promotion layer creates a permanent bond between PDMS and Au.[4,5] The thiolated end of the MPTMS molecule favors interactions with the gold surface, while the silane is able to bond with the PDMS network, as shown in **Figure 1**. Various methods of monolayer deposition including liquid immersion, spin coating, and vapor depositon were assessed and it was determined that functionalization of gold surfaces using a solution of MPTMS in ethanol provided the most consistent bonding.[5] We propose the use of a MPTMS self-assembled monolayer to securely bond PDMS microfluidic devices to the gold surface of SPRi slides and prisms. This combination of microfluidics and SPRi will allow for real-time, label-free imaging of bacterial susceptibility to multiple antibiotic mixtures and concentrations.

Materials and Methods:

Microfluidic devices were fabricated by curing PDMS (Sylguard 184) on top of a silicon master mold (base : curing agent = 5:1). The device design consists of a main channel with submicron constrictions branching off it and connecting to seven food chambers. Preliminary experiments to observe interactions of *Staphylococcus aureus* with varying concentrations of penicillin streptomycin were conducted in PDMS devices bonded to glass and observed with an optical microscope.

Gold-coated SPRi slides (HORIBA) were modified in a 25mM MPTMS solution in absolute ethanol for 2 hours, rinsed with absolute ethanol, and dried with argon. PDMS devices were treated with oxygen plasma (Anatech SP-100) and placed on the modified gold surface. The food channels were filled with 1000X and 200X dilution of penicillin streptomycin and the main channel was filled with a bacterial solution of *S. aureus* in lysogeny broth growth media.

Results and Discussion:

Filling of devices bonded to MPTMS functionalized gold showed that the MPTMS adhesion layer provided a leak free seal for the microfluidic network. Growth of *S. aureus* in new devices showed the that the toxicity of MPTMS did not have any affect on the growth of the bacteria.

In preliminary experiments with the constructed system, the effect of the penicillin/streptomycin (pen/strep) cocktail concentration was observed, as the growth of *S. aureus* was inhibited in food chambers containing higher concentrations of pen/strep.

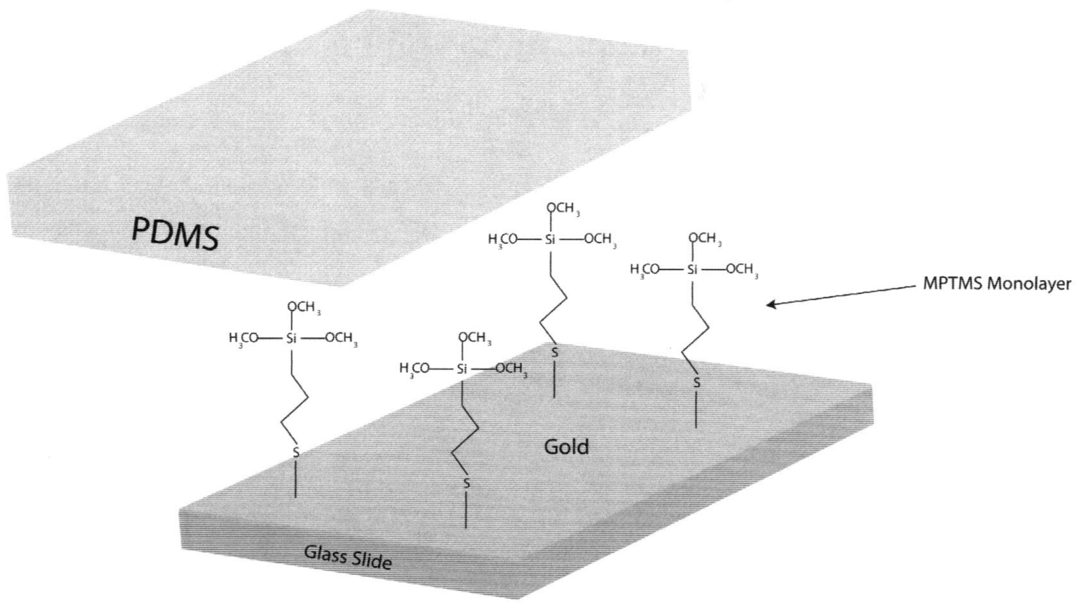

Figure 1: MPTMS Adhesion Layer on Gold

References

1. Abadian, P.N., Tandogan, N. Jamieson, J. J., & Goluch, E.D. "Using Surface Plasmon Resonance Imaging to Study Bacterial Biofilms." *Biomicrofluidics* **8**(2) 021804 (pp.1-11) (2014)

2. Xiong, L., Chen, P., & Zhou, Q.Adhesion promotion between PDMS and glass by oxygen plasma pre-treatment" *Journal of Adhesion Science and Technology* **28**(11) pp. 1046-10545 (2014)

3. Tandogan, N., Abadian, P. N., Epstein, S., Aoi, Y., & Goluch, E.D. "Isolation of Microorganism Using Sub-Micrometer Constricitons" *PLoS One* **9**(6) e101429 (pp. 1-7) (2014)

4. Byun, I., Coleman, A.W., & Kim, B. "Transfer of Thin Au Films to Polydimethylsiloxane (PDMS) with Reliable Bonding Using (3-mercaptopropyl)trimethoxysilane (MPTMS) as a Molecular Adhesive." *Journal of Micromechanics and Microengineering* **23**(8) 085016 (pp. 10) (2013)

5. Atmaja, B., Frommer, J., & Scott, J.C. "Atomically Flat Gold on Elastomeric Substrate." *Langmuir*
22(10) pp. 4734-740 (2006)

398767 Exploring Applications for Digital Microfluidics in the Realm of Medical Diagnostics and Environmental Sampling

Abstract:

Exploring Applications For Digital Microfluidics In The Realm Of Medical Diagnostics And Environmental Sampling

Carmen Gondhalekar and Branden Kusanto

Sandia National Laboratories and Oregon State University

Abstract:

The study of digital microfluidic technology is an emerging field in which scientists use a substrate patterned with electrodes and electrostatic forces to manipulate micro-sized volumes of solution and perform experiments in vitro. Droplet manipulation through the use of remote controlled devices allow for automated DMF systems which require little or no human handling. This technology has multiple implementations which can prove beneficial in occasions, such as emergency response, where portable low-cost point-of-care systems are needed to conduct assays to detect harmful agents. Sandia National Laboratories in Livermore California is implementing DMF technology in nuclear radiation detection and disease diagnostics. We worked together with a team of Sandia engineers and biologists to explore four different functions of Sandia's DMF system in the fields of radiation detection, disease biomarker detection, and viral RNA extraction from blood.

Automated, Tablet-Based Microscope for Imaging 3D Culture Models of Breast Epithelial Growth and Behavior

Sarah McFann[1], Benjamin Ricca[2], Mike D'Ambrosio[2], Johnathan Wang[2], and Daniel Fletcher[2]
[1]Department of Chemical and Biological Engineering, The University of Alabama
[2]Department of Bioengineering, The University of California, Berkeley

Mammary epithelial cells within the body reside in a chemically and mechanically complex 3D environment known as the extracellular matrix (ECM). Though the ECM has been shown to play an important role in both the development and maintenance of breast architecture, the role of external forces that deform the matrix surrounding the epithelial cells is still unclear. In this study, we applied a compressive strain that generated a transient force to a 3D malignant breast epithelial culture and found that single malignant cells could be "phenotypically reverted" to develop into growth-arrested, polarized structures. To further characterize the inputs that lead to this type of mechanically induced phenotypic reversion, the effectiveness of mechanically induced phenotypic reversion, and the pathway through which mechanically induced phenotypic reversion occurs, we needed a high-throughput method capable of imaging multiple developing 3D cell cultures in real-time. To address this need, we developed an inexpensive, tablet-based, automated microscope to collect time-lapse images of the growth and behavior of 3D breast epithelial models.

396377 Graphene Field Effect Transistors for Breast Cancer Diagnosis

Abstract:

We report the manufacturing of a device to detect the cancer indicative biomarker HER2 (human epidermal growth factor receptor 2) protein. The genetically engineered single chain variable fragment of the HER2 immunoglobulin, with high selectivity towards the HER2 protein, was attached to the graphene field effect transistors (GFETs) using 1-Pyrenebutyric acid N-hydroxysuccinimide ester as the linking molecule. Electron transport properties of the GFETs are measured before and after each chemical alteration. A shift in the current vs. gate voltage curve is observed after the addition of the biomarker that is dependent upon the concentration used. A Hill – Langmuir curve was then fit to the data in order to quantitatively track the response of the devices as a function of the HER2 concentration. Concentrations ranging from the pg/mL to μg/mL could be successfully detected and differentiated.

398798 Single Cell Encapsulation Using a Droplet-Microfluidic Array

Abstract:

The traditional approach of studying cell-behavior involves performing bulk analysis capable of only average measurements of how an entire population of cells responds to external treatments. Results obtained from such analysis can often be inaccurate because they do not account for the behavior of individual cells that often times respond differently from other cells in a heterogeneous population. Droplet microfluidics is one way of bypassing the complications that arise from such cellular heterogeneity. In this project, we sought to design, fabricate and optimize a two-phase droplet microfluidic device capable of isolating individual cells in picoliter-sized droplets. The underlying principle of the device is to encapsulate cells in discrete aqueous droplets in a continuous oil phase. The droplet generators were fabricated using poly(dimethyl siloxane) (PDMS) by established soft lithography and PDMS replication techniques. The fluidic channels were treated using an Aquapel solution to alter the surface properties of the PDMS to make it more hydrophobic. Succeeding error correction, the flow rates of both the inlet aqueous and oil streams were optimized to continuously produce droplets of reliable size. Cell encapsulation rates were performed using HeLa cells to ensure single cell trapping in each droplet. Finally, the system was modified to include a droplet trapping array downstream of the flow focusing junction. The array was designed to trap the aqueous droplets by taking advantage of the difference in fluid densities and the surface tension of the drops to perform the time-dependent analysis of cellular behavior, an aspect of droplet microfluidics that is currently limited. The results presented here are the first step in the development of a new microfluidic platform capable of quantifying intracellular kinetics at multiple time points in intact single cells in a high throughput manner.

398717 Microfluidic Device for Washing Frozen/Thawed Erythrocytes

Abstract:

Over 15 million units of erythrocytes (red blood cells) are transfused every year in the United States. Cryopreservation of erythrocytes would help prevent shortages and provide a larger bank of rare blood types to draw from for transfusion. Unfortunately, cryopreservation is accompanied by the time consuming process of deglycerolizing erythrocytes. Glycerol is a common cryoprotectant used to preserve erythrocytes as they are frozen. Its ability to permeate cell membranes allows protection from osmotic damage that plagues cells as they are being cryopreserved. Unfortunately, glycerol is toxic to cells after an extended exposure, so it must be removed after thawing of the cells. This is traditionally done in batch process in which it takes and hour or more to deglycerolize a single unit of blood. With a three-step process, the deglycerolization time has been shown to be reduced to around three minutes. In the three-step process, a 12% saline solution is introduced to the erythrocytes, which draws the glycerol from the cells through osmotic pressures. A 3.4% glycerol solution is then used to dilute the 12% saline solution and finally a 0.2% glycerol solution is used to return the cells to isotonic conditions. The success of the deglycerolization process is dependent upon the amount of time the erythrocytes spend in the 12% saline solution. Too short a time spent in the 12% saline solution, and when the 3.4% saline solution is introduced, water enters the erythrocytes rapidly and they lyse; too long spent in the 12% saline solution, the cells shrivel and die. This three-step batch process, however, requires a large volume of solutions. A microfluidic device, such as the one used in this project, decreases the volume of saline solutions needed to preform the three-step deglycerolization process and is able to run the process continually. The residence time within the device corresponds to the amount of time the erythrocytes spend in the 12% saline solution, and 3.4% saline is added within the device to dilute the 12% saline. The 0.2% glycerol solution is added manually as the last step of the three-step process. The goal of the project is to reinforce the idea that rapid deglycerolization is possible with low hemolysis and that the microfluidic device can preform the same function as the batch process. Ultimately, as residence time that allows for under a 20% hemolysis to match the data found in the previous study is to be consistently achieved. So far, the general trend of the previous study's hemolysis data has been replicated using the microfluidic device, but a hemolysis value under 20% has not been achieved. Working towards an ultimate goal of a 20% or less hemolysis value, more residence times must be evaluated and repeated trials to test consistency should be performed.

396280 The Effect of pH and Ionic Strength on the Filtration of Bacteriophage φX-174 through a Commercial Virus Removal Filter

Abstract:

Many mammalian cell lines used to produce recombinant protein products naturally produce viruses that can pose a serious threat to human health. Effective virus clearance is a crucial part of bioprocessing, and virus filtration membranes are a major component of the overall virus removal strategy. These membranes require high levels of virus retention and high transmission of the desired protein product. Virus filtration complements other virus clearance steps and is thought to provide high virus removal based on a size-exclusion principle. The pores of virus filtration membranes are designed to allow antibodies (~8 nm) to pass through the membrane, while the viruses (≥20 nm) are retained. However, there is some evidence that virus retention may also be a function of solution conditions, suggesting that adsorptive phenomena may also play a role in virus retention. This presentation examines the effects of solution pH and ionic strength on the virus retention characteristics of a commercial virus removal filter (Millipore's Viresolve NFP membrane). The data show that virus retention increases as pH decreases. The increase in retention at pH below the isoelectric point of the virus may be due to electrostatic attraction between the charged virus and the charged pores of the membrane. At higher pH, the membrane becomes more negatively charged, which may lead to a repulsive electrostatic interaction and a corresponding reduction in virus retention. Virus retention at low pH was also greater at lower ionic strength, consistent with the effects of electrostatic interactions on virus retention. These results provide important insights into the factors governing virus filtration membrane performance in bioprocessing applications.

Development of a Flow-Free Microfluidic Gradient Generator
Kelly O'Quinn, Michael Tullier, John Pojman, and Adam Melvin

Microfluidic technology has flourished since its emergence over thirty years ago, allowing for rapid scientific advances in a wide variety of fields. Microfluidics allows for precise spatial and temporal control of the chemical and physical microenvironment and has numerous applications ranging from DNA analysis to inkjet printing. In microfluidics, all fluid flow is laminar, forcing mixing to occur via length-scale diffusion. In the case of a microfluidic gradient generator, this allows for a chemical gradient to develop perpendicular to the direction of fluid flow. Applications of microfluidic gradient generators include quantifying cell migration in response to gradients chemical cues (chemotaxis) and culturing of cells for extended periods of time. However, many of the available devices require direct flow through the fluidic channels and are, thus, incompatible with non-adherent cells like bacteria and algae. To address this issue, we developed a double-layer microfluidic device consisting of a top PDMS-layer, imprinted with the fluidic channels, on top of a bottom layer of agarose, allowing for rapid diffusion of chemical cues. One disadvantage of this PDMS/agarose system is that the two layers do not readily adhere, requiring an external chamber to create a tight seal. In an effort to overcome this limitation, we fabricated devices with a PDMS-alternative polymer made using thiol-ene chemistry. Both devices were characterized using fluorescent tracers and numerical simulations to evaluate the mass transfer dynamics. As a proof-of-principle, the device was used in preliminary studies to culture the model algae *Chlamydomonas reinhardtii* for several days by continuous nutrient perfusion.

University of Puerto Rico
Mayagüez Campus
Chemical Engineering Department

Filter Integrity Test: Wetting & Procedures

The Pharmaceutical and Biotechnology sectors are growing areas in the field science, which affect most of our daily basics, like health, food, and safety. Filters made of membrane full of pores are an essential part of these sectors and their processes. When working with the production of different drugs use worldwide, or making food, the company reputation and revenues are at stake if any bacteria or virus make it to the essence of the process. The Food and Drug administration (FDA), set the standards needed for the safety in these sectors. Filters are uses in different stage of the process to ensure that only the material of interest is present alone. These filters are made by different companies with different materials and before using them, the company need to be sure that the filter is integral with not tear or imperfection.

The filter integrity test consist of using the bubble point and diffusion of the membrane to pass air through the filter, but before performing the test (FIT), the filter need to be wetted with a wetting agent to add resistance to the flow of air and this way measure the pressure the filter can hold before letting air pass through. There are minimum values that can be use from the vendor of the filters to analyze the data. There are many different procedures and variables for wetting a filter, and the objective of this project is to find the optimal set of back pressure, flow rate, time and type of wetting, to completely wet a filter before the FIT and get positive result if the filter is integral and not false failures.

Companies use filters and perform these tests, many times with a good wetting procedures, but improvement is always available, and false failures mean that the company need to retest the filter again or send them back to the vendor by FDA regulation, wasting capital and time, with a better wetting the filters could pass the FIT from the first time, without the need of performing again the test.

Johnny Enrique López Calero
Undergrad Student at UPRM
Chemical Engineering & Biotechnology

Contact Information:
Johnny.lopez1@upr.edu

1. Student Presenters
   Matthew J. Thomas & Katlyn A. Bramblett
   Chemical & Biomolecular Engineering
   College of Engineering

2. Mentor
   Kevin N. West
   Chemical & Biomolecular Engineering
   College of Engineering

3. Co-Authors
   None

4. Title: Thermophysical Properties and Gas Solubilities of Brominated Vegetable Oil

5. Abstract

Vegetable oils are versatile compounds that are used in a number of industrial settings in addition to their culinary applications. These areas include the medical and fine chemical industries and they are also used as lubricants in machining operations. We are interested in vegetable oils as their properties are similar to those of lipidic ionic liquids, non-volatile, ionic compounds with lipid-like solvent properties. As such, vegetable oils can be used as less expensive analogs for property screening and to understand structure property relationship for lipidic ionic liquids. Brominated vegetable oil (BVO) is one such compound of interest. BVO is simply a vegetable oil where the carbon-carbon double bonds have been reacted with Br_2, adding two bromines at each site.

BVO has been used by soft drink companies as an emulsifying agent to incorporate more organic compounds, which enhances the flavor of the beverage. For our research, BVO provides a means to examine how bromination would affect the properties of a lipidic ionic liquid, without the cost and time associated with synthesizing such a compound. However, little is known of the physical properties of this compound. The purpose of this research is to measure the physical properties of BVO and compare these values to those other vegetable oils. The physical properties being measured included density, viscosity, refractive index, heat capacity, and gas solubility as a function of temperature. The experimental data is being measured, correlated and compared with that of unsaturated natural oils.

Ribosomal protein L29 enhances protein expression in bacteria by increasing translated mRNA transcript

Sang Hyun Ju, Seung Hee Cho and Lydia M. Contreras

Among the four types of ribosomal exit tunnel proteins: L4, L22, L23, and L29, we focused on the non-essential L29 to enhance cellular expression levels of Fcγ receptors. L29 showed much stronger effects when compared to other fusion partner proteins such as MBP or TrxA, especially in difficult-to-express proteins. Among FcγIIa, FcγIIb, and FcγIIIa, only L29-FcγIIIa showed enhanced levels compared to wild type FcγIIIa. qRT-PCR confirmed that total mRNA levels did not change while actively translated mRNA amounts did increase. To understand the mechanism, we are currently in the process of testing affinity towards ribosomes in L29, TrxA, and MBP by expressing them side by side in two separate PET vectors.

Regulation of ABI3 Transcription via Cloning of Drought-Responsive Promoters in *Arabidopsis Thaliana*

Z. Bhuiyan[1], J. Wind[2], I. Ezcurra[2]
[1]Department of Chemical Engineering, City College of the City University of New York, New York, NY, USA
[2]Division of Industrial Biotechnology, Royal Institute of Technology (KTH), Solna, Stockholm, Sweden

The ABI3 (Abscisic Acid Insensitive 3) gene in *Arabidopsis thaliana* is one of several genes responsible for establishing the drought tolerance of the plant. ABI3 is transcribed to a high degree in germinating seeds in order to protect against drought-like conditions created when seeds become dormant. While the specific mechanisms are still under study, it is hypothesized that drought-protection is achieved by an increase in the storage of nutrients in the form of lipids as well as developing proteins that protect against desiccation. Recent studies have shown the involvement of several drought-responsive promoter genes in regulating the transcription of the ABI3 gene during seed germination and in controlling the ability of Arabidopsis to adapt to dehydrating environments. These promoters, however, are not nearly as active in mature plants and consequently, the drought-resistance properties of *Arabidopsis* are impaired. Thus, the objective of this project is to clone the drought-resistant promoters ATPUB19, ANAC032, HSFA6B and ATHB-12 into an artificial plasmid and transform mature *Arabidopsis* plants with them in order to observe the effect on drought-resistance.

In order to do this, the promoter genes will be amplified from genomic DNA using conventional PCR methods and inserted into the artificial p214 plasmid using Overlap Extension PCR followed by *E. coli* transformation. The modified plasmid will then be transferred to *A. tumifaciens*, a bacterium that is capable of infiltrating and transforming the Arabidopsis genome.

Up to this point there has been success in cloning and amplifying the drought responsive promoters from genomic DNA in order to produce a large quantity of pure promoter fragment. While insertion of these promoter fragments into the artificial p214 plasmid containing the ABI3 gene has been successful, transformation of *E. coli* with this modified plasmid is still underway. Future work will revolve around transfer of the ABI3-Promoter plasmid construct to *A. tumefaciens*. Ultimately, quantitative analysis of the nucleic acids present in transformed plants will indicate the effects of the promoters on the drought response.

Deglycerolization of Human Red Blood Cells in Microfluidic Devices

Audrey Dickinson, John Lahmann, Dr. Adam Higgins

School of Chemical, Biological, and Environmental Engineering, Oregon State University, Corvallis, Oregon 97330

Blood transfusions are one of the most common medical procedures with approximately 15 million units of blood transferred per year in the United States alone. Today, almost all red blood cell units are stored in a refrigerated state, which allows for a storage shelf life of up to 6 weeks. This storage shelf life can be increased to up to 10 years through a glycerolization process. Glycerol is added to the red blood cells and acts as a cryoprotectant. The glycerolized blood is then stored in a freezer at -80 degrees Celsius. One of the main reasons that glycerolization is less common than just refrigeration is that the post-thaw process to remove glycerol is time consuming, taking about 1 hour per unit of blood. The addition and removal of glycerol can cause large osmotic gradients in the red blood cells, which can damage the cells. This can be avoided by using a multistep process that allows for equilibration to remove the glycerol. The first step is to wash the cells with 12% saline solution, equilibrate, then a 3.4% saline solution wash and finally the addition of 0.2% glucose solution to bring the resulting concentration to isotonic conditions. A microfluidic device can be implemented in this procedure to standardize the equilibration times and to provide a continuous process. The microfluidic device used has an inlet for the glycerolized blood, two inlets for the 12% saline then a channel and at the end of the channel, two more inlets for the 3.4% saline, resulting in 5 inlets and one outlet that expelled the saline-glycerol-blood mixture. A 0.2% glucose solution would then be added to the solution by hand. The microfluidic device allows for smaller changes in the residence, and equilibrium time than batch dilutions. The microfluidic device was run at multiple residence times and the final solution was tested for hemolysis of red blood cells. The data shows that a residence time can be achieved that limits the mechanical damage to the blood while still removing the glycerol. Future developments would include a scale up of the device since the throughput will need to be much higher to process a unit of blood quickly in an emergency situation. There will also need to be a device used to separate the red blood cells from the saline, glycerol, glucose solution. A plasmapheresis or membrane device could be implemented.

Genome sequence, metabolic model reconstruction and ^{13}C-metabolic flux analysis of an extremely thermophilic, fast growing, xylose-consuming *Geobacillus* strain LC300

Lauren Cordova, Christopher P. Long, Keerthi P. Venkataramanan, Maciek R. Antoniewicz

Abstract

In this work, we have isolated and characterized a new extremely thermophilic microorganism that can efficiently utilize xylose for cell growth. A single colony of the microorganism was isolated through plating on agar plates and sequenced using PacBio. The organism was identified to be a new *Geobacillus* strain that we have termed LC300. The organism displays extremely fast growth rates on defined medium with xylose as a carbon source, a C5-sugar derived from cellulosic biomass. The genome sequence of *Geobacillus* LC300 was then annotated and biomass composition measured. Based on annotated genome a detailed metabolic model of central carbon metabolism was reconstructed. The model was then validated using ^{13}C-labeling experiments. Finally, we applied ^{13}C-metabolic flux analysis to quantify precise metabolic fluxes for this organism.

Lung cancer resulted in more deaths in the United States than colon, breast, and prostate cancer combined in 2012. One indicator of lung cancer is mitochondrial dysfunction such as: down regulated apoptosis and inefficient ATP production. This dysfunction makes mitochondria of cancer cells more permeable and structurally different from healthy cells. Therefore, mitochondria-targeted drug therapy has emerged as a means to target lung cancer.

PAMAM dendrimers, are 3D hyperbranched polymers with multifunctional surface groups that can be used to conjugate therapeutics, and are promising drug nanocarrier (DNC) systems. Delocalized lipophilic cations, such as triphenylphosphonium (TPP), have been shown to preferentially accumulate in the mitochondria due to its negatively charged inner membrane potential. The use of TPP targeting should thus increase the effectiveness of a therapeutic directed to the mitochondria. Polyethylene glycol (PEG) has been known to enhance the biodistribution of DNCs, decrease the toxicity of constructs, and be used a linker to targeting groups. Overall, the combination of lipophilic cations and PEG with drug nanocarriers may offer new opportunities to target mitochondria, and thus treat lung cancer.

In this work, a series of TPP and PEG-TPP modified Generation 4 PAMAM amine terminated dendrimer ($G4NH_2$) conjugates were synthesized, and characterized by ^1H NMR, MALDI, and DLS. We investigated the effect of TPP density and PEG-TPP density on the dendrimer internalization, mitochondrial targeting, and cytotoxicity in an alveolar adenocarcinoma cell line (A549). Using flow cytometry and confocal microscopy, the conjugates of TPP to dendrimer were found to be effectively internalized by the cell and to target the mitochondria, respectively, when compared to bare dendrimer. The addition of PEG linker was found to significantly decrease the cytotoxicity of the conjugate, while maintaining the same level of mitochondrial targeting despite decreased internalization. Combined, these results suggest that the conjugation of TPP to $G4NH_2$ via a PEG linker may be a potentially viable strategy to deliver anti-cancer therapeutics to the mitochondria for treating lung cancer.

395904 The Effects of 4-Hydroxynonenal on Reactive Oxygen Species Generation By Pressurized Macrophages

Abstract:

Previous studies demonstrated that exposure of macrophages to pressure fluctuations alters their capacity to generate reactive oxygen species (ROS). For example, it has been determined that *in vitro* exposure of macrophages to 20 mmHg above ambient pressure can lead to a decrease in ROS generation as well as parallel changes in F-actin distribution. Different chemicals exhibiting concentration dependent pro-oxidant characteristics, such as quercetin and trolox, have previously been explored as a means to counteract the anti-oxidant effects of pressure on ROS generation. Interestingly, these chemicals at concentrations previously shown to stimulate ROS generation failed to recover ROS in pressurized macrophages. In the present study, we used 4-Hydroxynonenal (4-HNE) as an alternative approach, as it has previously been demonstrated to stimulate ROS generation in macrophages. Macrophages were exposed to 20 mmHg above ambient pressures for 2 hours and assessed for their ROS production and F-actin organization. We determined that although 4-HNE may be used in order to recover basal ROS levels in pressurized cells, it was incapable of counteracting the suppressive effects of pressure. We also observed that pressure-sensitive reorganization of F-actin was unaffected by 4-HNE. Overall, our results suggest that the effects of pressure on macrophage ROS generation and F-actin distribution are independent of 4-HNE.

397791 CHO Platform Evaluation for Biopharmaceuticals Production

Abstract:

There is a high demand to develop an efficient platform for biopharmaceutical production, especially considering the fast growth of biosimilar. Chinese hamster ovary (CHO) is the most popular expression system for therapeutic proteins in the biopharmaceutical industry. However, the lack of knowledge of a systematic evaluation and comparison of various CHO systems might hamper the production of recombinant proteins. In this study, we used biosimilar as a model protein to evaluate different expression vectors (dhfr and GS), CHO host cells (i.e., CHO K1, CHO DG44, and CHO S), and commercial CHO media. The stable cell line development using different systems was also compared, including transfection, amplification and single cell cloning. In addition, advanced Omics technology was applied to identify the regulator candidates, which are being used to construct a novel engineered CHO host cell. In the future, an efficient CHO platform will be created to achieve high production, high quality and high stability in biopharmaceuticals production.

398177 E. coli Pilot-Scale One-Step Purification of Amine Dehydrogenases Employing an Aqueous Two-Phase Extraction System

Abstract:

This project explores for the first time the use of a large-scale aqueous two-phase extraction system (ATPS) to purify dehydrogenases. Dehydrogenases are an important family of enzymes that generates chiral alcohols and amines for pharmaceutical and crop protection agents. This work first explores the purification of popular dehydrogenases (GDH, FDH) as well as the newly developed amine dehydrogenase family (L-AmDH, F-AmDH and cFL1-AmDH) in an ATPS. In aqueous two-phase systems (ATPS), when polymers and salts are added to water, the water forms two phases. The ATPS consists of a top polymer layer and a bottom salt layer. The goal is to selectively partition the dehydrogenase into the top phase to achieve pure protein, while the cell lysate and debris will remain in the bottom phase. Each enzyme will require different purification conditions (PEG, salt, temperature, etc.) and this work investigates these conditions and develops a general purification protocol.

Microalgae Production in an Open-Raceway-Pond: Analyzing the Paddle Wheel Velocity Effect on the L/D Frequency

Soo Hyun Park[1], Jay H. Lee[1*]
[1] Dept. of Chemical and Biomolecular Engineering, Korea Advanced Institute of Science and Technology, 291 Daehak-ro (373-1 Guseong-dong), Yuseong-gu, Daejeon 305-701, South Korea

Tel: 82-42-350-3926, E-mail: jayhlee@kaist.ac.kr

Microalgae have received much attention as potential feedstock for the third-generation biofuel. Considering the scale of production needed and the economics associated, an open-raceway-pond cultivation system is likely to be the choice for the commercial production. Hence, a model that can be used to determine an optimal balance between biomass production and operating cost can be highly valuable on the way to a successful commercialization of microalgal biofuel.

In this study, we consider light as one of the dominant factors for the biomass production. Even though light is imperative for microalgae to grow, if exposed for too long, microalgae can go into a photo-inhibited state, resulting in a reduced biomass production. Hence, the length of exposure is a crucial parameter for the biomass production. An integrated model is constructed by combining the Droop model to describe the biomass production and the Han model to illustrate the physiological property of microalgae with respect to light. The effect of light on the biomass production is observed through mixing. Two simulations are conducted: With mixing and without. From the simulations, it is seen that 1.5 times higher biomass production is achieved by applying mixing. To link the effect of light to a raceway pond velocity varied by mixing, a term called 'Light/Dark frequency (L/D frequency)' that describes a dynamic pattern of light is designated. It is postulated that higher biomass production can be reached because the velocity and the L/D frequency are directly proportional, however, beyond a certain velocity region (above 0.3m/sec), the effect of velocity becomes insignificant to the biomass production. As the pumping accounts for a major portion of the operation cost, by connecting these results with power consumption, the optimal mixing condition for the biomass production can be found.

Acknowledgment

This study was financially assisted by Undergraduate Research Participation (URP) program of Korea Advanced Institute of Science and Technology (KAIST), South Korea.

Purification of Invertebrate Hemoglobins for use as an Alternative to Donated Blood

Jack A. Dienes, Devon C. Zimmerman, Jacob J. Elmer
Villanova University, 800 E Lancaster Ave, Villanova, PA 19085

Although donated blood will always be the preferred material for transfusion, its limited availability has motivated the development of hemoglobin based oxygen carriers (HBOC's) that can be used when donated blood is unavailable. For example, donated blood is often unavailable in military situations since it requires refrigeration facilities that are impractical on the battlefield. Our goal is to develop an HBOC that does not require refrigeration, thereby satisfying this vital need. In this work, we purified and characterized the hemoglobin from several different invertebrates for use as HBOC's, including earthworms (*Lumbricus terrestris* and *Eisenia fetida*), bloodworms (*Glycera dibranchiata*), and sandworms (*Nereis virens*), as well as the novel hemoglobins of ribbon leeches (*Nephelopsis obscura*) and Alabama jumper worms (*Amynthas gracilis*). Each of the hemoglobins was highly purified using a combination of metal (zinc) precipitation and tangential flow filtration. The purity and molecular weight of each hemoglobin were then analyzed with native and denaturing PAGE gels. The oxidative stability and oxygen transport characteristics of the hemoglobin samples were then compared and used to identify the best candidate for future animal studies. The hemoglobin from *L.terrestris* and *G.dibranchiata* were identified as lead candidates, since they oxidized the least during purification (5.3% for LtEc, 8.5% for GdHb). These hemoglobins also have oxygen affinities (26.50mmHg for LtEc in Hemox buffer, 22.95mmHg for GdHb in TRIS) that are highly similar to human red blood cells (28mmHg in Hemox buffer) at body temperature (37°C). We are currently comparing the oxidation rates of LtEc and NoEc to see which is more stable, and we are also comparing the oxidation rate of LtEc in TRIS buffer against the oxidation rate of LtEc in Ringer's Solution. Overall, our results show that invertebrate hemoglobins may be promising HBOC's, which further motivates the investigation of even more diverse invertebrate hemoglobins in the future.

398693 Purification of Maltose Binding Protein By Starch Precipitation

Abstract:

A central step in the production of high purity proteins is their recovery and purification. The maltose binding protein (MBP) affinity tag is commonly used in this process, since it has the ability to selectively bind to immobilized amylose. I am exploring the possibility of using a commercial grade cationic starch, which contains large amounts of amylose as part of its polymer backbone, and a salting out method to precipitate an MBP-tagged protein. Different salts including sodium chloride, ammonium sulfate, sodium sulfate and magnesium sulfate can selectively precipitate starch, which will be bound to the MBP-tagged protein, thus co-purifying it. Varying the salt solutions will allow the ideal composition and concentration of the salt solution for this purification method to be determined. The effectiveness of the approach will be measured through gel electrophoresis, allowing the recovery and yield of the purified proteins to be quantified. Because starch and salts are relatively cheap, the success of this approach will lead to a new, feasible option for mass purification and production of proteins. At this time, the expression and starch-salt purification of the MBP-tagged protein is still in progress. The data includes analysis of electrophoresis gels and a comparison between this starch-salt method and other, standard purification methods.

Abstract

GDP-fucose: There have been more interest in the human milk oligosaccharides especially fucosyl oligosaccharides recently because they can prevent infant from enteric and other pathogens. Trisaccharide 2'-fucosyllactose (2'-FL) is one of those important oligosaccharides found from human milk, and because of its prebiotic and anti-infective properties, 2'-FL is discussed as nutritional additive for infant formula. Although chemical synthesis of fucosylated oligosaccharides, like Lewis blood group antigen, has been achieved for a long time, it is very expensive and laborious. It is very unlikely that chemical synthesis of fucosylated oligosaccharides can be applied in industrial production. The alternative way, enzymatic synthesis of fucosylated oligosaccharides, is being used because it provides more efficiency by the specificity of glycosyltransferases. So the most promising way to synthesize a large amount of 2'-FL is using the whole cell biosynthesis in *Escherichia coli* (*E.coli*) by intracellular synthesis of GDP-L-fucose. GDP-L-fucose is a sugar nucleotide, and it can be regarded as an activated form of sugar molecules. It is the donor substrate of fucosyltransferases to synthesize fucosylated antigens (like 2'-FL).

E.coli: *E.coli* is chosen as the host organism because of these four reasons: (i) it grows really fast. (ii) It is easy to reach high cell density. (iii) The media for cell growth can be made from readily available and inexpensive components. (iv) Transformation with exogenous DNA is easy and fast. *E.coli* JM109 is picked and its chromosome is modified. As shown in the picture below, the strain is constructed by integrating gene CDT and CBP into the poxB gene location and transformation of gene pQFKP which was from pQE80L expression vector. The CDT gene allows the *E.coli* to take cellobiose as carbon source, and CBP can transform cellobiose into glucose which provides energy to *E.coli.* The L-fucokinase-GDP-L-fucose putophophorylase, or FKP, can catalyze the transformation of L-fucose to GDP-fucose. Also, as shown in the picture, P_{T5} is the promoter for FKP gene experssion through IPTG induction. IPTG, or isopropyl beta-thiogalactoside, is a non-metabolized inducer to induce the synthesis of FKP from the recombinant *E.coli* JM109. Because when *E.coli* is exposed to enough lactose, lactose will bind to the repressor proteins and will cause the induction of the production of two different proteins, permease and β-galactosidase. Permease is used to transport carbohydrate and β-galactosidase can hydrolyze the disaccharide lactose to the monosaccharides, so that *E.coli* can take these monosaccarides as food and grow. Here these two genes are replaced with FKP gene so that when *E.coli*'s non-metabolized inducer(IPTG) is added, it will be tricked to produce the FKP protein rather than permease and β-galactosidase.

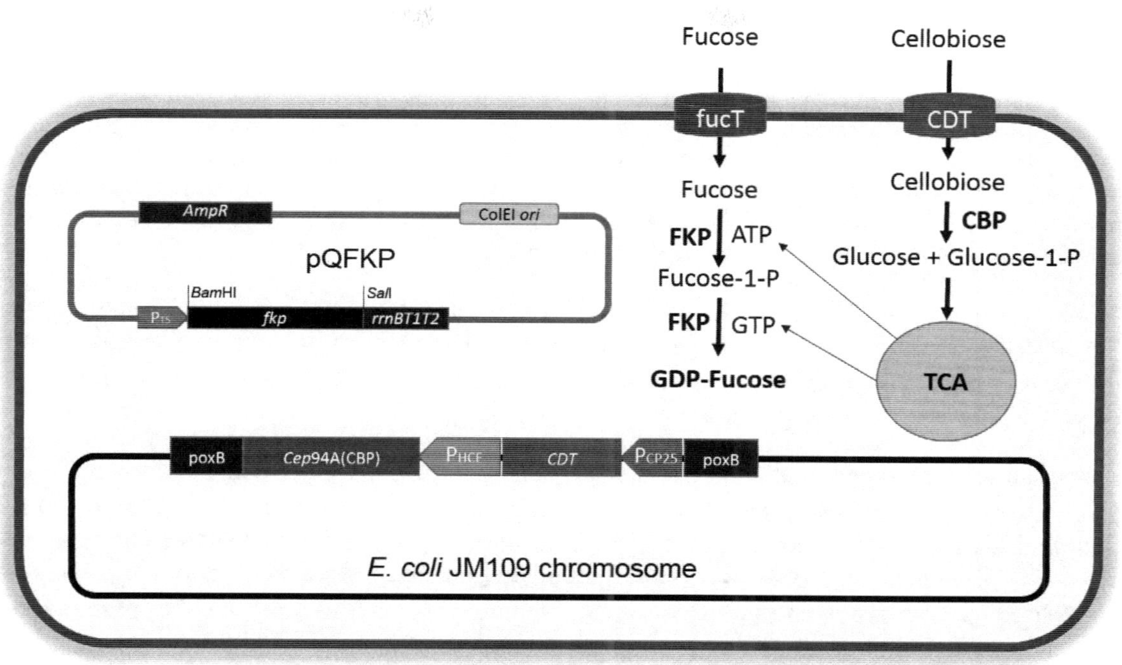

Cellobiose: Cellobiose is our interest carbon source for the GDP-fucose synthesis because cellobiose is cellulose, one of the most abundant carbon source in the world. There are two mechanisms of cellobiose assimilations, hydrolysis and phosphorolysis. Phosphorolysis is preferred because: phosphorolysis allows saving of one ATP in phosphorylation, representing a net gain of 7 ATP for each cellobiose molecule metabolized, compared to a net gain of 6 ATP for the hydrolysis route. This one net gain of ATP for phosphorolysis allows more tolerance for the recombinant *E.coli* to grow under the effect of inhibitor. In the previous experiments we found that if cells are cultivated in 5% sodium acetate (inhibitor) media, the final cell density of cellobiose phosphorolysis *E.coli*, or CBP, was only slightly lower than what achieved at lower acetate concentration. But for cellobiose hydrolysis *E.coli*, or BCL, cells did not show any sign of growth during the 48h of cultivation. So phosphorolysis is definitely the better option.

This study investigates the advantage of using cellobiose as carbon source for *E.coli* cell growth and the amount of GDP-L-fucose synthesized, with cellobiose phosphorylase. We compared the result of using cellobiose with those using glycerol and glucose. The comparisons showed that the cell growth in cellobiose media was the most, and the yield of GDP-L-fucose synthesized in cellobiose media was around 67.14% higher than that synthesized in glycerol media, and around 235.6% higher than that synthesized in glucose media.

398821 Role of Vibrational Motions on Thymidylate Synthase Catalysis

Abstract:

Thymidylate Synthase (TSase) catalyzes the last committed step in thymine biosynthesis, i.e., the conversion of deoxyuridine monophosphate (dUMP) to thymidine monophosphate (dTMP), using 5,10-methylenetetrahydrafolate as both the methylene and hydride donor. dTMP is an essential DNA building block, thus TSase activity is critical is fast proliferating cells and is thus a target for chemotherapeutic drugs and antibiotics.

Recent studies investigated the role of vibrational motions of proteins in enzyme catalyzed reactions. Studies isolating effects of fast protein vibrations in the catalyzed reaction are conducted by comparing light (natural distribution of isotopes) and heavy (^{13}C, ^{15}N, ^{2}H labeled) TSase. A broad spectrum of kinetic properties and activation parameters have been measured to compare these two isozymes.

Initial velocity data were analyzed to obtain the k_{cat} value for each enzyme at four temperatures (5°C, 25°C, 30°C, 40°C). Rates for both light and Heavy TSases were measured at different dUMP concentrations as a function of 5,10-methylenetetrahydrafolate concentration and vise versa. Rates were fit to the Michaelis-Menten equation, with or without substrate inhibition, as 5,10-methylenetetrahydrafolate showed substrate inhibition. Parameters K_M and K_{cat} were compared for both heavy and light TSases at all temperatures and were identical (within the experimental error) at 20°C, but varied at all other temperatures. To further assess whether or not the mass difference within the enzyme changes the rate-limiting step for the specific reaction, the kinetic isotope effect was measured on k_{cat}. Intrinsic isotope effects were also measured and findings indicate that k_{cat} in the heavy TSase was no longer rate limited by the hydride transfer. The temperature independence of the isotope effects for the light enzyme indicated ideal enzyme dynamics. A large temperature dependency for heavy TSase, on the other hand, indicated that those fast vibrations of the light TSase are altered by the slower dynamics in the heavy TSase enzyme. Future work will include analysis and further study on the role of vibrational motions versus the role of binding between the heavy and light TSase through the use of isothermal titration calorimetry experiments.

398344 Process Engineering of Clostridium Tyrobutyricum to Improve Butyric Acid Production

Abstract:

Butyric acid production by *Clostridium tyrobutyricum* has drawn increasing attention because it is widely used in food and pharmaceutical industries. *Clostridium tyrobutyricum* mutant PAK-Em with inactivated *ack* gene, encoding acetate kinase, was used in fermentations to produce butyric acid from glucose. A fibrous-bed bioreactor (FBB) was used to immobilize and adapt PAK-Em cells, which boosted the butyric acid yield to 0.38 g/g at pH 5.5 and 0.42 g/g at pH 6.5. PAK-EM produced much more butyric acid (23.2 g/L vs. 17.2 g/L at pH 5.5 and 63.0 g/L vs. 15.7 g/L at pH 6.5) than the wild type, but had a lower specific growth rate (0.08 h^{-1} vs. 0.11 h^{-1} at pH 5.5 and 0.09h^{-1} vs. 0.10 h^{-1} at pH 6.5) in the FBB. The integration of metabolic engineering and process parameter optimization increased butyric acid production significantly.

Investigating a Three Step Deglycerolization Method for Cryopreserved Red Blood Cells

Jolynn Meza Wynkoop, John Lahmann, Dr. Adam Higgins
Biological Engineering Lab, School of Chemical, Biological and Environmental Engineering
Oregon State University
Corvallis, Oregon 97330

Due to the high clinical demand for red blood cells (RBCs), there is a need to actively preserve cells in a manner that lengthens the shelf life while protecting RBC function. Red blood cell units stored in the refrigerated state have a short shelf life of 6 weeks. Comparatively, cryopreservation offers red blood cells an FDA approved shelf life of 10 years. The time consuming wash step of approximately one hour makes cryopreservation a less appealing option for blood banks. Due to the many time constraints that clinicians may be under in an emergency situation, a faster method for removal of cryoprotectants is needed. To rapidly remove cryoprotectants from RBCs, a multi-step deglycerolization process can be performed through the addition of different concentrated saline and glucose solutions. Rapid removal of the cryoprotectant, glycerol, can result in large osmotic gradients within the cell, so it is important to make predictions involving cell membrane transport to ensure that cells stay intact. In our experiments, cells have shown to undergo hemolysis, or lysing of RBCs, in two situations. The first situation involves the cells swelling beyond their maximum volume. Second, the RBCs could also lyse if they shrink past their lower osmotic tolerance limit and are then returned again to isotonic conditions through the addition of dilute glucose solution. This work intends to develop a method that results in the least amount of hemolysis while removing glycerol in as little as several minutes. This process was modeled mathematically using a three step procedure. Due to the mathematically optimized three-step deglycerolization method, saline and glucose concentrations were determined for which would allow for glycerol removal from red blood cells while keeping cells within their osmotic tolerance limits. Saline is used because it maximizes the concentration driving force for glycerol removal, allowing the procedure to occur rapidly. Another factor that allows for glycerol removal is increased temperature. This study was conducted at two different temperature, 37 °C and 45 °C. The goal of these experiments was to increase the rate at which glycerol could be removed from RBCs and cause little to no hemolysis. Experiments were performed by first exposing glycerolized blood to a 12% saline solution under given time frames. The ratio of blood to 12% saline was varied and experiments were conducted at 2-fold, 4-fold, and 10-fold dilutions, blood to 12% saline. After the initial dilution with 12% saline, there was a varying waiting period between 5 and 120 seconds while the solution stayed on the horizontal shaker. Then the next solution of 3.4% saline was added to the mixture and another wait period of 120 seconds followed. Lastly, to bring the cells back to isotonic conditions, a 0.2% glucose solution was added. All solutions were held at the given temperature by being stored in a closed-lid water bath throughout the process. The results of these experiments yielded higher than expected hemolysis with the lowest hemolysis value being around 40% hemolysis for the four-fold dilution at 37°C and with waiting 10 seconds between the first two additions of saline solutions. Continuing research is being done to investigate the mathematical model to better predict the solution concentrations of saline needed for the rapid removal of glucose from frozen-thawed RBCs.

398361 Biodegradable, Selectively Permeable Membranes for Crosslinking of Polypeptides

Abstract:

Elastin-like polypeptides (ELPs) are bioinspired stimuli-responsive biopolymers; these polypeptides undergo a lower critical solution temperature phase transition in water. We aim to utilize this phase transition to create new particle structures at the microscale by crosslinking multiphase liquid states inside droplets. To do so, we have engineered microcapsules that contain ELPs surrounded by a nanoporous biodegradable polymeric membrane. The porous microcapsule acts as a semi-permeable membrane, maintaining the larger protein molecules (MW ≈ 30 kDa) inside the capsule while allowing smaller crosslinking molecules (MW ≈ 500 Da) to diffuse into the capsule through the membrane pores.

To fabricate the capsules, we use controlled biphasic flow within a capillary microfluidic device to form template liquid water-in-oil-in-water double emulsions; the inner drop consists of ELPs in an aqueous buffered solution, while the middle layer is a solution of polylactic acid, a biodegradable polymer, dissolved in toluene. The outer or continuous phase is a water solution containing a surfactant to promote capsule stability. Solvent evaporation is used to consolidate the ultra thin (500 nm – 2 μm) polymer solution middle layer to produce a biodegradable nanoporous microcapsule.

Following small molecule delivery, the dialysis membrane can be removed through an esterase enzymatic reaction that does not disrupt the inner ELP particles. ELP colloids can be further purified through multiple centrifugation and washing cycles. Liberated ELP microparticles can be used downstream in a variety of biomedical applications, from drug delivery to compartmentalized micro-reactions.

Abstract:

The ability to maintain the cleanliness of healthcare products, instruments and other surfaces is undoubtedly a serious hygiene matter. Infections developed in medical and retirement facilities constitute a serious and growing health concern.

One strategy for reducing bacterial contamination utilizes the Oligodynamic effect, which refers to the toxic effect of metal ions on living cells, such as viruses, prokaryotic and eukaryotic microorganisms, even when present at relatively low concentrations. This antimicrobial effect is exemplified by the presence of ions derived from silver.

Introduction of silver ions (Ag^+) within the matrix of a midblock-sulfonated pentablock copolymer is demonstrated here to be an effective bactericide, causing changes in microbial growth patterns by penetrating membranes and disrupting the metabolism of several species of bacteria. The sulfonated pentablock copolymer used here is capable of absorbing remarkably high solvent levels and forming highly swollen gel networks in polar media, thereby making them suitable for use in hygiene and healthcare applications.

The copolymers have been impregnated with silver ions (Ag^+)from $AgN0_3$ solutions. Copolymers were exposed to different concentration levels of $AgNO_3$ (via the Ag^+ absorption step). Evidence of inhibition of microbial growth was examined after diluting the liquid bacterial culture onto LB agar plates. *E. coli* (gram-negative) was the microorganism that was specifically examined.

Hydrogels in Spinal Disk Replacement

Author: Kylee Mockler Martens

Abstract

The goal of this research project is to test gels and gel-infused foams to compare to the human *nucleus pulposus* as a replacement for ruptured spinal disks. This research project took place at Oregon State University under the direction of the School of Chemical, Biological, and Environmental Engineering professor Dr. Skip Rochefort. This research is extremely important because herniated disks and other related back problems are common and difficult to prevent. Spinal disk replacement techniques have been undergoing research and development for over fifty years, although none have been able to completely solve the problem. The current surgical treatments can be costly and don't completely resolve the problem. It is important that the gel or gel-infused foam to be infused into the body mimics the functionality of the human *nucleus pulposus*. However, it is believed that an even better improvement can be used as a spinal disk replacement. A herniated disk is caused from excessive load on the spinal disk causing it to bulge and put pressure on the spine causing severe pain and creating numbness in the legs.

The materials used for this research project for the composite gels were Agarose, Gelzan, and Kelcogel LT100. There were two composites that were tested with all three substances. One was composed of 1.5% Agarose, 1% Gelzan, and 0.5% Kelcogel LT100. The second composite tested was 0.75% Agarose, 0.5% Gelzan, and 0.25% Kelcogel LT100. Many experiments were also conducted testing 1% Agarose, 2% Agarose, and 3% Agarose. There were three foams that were tested for this project so far. There was a white foam, a black foam, and a gray foam. The white foam was the least stiff with the gray being the stiffest. With the functionality of the foam in mind, the gray foam could be eliminated because the foam would be too stiff to be used as a spinal disk replacement.

The tests that were conducted were a compression test, oscillation test, steady shear rate, and a frequency test. The tests were performed on the AR 200ex rheometer and the Bose EnduraTec 3200 rheometer. Data was compiled to determine the modulus. The modulus is determined from the slope of the linear region of the stress versus strain graph. Tests were performed to calculate G' and G", the elastic and viscous modulus. Viscosity is the ability of a substance to resist deformation. Viscosity can also be determined from shear stress over strain. Elasticity is the reverse process when a fluid can return to its original form.

From many tests and trials, the gray foam has the highest modulus, meaning it is the stiffest. Through many trials of different composites including Agarose, Gelzan, and Kelcogel LT100, it seems that a gel with just Agarose does just as good of a job. It is unnecessary to make as complicated of a gel when one with a simpler structure is easier. The data also shows that the gel-infused foams have a higher modulus, meaning they can endure more stress before breaking. The function of the artificial foam will also improve the longevity of a replaced spinal disk because it will counteract the compression of the gel. The foam serves the purpose of maintaining the shape of the disk. It can be concluded that the gray foam is not a potential candidate and should not be tested in future research. Future research should conclude of other foams or more tests with the white and black foam, infused with varying percents of Agarose.

Effects of Cross Linking Time and Temperature on Alginate Beads Dissolution

Ai Lin Chin*, and Dr. Robby Sanders

Department of Chemical Engineering, Tennessee Technological University, Cookeville, TN 38505-0001

Alginate, a compound extracted from the cell-wall of brown algae, is made up of mannuronic acid and guluronic acid, structures of which are shown in Figure 1. Being naturally derived, it is used widely in the food and textile printing industries as a stabilizer or emulsifier or both. When it exists in the form of sodium alginate, it can be dropped into a calcium chloride bath to form beads of calcium alginate as reported by Ferrell and Vernengo. These beads are formed relying on the cross linking action between the sodium alginate and the calcium chloride solution.

Figure 1. Mannuronic acid and guluronic acid found in alginate.

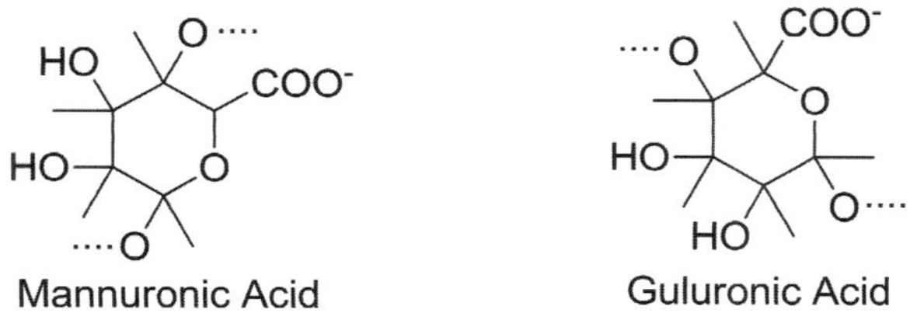

Current efforts in our lab have suggested a dissolution of the beads when placed in water. Thus, research is being conducted with the goals of developing a better understanding of bead volume with respect to time and ultimately incorporating this information into compartmental models to describe the release of drug simulants from the beads.

397353 PEGDA Cross-Linked Pnipam Microgel Particles for Protein Delivery to Macrophages

Abstract:

Macrophages phenotypically exist on a spectrum, with M1 classically activated macrophages and M2 alternatively activated macrophages being the ends of this spectrum. Tumor associated macrophages (TAMs) are M2-like and promote tumors. Interleukin-12 (IL-12) has been shown to reprogram M2 macrophages to M1 macrophages, which are tumor regressive. Local delivery of this protein to the tumor microenvironment will lead to repolarization of TAMs to tumor-toxic M1 macrophages. Poly(N-isopropylacrylamide)-co-acrylic acid-co-poly(ethyleneglycol) diacrylate (pNIPAm-co-AAc-co-PEGDA) mcrogel particles were synthesized to be used as a drug delivery vehicle for IL-12. LPS stimulated macrophages are used to model M1 macrophages *in vitro*. IL-4 stimulate macrophages are used to model M2-like macrophages *in vitro.* The cytotoxicity, degradative properties, macrophage cytokine secretion profiles, particle internalization and FITC release was studied. Viability of polarized macrophages in the presence of these particles was found to be above 70%. These particles showed evidence of degradation after 2 days. Internalization of these particles by RAW 264.7 macrophages has been shown. Results from this work will be applied to the design of a drug delivery vehicle that will be used to increase the efficacy of drug-based tumor cytotoxicity.

397007 Evaluation of Porous Polycaprolactone Microspheres for Extended-Release Drug Delivery

Abstract:

This study evaluated the suitability of porous polycaprolactone (PCL) microspheres to act as a 7-14 day extended-release drug delivery system for a hydrophobic compound, modeled using rhodamine 123 (Rh123), to the retina. At this time, the targeted application of the actual drug has not been published and is not permitted for discussion. Porous PCL microspheres were formed via a combination ink-jetting and thermally-induced phase separation methods. Microspheres were characterized with scanning electron and light microscopies which displayed a rough surface morphology and a mean diameter of 117.0 ± 38.6 micrometers. Degradation testing of the microspheres utilizing gel permeation chromatography showed no change in microsphere molecular weight over the course of 23 days. Rh123 was loaded in two different ways, one pre-formation and one post-formation of the microspheres. Fluorescence spectroscopy measured drug release of each loading method from the microspheres, revealing low drug release and/or loading efficacy. Further research recommended to determine the suitability of the proposed drug delivery system include polymer blending to increase speed of degradation and use of a dialysis bag for drug release studies to increase accuracy and precision.

Title: Developing Correlations Between Bulk Powder Properties and DEM Contact Model Parameters

Authors: Charles Foster: charles.foster@rutgers.edu, Amanda Rogers: amandarog@gmail.com, Dr. Marianthi Ierapetritou: marianth@soemail.rutgers.edu

Keywords: cohesion, powder mixing, granular solids, rotary cylinder, axial dispersion, Discrete Element Method, DEM

The majority of pharmaceutical tablets produced for consumption are made through processing of granular solids. To ensure product quality, it is important that the exact concentration of the active ingredient maintained during manufacturing, and it is for this reason that the mixing of granular solids is of particular interest to chemical engineers.

Many experiments have been conducted to both qualitatively and quantitatively model granular mixing behavior. A standard tool for simulating the mixing of granular solids is the Discrete Element Method (DEM) (Sykut et al., 2007). The majority of research done in this field has been aimed at describing the behavior of cohessionless granular solids in different mixing devices. Processes that have been modeled using DEM simulations based on cohesionless granular solids include V-blender (Moakher et al., 2000, Lemieux et al. 2007), tote blender (Lemieux et al.,2007), double cone (Moakher et al., 2000), tumbling ball (Cleary and Sawley, 1999), ribbon mixer (Kaneko et al., 2000), and flat blade mixer (Stewart et al., 2001).

Many pharmaceutical powders, however, exhibit cohesive behavior. Cohesion is expressed in terms of energy and is defined as the shear stress of a powder when no normal stress is applied to the plane of shear (Orband and Geldard, 1997). Substantially less effort has been put into simulating the mixing of cohesive granular solids, and most research in the field has been qualitative, aimed at matching experimentally observed behavior with simulation behavior in order to tune DEM parameters (Mishra et al., 2002; Alexander et al., 2006; Chaudhuri et al., 2006). The focus of this study is to examine axial dispersion in a rotating cylinder through both experiments and DEM simulation. The goal of this research is to develop an empirical correlation between Discrete Element Method (DEM) contact model parameters and the measurable bulk powder properties of two cohesive granular solids. Contact model parameters for the DEM simulation will be tuned based on comparison to performance data from an experimental system.

The components that are used in this experiment are Acetaminophen (APAP) and a fluid cracking catalyst (FCC). APAP exhibits strong cohesion when compared to other powders with similar particle size and compressibility, while FCC exhibits weak cohesion when compared to other powders with similar particle size and compressibility. It is the hope that through this research engineers will have a better understanding of how DEM simulation parameters and cohesion are related, and in the future be able to use DEM simulation software to accurately model pharmaceutical manufacturing processes involving cohesive powders.

References:

Alexander, A.W. (1), et al. "Avalanching Flow Of Cohesive Powders." *Powder Technology* 164.1 (2006): 13-21. *Scopus®*. Web. 1 Sept. 2014.

Chaudhuri, B., et al. "Cohesive Effects In Powder Mixing In A Tumbling Blender." *Powder Technology* 165.2 (2006): 105-114. *Scopus®*. Web. 1 Sept. 2014.

Cleary, P.W., Sawley, M.L. "Three-dimensional modelling of industrial granular flows." *Second International Conference on CFD in the Minerals and Process Industries* (1999): 95-100. *CSIRO*. Web. 1 Sept. 2014.

Kaneko, Y. (1), T. (1) Shiojima, and M. (2) Horio. "Numerical Analysis Of Particle Mixing Characteristics In A Single Helical Ribbon Agitator Using DEM Simulation." *Powder Technology* 108.1 (2000): 55-64. *Scopus®*. Web. 1 Sept. 2014.

Lemieux, M. (1), et al. "Comparative Study Of The Mixing Of Free-Flowing Particles In A V-Blender And A Bin-Blender." *Chemical Engineering Science* 62.6 (2007): 1783-1802. *Scopus®*. Web. 1 Sept. 2014.

Mishra, B.K. (1), C. (2) Thornton, and D. (2) Bhimji. "A Preliminary Numerical Investigation Of Agglomeration In A Rotary Drum." *Minerals Engineering* 15.1-2 (2002): 27-33. *Scopus®*. Web. 1 Sept. 2014.

Moakher, M., T. Shinbrot, and F.J. Muzzio. "Experimentally Validated Computations Of Flow, Mixing And Segregation Of Non-Cohesive Grains In 3D Tumbling Blenders." *Powder Technology* 109.1-3 (2000): 58-71. *Scopus®*. Web. 1 Sept. 2014.

Orband, J.L.R., and D. Geldart. "Direct Measurement Of Powder Cohesion Using A Torsional Device." *Powder Technology* 92.1 (1997): 25-33. *Scopus®*. Web. 23 Sept. 2014.

Stewart, R.L. (1), et al. "Simulated And Measured Flow Of Granules In A Bladed Mixer- A Detailed Comparison." *Chemical Engineering Science* 56.19 (2001): 5457-5471. *Scopus®*. Web. 1 Sept. 2014.

Sykut, J, M. Molenda, and J. Horabik. "Discrete Element Method (DEM) As A Tool For Investigating Properties Of Granular Materials." *Polish Journal of Food and Science* 52.2 (2007). 169-173. Web. 24 Sept. 2014.

Understanding the Effects of Nanotopography on Epithelial-to-Mesenchymal Transition

Akili Smith-Washington, Tuskegee University
2014 University of California, San Francisco Summer Research Training Program
Tejal Desai, PhD, Jessica Walsh, PhD
Department of Bioengineering and Therapeutic Sciences

EMT (Epithelial-to-mesenchymal-transition) is a biological process where polarized epithelial cells make a phenotypic transition to motile mesenchymal cells. This change is seen in 3 different situations: embryogenesis, wound healing and fibrosis, and metastasis in cancer. Studies of this phenomenon illuminate that EMT may be induced artificially. Previous studies by Kimberly Kam and Laura Walsh (NanoLetters 2012) demonstrate that nanotopography increases permeability in epithelial monolayers of human intestinal Caco-2 cells. The purpose of this project was to examine the effects nanotopographic films on EMT in human intestinal Caco-2 cells and MDCK cells expressing E-cadherin GFP by detecting TGF-β, Smad2, claudin, occludin, and E-Cadherin-proteins known to be expressed in cells that have undergone EMT using RT-PCR and immunofluoroescent staining.

MEASUREMENT OF ELASTIN EXPRESSION AFTER *IN VITRO* APPLICATION OF POTASSIUM CHANNEL OPENERS IN RFL6 CELLS

Parulekar, Jaya[1], Wagenseil, J.[2], Staiculescu, C[2].

[1]Illinois Institute of Technology, [2]Washington University in St. Louis

Abstract: The goal of this study is to develop an *in vitro* assay for elastin expression using RFL6 cells, which were derived from fibroblasts present in rat lung tissue, and to test different doses of minoxidil and diazoxide, potent vasodilators and potassium channel openers. Elastin amounts in response to drug treatments are assessed using a commercial Fastin Elastin assay. Our project will shed light on optimal pharmaceutical doses for increasing elastin expression and also provide a relatively high throughput method for testing additional anti-hypertensive drugs that may affect elastin expression and subsequently decrease arterial stiffness and blood pressure.

Introduction: Hypertension is a chronic medical condition resulting in increased pressure in the arteries and increased work load on the heart. Large arteries normally exhibit a significant degree of elasticity due to elastin, a protein that aids connective tissue in maintaining its elasticity. With increasing age, degradation of elastin occurs, resulting in increased arterial stiffness. Traditional anti-hypertensive medications decrease blood pressure, but do not address changes in elastin content or arterial stiffness. Vasodilating pharmaceuticals that encourage production of elastin and helper proteins by smooth muscle cells in the arterial wall, particularly potassium channel openers, may increase functional elastic fibers, decrease vessel stiffness, and be a viable treatment option for hypertension.

Materials and Methods: RFL6 cells were cultured to confluency at the same passage in T-75 flasks for 48 hours. The contents of these flasks were then split into 6-well plates containing Fetal Bovine Serum media for 48 hours. The cells were then placed in serum-free media for 48 hours, after which the wells were washed out with PBS before addition of the minoxidil or diazoxide. The drugs were administered separately, dissolved in dimethyl sulfoxide, and diluted with serum-free media before being added to the cells. Concentrations of drug administered will be 0.5, 5, 50, and 500 μM and will be added to 4 of the 6 wells. The other two wells will consist of media with dissolved DMSO and media with no drug or DMSO. These two wells represent the controls of this experiment, with the well containing no added drug or solvent representing a baseline reference for elastin production in untreated RFL6 cells. Cells were incubated for 48 hours, and elastin content was quantified using the Fastin Elastin Assay (Biocolor Ltd.), a quantitative dye-binding method for analysis of both soluble and insoluble elastin released into tissue culture media. The sample absorbance readings were measured with a spectrophotometer and were then fit to the standard curve produced to determine the quantitative amounts of elastin produced by the drugs.

Results: Four sets of cells were cultured in six well plates and administered with DMSO control, 0, 0.5, 5, 50, and 500 μM concentrations of minoxidil and diazoxide separately. For minoxidil, DMSO alone seemed to increase elastin content within the cells. As drug concentration increased, elastin content correspondingly increases (Figure 1a). The greatest average elastin content is observed at 50 μM. At 500 μM of diazoxide, elastin content decreased slightly, indicating that the drug dosage exhibited saturation. For minoxidil, as drug concentration increased, elastin content also increased, except at 0.5 μM concentration (Figure 1b). The greatest average elastin content is observed at 50 μM, significant with respect to DMSO and 0 (p-value equal to 0.015 and 0.018, respectively). At a concentration of 500 μM of minoxidil, elastin content decreased slightly, making similar indications as those from diazoxide.

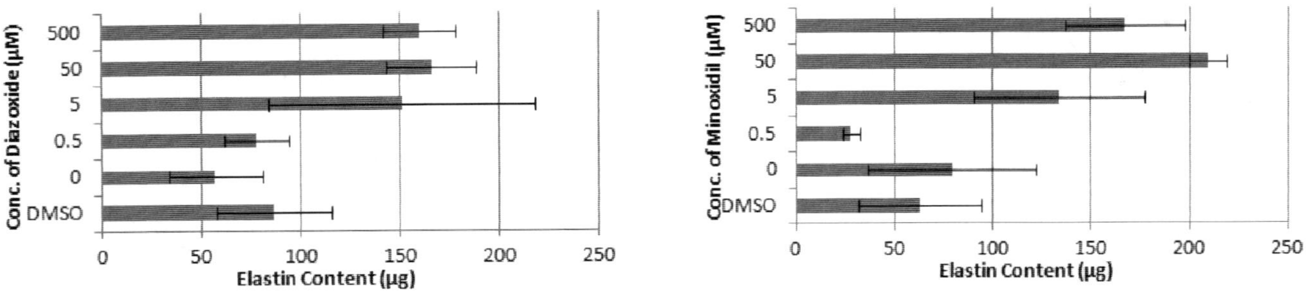

Figures 1a and 1b: Extraction of Elastin from RFL6 Cells Treated with Diazoxide (a) and Minoxidil (b)

Conclusion: From the results obtained, it can be interpreted that diazoxide and minoxidil exhibit changes in elastin content as drug concentration increases, with an especially significant elastin content level at 50 µM concentrations for both drugs. Seeing as these drugs exhibit an effect on increasing elastin content compared to the control cell, future work involves transitioning from application of drug to cell line to application of drug to primary cells. Applying minoxidil and diazoxide to arterial cross sections, for example, would be an even more appropriate model for elastin production in blood vessel walls.

Acknowledgements: Funding by the AMGEN Scholars Program, NIH grants HL105314 and HL115560 and Washington University in St. Louis are gratefully acknowledged.

In a variety of biomedical research fields, cultured mammalian cells are often used as models to study a variety of intracellular or extracellular responses, perturbations, and features. Flow cytometers are powerful tools in this endeavor. A flow cytometer works by passing cells through a focused laser beam, and as the cell passes through the laser, the flow cytometer will collect scattered and fluorescent light that is emitted. The amount of fluorescence and scattered light intensity is used to determine features such as cell morphology, size, and biochemical and molecular changes. Cytometers are prevalent in clinics where peripheral blood cell (WBC) counting is vital for the prognosis of various immune disorders. The count of a person's WBCs, or leukocytes provides an assessment of the body's response to infection. Because there are a number of different WBCs , the standard cell counting assay requires a certain level of complexity whereby multiple fluorescent molecules are tagged to the population of leukocytes. This will allow the cytometer to differentiate among each WBC type because each will emit a particular fluorescence color. In order to alleviate challenges that occur in cytometry when multiple fluorescent species are used (e.g. spectral overlap and spillover) our laboratory is leading efforts in innovative uses for fluorescence lifetime-dependent cell counting. The fluorescent lifetime is defined as the average time for an excited molecule to return to its original ground state and emit fluorescent photons. The fluorescence lifetime is independent of fluorophore concentration and therefore will eliminate issues with spectral overlap. In this study we compare standard intensity-based cytometry with fluorescence lifetime-dependent cytometry by replicating the WBC count assay. We use mouse peripheral blood and label the leukocytes with fluorescently labeled antibodies. We count the WBC population (CD45+, CD8a+, CD3+, and CD4+) and determine if the fluorescence lifetime parameter provides more quantitative WBC counts.

Our laboratory's objective is to show that fluorescence lifetime is a novel way to improve the accuracy of clinical counting of leukocytes. Because leukocytes are essential to the body's immune defense having an accurate and timely way to evaluate the amount and type of leukocytes in your body's blood stream could prove significant in the prognosis of immune disorders.

In order to accomplish this objective Peripheral mouse blood was purchased from Bioreclamation. The blood was then lysed using a solution of NH4Cl, EDTA, Sodium bicarbonate dissolve in distilled water in a protocol published by Santa Cruz Biotech. The peripheral mouse blood is then added to the solution of the lysing agent incubated at room temperature and then centrifuged. After the blood has been lysed the white blood cells that are left are re-suspended in PBS and then centrifuged again. After centrifugation different fluorescent antibodies were added to different cluster of differentions (CD) within the cells. CD3+, CD4+, CD8a+ and CD45+ were labeled with different fluorescent antibodies. CD3+ was labeled with FITC, CD4+ was labeled with PE, CD8a+ was labeled with PE Cy5 and CD45+ was labeled with PE Cy7. The fluorescent cells were then run on a flow cytometer called an Accuri, which is similar to clinical flow cytometers, and compared to published results. Once the protocol was consistent with the publications the fluorescent antibodies were run on a Vantage Flow Cytometer modified to measure fluorescence lifetime. A data analysis technique called a Phasor Plot was used to interpret the lifetime data.

Undergraduate Researcher: Daniel Puleri

Graduate Mentor: Charles Zhao

Faculty Advisor: Prof. Hang Lu

Affiliation: Georgia Institute of Technology

Topical Area: Food, Pharmaceutical, and Biotechnology

Image alignment and segmentation algorithms for image-based screens of subtle features using *C. elegans*

Introduction

The quantitative phenotyping of subtle traits is crucial to the genetic mapping of currently unexplored regions of the genome. By using high-throughput microfluidics and automated image processing, we show that we can acquire phenotypic features rapidly and quantitatively from a model system. With high-volume, high-information feature data, one can statistically map genomic regions with phenotypes; alternatively, one can search for genes of interest in a forward genetic screen. Forward genetics typically uses mutagenization to create an abundance of phenotypes and then determines which genetic mutations correlate with a change in phenotype by isolating the differing phenotypes and comparing changes in DNA. Determining the effect on one particular gene by comparing populations with different genetic backgrounds could shed light on epistatic effects—an important topic in modern biology. Finding which genes affect phenotypes, such as those for diseases, can then be used for future downstream medical research with more direct clinical applicability.

When studying small structures such as neurons, microscopy is a vital tool for observing phenotypes. The images that one gathers from conventional fluorescent microscopy are typically information-rich; however, the relevant data is obfuscated by noise and false positives.

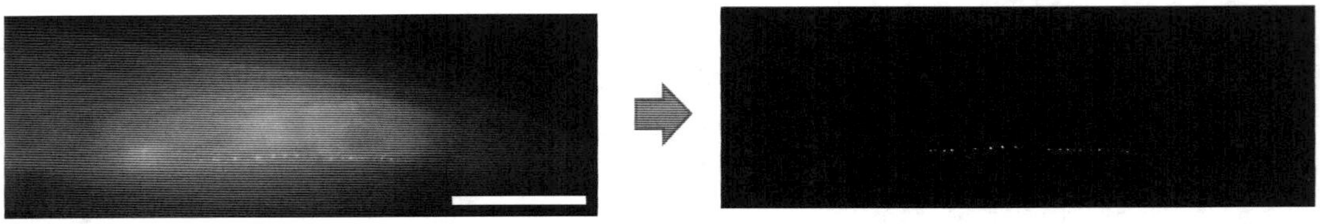

Figure 1. The synapse detection algorithm takes in a maximum projection of a set of images in the z-plane (left) and generates an estimate of the pixels in the image that are part of a synapse (right). Scale bar approximately 30μm.

Using automated image processing, one is able to phenotype large populations of organisms characterized via microscopy in an impartial and high throughput manner that allows human intervention to be minimized. I have used a combination of statistical and heuristic techniques to determine which features of an image are important enough for extraction. For example, we characterize features that would be hard for a human to determine, such as total synaptic intensity and major axis length—these quantitative phenotypes allow us to make more specific claims about genetic effects, when compared with more qualitative observations.

For our studies of the synaptic domain we use *Caenorhabditis elegans* (*C. elegans*), a soil nematode that is widely used in biology due to the breadth of knowledge about the species and its ease of handling. Using traditional techniques to handle and image *C. elegans* is both protracted and cumbersome. The Lu Lab uses microfluidic devices to augment the handling of *C. elegans* and enable high throughput imaging and sorting of worms on-chip[1]. By using microfluidic devices, we are able to gather the large amounts of data needed to perform statistical mapping of genes.

[1] Chung, K., Crane, M. M. & Lu, H. Automated on-chip rapid microscopy, phenotyping and sorting of C. elegans. *Nat. Methods* **5,** 637–643 (2008).

More specifically, we study the synaptic morphology of *C. elegans* which is visible due to the use of an integrated fluorescent protein (*Pmig-13::snb-1::yfp*) that localizes to the synapse. Synaptic morphology is an important trait that has been linked to diseases ranging from schizophrenia to autism spectrum disorder (ASD)[2,3].

Background

Caenorhabditis elegans is a non-parasitic nematode that has been used as a model organism in the laboratory since the 1960s. The organism is about 1mm in length and is transparent. *C. elegans* is also one of the only organisms to have its connectome fully mapped—a description of how its neurons are physically laid out.

The model organism *C. elegans* is well-suited to study the subtle phenotypes of multigenic effects such as in synaptic morphology—which may play an important role in neuronal disease such as autism. Current understanding of synaptic morphology and aging, given the large amount of quantitative data needed to characterize these traits, is relatively incomplete even in the oft-studied model organism *C. elegans.* The need to generate mutants, rapidly characterize worms, and sort populations using traditional methods is tedious and very labor intensive. Using a novel high-throughput microfluidic imaging platform developed in the Lu Lab, these concerns can be mitigated and automated quantitative characterizations of synaptic domains can be performed.

Overall, *C. elegans* is easy to work with due to its rapid mating and transparency under the microscope, which makes it ideal for usage in high magnification experiments. It is easy to house, has a rapid reproductive cycle, and can lay as many as 300 eggs per mother.

Methods/Techniques used

The microfluidic device that is being used was designed by Dr. Adrianna San Miguel for the imaging of *C. elegans.* The device allows for the high volume imaging of worms in a "drive-through"-like process where worms are pushed up to the imaging channel and then directed through one of two exits based on phenotype. The camera is aimed to image worms in a tail-first orientation, while discarding worms in the incorrect orientation.

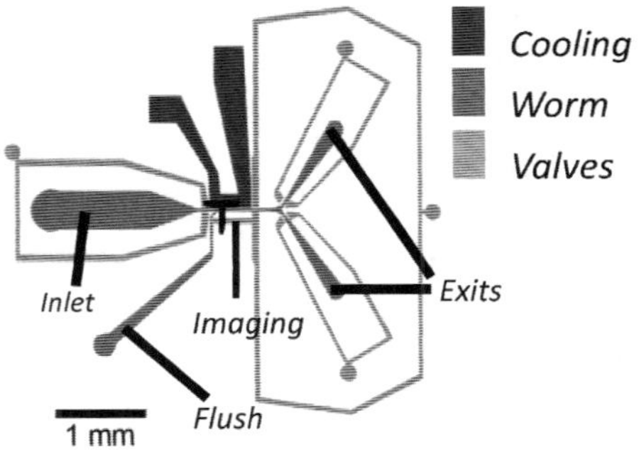

Figure 2. A diagram of the microfluidic device used for imaging and sorting.

The microfluidic devices are fabricated by curing polydimethylsiloxane (PDMS) over a silicon or polyurethane master. The features of the microfluidic devices arise from the detailed height variations of the silicon master. The devices are bored into to create the holes for the inlet, outlet, and valve tubes. Afterwards, the device is plasma bonded to a glass slide which acts as the transparent floor which rests upon the microscope objective.

Worms are imaged at regular points in their lifecycle using an upright fluorescent microscope at 40x.The images are then captured and analyzed in MATLAB®. To distinguish the synapses from background noise, local background removal and thresholding identify synapses and other bright objects. Synapses are then identified from other points of high intensity, mostly autofluorescent lipids, by using DBSCAN, a clustering algorithm. The clusters identified are then checked for their angle to only allow pixel groups in the horizontal dimension. Clusters in the same vicinity are then connected so that the longer axon can be

[2] Shinoda, Y., Sadakata, T. & Furuichi, T. Animal models of autism spectrum disorder (ASD): a synaptic-level approach to autistic-like behavior in mice. *Exp. Anim. Jpn. Assoc. Lab. Anim. Sci.* **62,** 71–78 (2013).

[3] Calahorro, F. & Ruiz-Rubio, M. Caenorhabditis elegans as an experimental tool for the study of complex neurological diseases: Parkinson's disease, Alzheimer's disease and autism spectrum disorder. *Invertebr. Neurosci. IN* **11,** 73–83 (2011).

reconstructed. The features of the worms are extracted and compared to other populations based on a feature set that characterizes, among other things: synapse size, shape, intensity, and domain length.

Due to the length of the worms, some images had to be taken off-axis to facilitate the entire worm in the image. Using a linear regression, the slope of the worm was determined and then it was aligned onto the horizontal axis. The centroid was used to center the worm so that all other processing could be done with a regular reference frame. Post-processing is then performed on the worm using robust principal component analysis (RPCA)[4].

Results

Linear regression to straighten images, regression to determine cluster linearity, and vertical line detection are some of the many heuristics we employ to "clean up" images so that they are more amenable to further processing and analysis. Moreover, I implemented an additional heuristic that connects separated clusters of synapses so that the synapses, if they are in a certain vicinity, can be reconnected to form the complete chain of synapses. Seemingly simple heuristics like this need to be used to correct slight defects that are not picked up by major pieces of our algorithm.

When analyzing images and comparing their 29 features to sort worms into respective populations, we quickly ran into a multiple comparisons problem. In addition, there are underlying issues such as inherent relations between image features that share similar properties like synaptic domain length and number of synapses. To resolve these issues, we propose the use of principal component analysis (PCA). Comparing strains by their strongest component, the component with the eigenvalue of the largest magnitude, we can more successfully segment strains into separate populations. However, this comes with some drawbacks. The PCA is sensitive to large data sets that may contain correspondingly large amounts of noise. To combat this, we have implemented a newer technique called robust principle component analysis (RPCA). Using this technique in a novel setting, we are able to mitigate the aforementioned issues and I was able to optimize it for use with our large datasets which helps yield which features are most important in distinguishing populations. This reduced the memory load from several hundred gigabytes to less than four—a number that is manageable by a consumer laptop.

We were then able to uncover that the genetic background of the worm does affect key features of synaptic morphology. Laboratory adapted wild type strains from different original natural environments showed differing effects on the particular phenotypes. This shows that synaptic morphology is not only controlled by specific alleles; but, is also affected, albeit subtly, by the organism's underlying genetic background.

[4] Cand, E. J., et al. (2011). "Robust principal component analysis?" J. ACM 58(3): 1-37.

398426 The Effects of Y-27632 on the Propagation of Glioblastoma Stem Cells

Abstract:

Glioblastoma is the deadliest form of brain cancer. Patients diagnosed with glioblastoma have poor prognosis, and their median survival rate is approximately one year. Recent research shows evidence for a specialized subpopulation of glioblastoma cells called glioblastoma stem cells (GSCs). GSCs have the ability to self renew and differentiate into the heterogeneous tumor cells that constitute the entire tumor. GSCs are also thought to be resistant to current treatment techniques such as chemotherapy and radiation. For these reasons, the study of GSCs is an area of extreme interest in current glioblastoma research; however, very few GSCs exist in glioblastoma cell lines grown in conventional *in vitro* conditions. When glioblastoma cells are grown in serum-free media, they form tumorspheres that imitate the *in vivo* tumor niche and are enriched for GSCs. We have shown that the addition of the Rho-Kinase (ROCK) inhibitor Y-27632 to serum-free, cancer stem cell media increases the total number of cells and the number of GSCs that can be grown *in vitro*. Y-27632 has been implicated in suppressing apoptosis, and it was hypothesized that Y-27632 would increase the number of glioblastoma stem cells (GSCs) grown *in vitro* and improve cloning efficiencies. Indeed, our data showed that supplementing the GSC media with Y-27632 inhibits apoptosis. Furthermore, we demonstrated that Y-27632 positively affects the cells' ability to form spheres using size analysis and limiting dilution assays. We also characterized the expression of known stem cell markers in cells grown with the inhibitor using various molecular and cell biology techniques. Our data therefore showed that Y-27632 has an overall positive effect on tumorsphere formation and is poised to have a promising application for cancer stem cell culture in the future.

Semi-quantitative Readout in a Paper-based Test for the Home Monitoring of Phenylalanine Levels in Phenylketonuria Patients

Liam Wong and Elain Fu
Chemical, Biological, and Environmental Engineering
Oregon State University

Phenylketonuria (PKU) is an inherited genetic disorder that results in increased levels of an amino acid called phenylalanine (Phe) in the blood. If PKU is not diagnosed and treated, the Phe build up can lead to mental disabilities and other serious health problems. The most common therapy for individuals with PKU is a strict diet to achieve target Phe levels in the blood. Thus, monitoring Phe levels in the blood of these individuals is a critical component of therapy. Existing tests for PKU require high-resource laboratories, which are not available to many populations and cannot provide the real-time feedback required for effective therapy monitoring. In many regions of the world, the level of resources is too low to support testing that requires any instrumentation. It would be ideal to develop an instrument-free, rapid and simple test for PKU patients to monitor their Phe levels. We are developing a paper-based test that provides a semi-quantitative output of the concentration of Phe from a sample of blood. Users will be able to measure the concentration of Phe in a blood sample by looking at the simple pattern or intensity of color produced and matching it to the expected result for "low", "medium", or "high" levels of Phe. The focus of this presentation is on the demonstration of the use of colorimetric indicators to achieve distinct visible output for clinically relevant concentration ranges of Phe in a sample. Specifically we are optimizing reagent concentrations to produce the most robust signal and evaluating different colorimetric indicators to determine which indicators produce the ideal response curves (signal vs. Phe concentration) for quantitative readout. The ideal intensity-based response curve should have a steep slope to produce the greatest difference for a specific difference in Phe levels. In addition to the presentation of this data, future plans for the incorporation of the readout developed into a simple disposable test will be described.

398433 The Investigation of Reovirus Induced Thymic Stromal Lymphoietin Expression through the Analysis of NF-Kb and Its Link in the Signaling Pathway

Abstract:

A cytokine known as thymic stromal lymphoietin (TSLP) is produced by non-hematopoietic cells in the body by the epithelial cells to promote the maturation of dendritic cells. Dendritic cells are major players in both innate and adaptive immune responses by mediating the responses of T-cells, particularly Th2 responses. Preliminary experiments demonstrated that reovirus, a virus that commonly infects the intestine, could stimulate intestinal epithelial cells to produce TSLP in vitro. We hypothesize that reovirus induces TSLP production through activation of the nuclear factor-κB (NF-κB) signaling pathway, where NF-κB is a protein complex that controls DNA transcription in cellular systems. To test this hypothesis, we assess TSLP production in reovirus-stimulated intestinal epithelial cell lines including iCCL2 and MODE-K, while disrupting NF-κB signaling by genetic and chemical methods including introduction of silencing RNA (siRNA) and specific chemical inhibitors of NF-κB activation such as CAPE and IKK inhibitor-VII (IKKi-VII). Our studies aim to delineate mechanisms by which epithelial cells respond to virus infection and initiate protective immune responses.

Development of a thermostable microneedle patch for influenza vaccination

Miraj Desai, Matt Mistilis, Andreas Bommarius, Mark Prausnitz

School of Chemical and Biomolecular Engineering

Georgia Institute of Technology, 311 Ferst Drive, Atlanta, GA 30332

The goal of this study is to optimize thermostable formulations of influenza vaccine to be used and delivered with microneedle patches. Thermostable formulations of vaccine are less reliant on cold chain transport and storage and will lead to cost and energy savings. Fabrication of influenza vaccine microneedle patches involves drying of vaccine in the patches, which can alter the molecular structure of the vaccine and lead to decreased activity. Before drying, replacement of phosphate buffered saline with an ammonium acetate buffer as well as the addition of other excipients such as polysaccharides and amino acids increased stability and helped retain vaccine activity. It was also found that drying the vaccines on PDMS resulted in greater stability as opposed to drying on stainless steel, while drying in the presence of surfactants was detrimental to stability and rendered the vaccine less effective. Lastly, microneedle patches formulated with a combination of calcium heptagluconate and arginine showed no loss in stability, even after drying and storage of the patches for six months at room temperature. The best retention of vaccine activity and stability resulted from combinations of various stabilizers, and this holds true even when stored for long periods of time. By combining stabilizers in the proper amounts, it is possible to drastically increase the thermostability of vaccine delivered by microneedle patches. This will allow vaccines to retain function even after drying and exposure to a wide range of temperatures and external conditions. Effectively implementing this concept can lead to elimination of cold storage and transport while maintaining optimal vaccine activity.

398572 Construction of Gene Deletion Mutants in *Teredinibacter Turnerae* to Determine Gene Function

Abstract:

Abstract

Teredinibacter turnerae is a bacterial endosymbiont of marine bivalve mollusks of the family Teredinidae[1]. *T. turnerae* has the ability to fix nitrogen and to use cellulose as a carbon source, an asset which it confers to its host as well[1]. This bacterium also produces anti-microbial compounds to inhibit the growth of transient microbes in shipworms[2]. Sequencing of the *T. turnerae* (strain T7901) genome was recently completed, and has led to the identification of nine polyketide synthesis or non-ribosomal peptide synthesis biosynthesis regions potentially encoding bioactive compounds[1]. So far, only two regions are characterized. Region two is known to code for production of tartralon (an anti-microbial compound), while region seven is known to code for turnerbactin (a triscatecholate siderophore) [1]. In this study, I construct two separate single gene deletion mutants (*ΔTERTU_1990* and *DTERTU_2288,* located in regions one and three respectively) to aid in the characterization of the genes.

References

1. Yang, J. C. *et al.* The complete genome of Teredinibacter turnerae T7901: an intracellular endosymbiont of marine wood-boring bivalves (shipworms). *PloS One* **4,**e6085 (2009).

2. Trindade-Silva, A. E. *et al.* Physiological traits of the symbiotic bacterium Teredinibacter turnerae isolated from the mangrove shipworm Neoteredo reynei. *Genet. Mol. Biol.* **32,** 572–581 (2009).

398620 Correlation of Concentration Versus Pressure: A Step in the Direction Towards a New Method for Measuring Endotoxin

Abstract:

Endotoxins are complex lipopolysaccharides that are found on the outer cell membrane of gram-negative bacteria. Bacteria will secrete endotoxins in large amounts throughout their life cycle. Endotoxins have an abnormally high heat resistance making it nearly impossible to destroy them under normal sterilizing conditions. Endotoxins affect both *in vitro* and *in vivo* cell growth and functions. In vitro, endotoxins have been known to affect cell development while *in vivo* endotoxins are known to elicit an inflammatory response, fevers, chills and even fatal septic shock in mammals. Endotoxins are typically measured in Endotoxin Units per milliliter (EU/mL), and the US Food and Drug Administration (FDA) sets limits on the concentration of endotoxin for medical devices and parenteral drugs.

A common way to test for the amount of endotoxin is to perform a "gel clot" assay that utilizes a Limulus Amebocyte Lysate (LAL) obtained from horseshoe crab blood that was shown to clot in the presence of bacteria. Specifically, the use of the LAL reagent in a reaction containing endotoxins will produce a gel clot under certain reaction conditions. Ultimately, the LAL reagent can be used to detect endotoxin down to specific levels associated with the sensitivity of the lysate. When a clot forms, the endotoxin concentration is then known to be at a level that is equal to or higher than the sensitivity level of the LAL. A slurry or no clot will be produced at lower levels of endotoxin for a corresponding LAL sensitivity.

The procedure to produce a gel clot takes 1 hour at 37 degree C and can become expensive. Further, testing samples in the field can require an iterative process that involves multiple wells, LAL, and endotoxin controls. While chromogenic and kinetic-based assays also exist, new methods for endotoxin detection could possibly result in more cost effective, high throughput approaches. In this work, our efforts to develop a new methodology for endotoxin testing that exploits mechanical properties of the gels will be discussed. The research might allow for quicker in house testing in many fields ranging from processing plants, hospitals, and laboratories. Further, such methods might result in reductions in the amount of LAL needed and cut down on the destruction of horseshoe crabs population and their environment.

398635 Measurement of Tag Translocation Efficiency in Saccharomyces Cerevisiae Via Confocal Microscopy

Abstract:

Low bioproduction of desired final products is a common problem in synthetic biology. Several technologies exist that can improve bioproduction. Compartmentalization in organelles via signal tagging is one such technology that has been shown to function effectively in *Saccharomyces cerevisiae*. Compartmentalizing metabolic pathways in organelles can increase bioproduction in several ways, such as improving the proximity of the enzyme to key metabolites, providing a physiochemical environment more ideal for efficient enzyme activity, and increasing the local concentration of metabolic intermediates. Despite its uses in synthetic biology, the translocation efficiencies of tags have never been quantified. This is the goal of the research detailed here. Several tags commonly used in the literature were chosen for the mitochondria (four), the peroxisome (three), and the vacuole (three) and compared with one another. Green fluorescent protein (GFP) was used to visualize translocation under a confocal microscope.

Roman Yakobov, Joshua Abrams, Jimmy Toussaint, Teresa Liberto, Maria Cortes, Ruven Pinchasov, Chirag Raval and David Rumschitzki

Early Atherogenesis: Aquaporin-1 & Pressure Affect Intima Thickness and Water Transport Across Rat Aorta Walls

ABSTRACT

Early *atherogenesis*, atheroma or plaque formation, is believed to begin with the blood-pressure-driven advection of the large *Low Density Lipoprotein* (LDL) cholesterol particles from the blood across the endothelium into the next vessel wall layer, the subendothelial intima (SI). LDL advects around rare endothelial cells (ECs) whose intercellular junctions are temporarily big enough for LDL passage. LDL can get stuck in the SI by binding to SI extracellular matrix (ECM). The rate of such binding presumably depends on LDL's local SI concentration. A slow, blood-pressure-driven flow of water and small molecules from the blood across the entire endothelium should dilute this concentration and slow LDL binding.

To explain the variation of aortic wall hydraulic conductivity with pressure, Y. Huang et al. (1997, 1998) proposed a hypothesis and later confirmed experimentally that pressure can compress the sparse SI, causing the ECs to partially block the internal elastic lamina's *fenestra* holes, water's putative entrance into deeper wall layers. Recent further work in our laboratory (Toussaint, Nguyen and Xue) found that endothelial cell (EC) expresses a water channel protein called *aquaporin-1* (AQP1) that allows some of this water flow to traverse *through* the ECs, as well as around them via *tight junctions*. They found that blocking or reducing functioning AQP1 numbers lowers the critical blood pressure to compress the SI and reduce the flow. Conversely, a theoretical model developed by another former student (Joshi) in our group predicted that increasing EC AQP1 expression shifts this critical pressure to higher values and thus creates a pressure regime in which increased AQP1 increases the trans-endothelial water flow by decompressing the SI. Raval in our lab experimentally demonstrated that increasing AQP1 indeed increases this flow. Our goal in this work is to show that this flow increase results from increased AQP1 decompressing the SI. If that is the case, it would suggest that such an increase might noticeably slow early plaque formation.

We utilize 10 μM Forskolin, a cyclic-AMP upregulator known to increase the expression of AQP1 in the membranes of some cell (Raval showed this for ECs) to raise the water permeability of ECs. We fix, section and stain rat aortas treated either with 10 μM Forskolin in $8.5*10^{-4}$ % of DMSO solution in water or $8.5*10^{-4}$ % of DMSO solution in water as a blank. Fixation is done at four different pressures, including at least one that should be in the regime of interest. Analyzing transmission electron micrographs, we assess the mean SI thickness as a function of treatment and transmural pressure.

Preliminary results suggest SI thicknesses vs. blood pressure for both Forskolin and blank treatments follow theoretically predicted trends. The SI thickness for a sample treated with Forskolin at 50 mmHg exhibits a ~135 nm expanded SI. As the theory predicts, at higher pressures (e.g 100 mmHg), Forskolin only expands the SI ~20 nm, since at 100 mmHg, pressure alone can compress the SI. We are currently obtaining more data to draw firmer conclusions.

398703 Towards Inducing hMSC Chondrogenesis in Hydrogels through Treatment with Kartogenin

Abstract:

Osteoarthritis, a disease involving degradation of joints, affects over 70% of Americans over the age of 55. Current treatments are limited to pain management and surgery, neither of which restore the lost cartilage. Stem cells, a multipotent cell source, have the potential to restore lost cartilage, thus improving function while also alleviating symptoms.

Kartogenin, a small, hydrophobic molecule discovered in 2012, has been shown to induce chondrogenesis in bone-marrow derived human mesenchymal stem cells (hMSCs). Kartogenin works by activating a variety of signaling pathways, including the TGF-β1 signaling pathway. We hypothesize that cyclodextrins can be used to increase the solubility of kartogenin, allowing it to be readily incorporated into injectable hydrogel scaffolds for stem cell delivery and differentiation. We are currently treating hMSCs in micromass culture with kartogenin. We are assessing chondrogenesis by using alcian blue to stain glycosaminoglycans, which are part of the extracellular matrix of chondrocytes.

We are also examining the encapsulation of kartogenin within cyclodextrins and the release of kartogenin from cyclodextrins. Following this work, we will treat micromass cultures with kartogenin-containing cyclodextrins and examine chondrogenesis through alcian blue staining and immunohistochemistry for collagen II. Future work will examine the release of kartogenin from cyclodextrins when encapsulated within gellan hydrogels with and without hMSCs..

INTRODUCTION:

Type 1 Diabetes is an autoimmune disease affecting millions of people worldwide, resulting in the loss of insulin producing beta cells in the pancreas, causing insulin deficiency. Insulin injection is the predominant treatment for type 1 diabetes, however; cell based therapy by the transplantation of insulin producing islets will provide a permanent treatment for diabetes. Alginate encapsulation has been used extensively for this process. However, there is a lack of sufficient donor pancreas and this has generated a critical need for a renewable source of insulin producing cells. Human embryonic stem cells (hESC) have been distinguished by their unique ability to differentiate to any cell type in the body, and our work focuses on pancreatic differentiation of hESCs. While hESC differentiation is largely dominated by chemical modulation of the cell microenvironment, recent reports have indicated the importance of the physical microenvironment in affecting cell fate. A previous project in our lab demonstrated the feasibility of encapsulating hESCs calcium alginate and deriving pancreatic islet-like cells under encapsulation. Interestingly, encapsulation enhanced pancreatic differentiation of hESCs compared to differentiation on tissue culture plastic. The objective of this work is thus to investigate the effect of physical and biochemical properties of the capsule on the growth and differentiation of encapsulated hESCs. This was done by varying the type and concentration of cation used. Our results demonstrate that cells encapsulated in softer gels, enhanced pancreatic gene expression. As the gel stiffness was increased, the cell growth is restricted and pancreatic gene expression was suppressed, even in the presence of adequate induction by growth factors.

METHODS:

Alginate encapsulation was achieved by suspension of the cells in an alginate solution comprised of filtered low viscosity alginate with gelatin. This solution was then added drop wise to a cation solution of 100mM $CaCl_2$, 10mM $BaCl_2$, or 20mM $BaCl_2$. DE stage was induced using ActivinA with Wnt3A for 4 days. Afterwards, PP stage was induced with KAAD-cyclopamine for 2 days and retinoic acid for 2 days. Nutrient diffusion was measured through a BSA diffusion assay over 24 hours. Cell viability was assessed at the last day of each differentiation stage using LIVE/DEAD assay. Individual colony area and aspect ratio were determined by undergoing image processing. This was done by taking the stock LIVE image and applying a thresholding and binarizing process. The binary image was then used to form an outline of the individual cell colonies. Cell proliferation was measured using the AlamarBlue assay and cell cytotoxicity was measured using the LDH assay. Differentiation was measured by PDX1 gene and protein expression.

RESULTS:

hESCs were encapsulated in alginate capsules and the biochemical and physical cues were changed by varying the crosslinker type and concentration. After encapsulation, we quantified the dynamics of cell growth in the capsules. Gel stiffness of encapsulated hESC at concentrations of 100mM $CaCl_2$, 10mM $BaCl_2$, and 20mM $BaCl_2$ were measured by atomic force microscopy. There was no significant difference between the stiffness of the 10 mM $BaCl_2$ gel and the 100 mM $CaCl_2$ gel. This stiffness measurement shows a relatively soft gel. The 20 mM $BaCl_2$ was significantly stiffer than the 10mM $BaCl_2$ gel. Nutrient diffusion was analyzed between the various stiffness gels, through timed BSA diffusion. It was observed there was no significant difference of BSA present in the supernatant in the gels of varying stiffness. We analyzed the cell proliferation of encapsulated hESC using the AlamarBlue assay during the stage-wise differentiation protocol. The 10 mM $BaCl_2$ gel and the 100 mM $CaCl_2$ gel conditions showed increased proliferation through differentiation. The stiffer 20 mM $BaCl_2$ conditions showed suppressed proliferation throughout differentiation to the PP stage. Cytotoxicity by the LDH assay showed minimal cell death throughout differentiation for all conditions. Cell viability was measured using the LIVE/DEAD assay. On day 1, there was little to no difference in viability between each alginate capsule conditions. By day 14, cell colonies with in each gel conditions significantly increased in size and showed good viability. The colonies in the 100 mM $CaCl_2$ condition were mostly elongated and asymmetrical, while the 10 mM $BaCl_2$ and 20 mM $BaCl_2$ kept a generally spherical and symmetrical shape. The average cell colony area for the 100 mM $CaCl_2$ increased throughout differentiation to the pancreatic progenitor stage. The softer 10 mM $BaCl_2$ condition shows an increased colony area in relation to the stiffer 20 mM $BaCl_2$ condition with a smaller average colony area. The aspect ratio is calculated by taking the minimum diameter of a colony divided by the maximum diameter, where 1 is perfectly spherical and elongates as it approaches 0. On day 1 of differentiation, all of the conditions showed colonies that were spherical ranging from approximately 0.85-1.00. Colonies in the $CaCl_2$ gel had a broad distribution of aspect ratios, while both the barium gels had colonies that are more spherical. Differentiation of the encapsulated cells was analyzed by measuring the gene expression of PDX1 by qRT-PCR. PDX1 is a transcription factor necessary for β-cell maturation and is hallmark indicator of pancreatic differentiation. The PDX1 gene expression of the 100 mM $CaCl_2$ and 10 mM $BaCl_2$ capsule was highly upregulated. The PDX1 gene expression for the stiffer, 20 mM $BaCl_2$ capsule was suppressed. Thus, alginate capsules made with higher $BaCl_2$ concentration inhibit differentiation compared to capsules made with $CaCl_2$ and low $BaCl_2$ concentrations.

DISCUSSION:

Our results demonstrated changes in gel stiffness have a clear effect on cell growth and differentiation within the alginate gel. There was no restriction of nutrient diffusion between the gels of varying stiffness. On day one post-encapsulation, all gel conditions elicited similar viability and proliferation of encapsulated cells. When the experiment proceeded to day 14, there was noticeable increase in proliferation of the conditions with similar stiffness, and suppressed proliferation in the stiffer 20 mM $BaCl_2$ gel. This is due to softer gels allowing for more rapid growth from the less restrictive gel. Our results also demonstrated that as the gel stiffness increased, the cell growth is restricted and PDX1 gene expression decreases. As alginate gel stiffness increase, pancreatic differentiation of hESC was suppressed. The alginate capsule made with 100 mM $CaCl_2$ was the softest of the gels we measured, and consequently showed the highest differentiation. The alginate capsule made using 20 mM $BaCl_2$, resulted in significantly lower PDX1 gene expression. When we decreased the barium concentration to 10 mM, we observed a gel stiffness very similar to the capsule made using $CaCl_2$. Using the lower $BaCl_2$ concentration rescued the PDX1 gene expression, and thus we observed very similar levels of gene expression as with the capsule crosslinked with calcium. This is of significant importance, since it indicates that even in the presence of conducive growth factor stimulation, non-conducive biomechanical environment can suppress differentiation. With the three alginate gel compositions tested in this project, both cell growth and the PDX1 gene expression was found to be inversely proportional to gel stiffness.

398889 The Development of Injectable Hydrogels for the Management of Chronic Pain

Abstract:

Pain medication is typically administered in one of three forms: bolus injection, tablet, or intravenously (IV). Some of these methods of introducing the drug into the body can cause dangerously high increases of concentration of the drug in the bloodstream and may be only effective for short windows time during which the patient is actually relieved of their pain. There is a need to develop controlled sustained release systems that may overcome these issues. One such method of controlled drug delivery is in the form of a hydrogel that is filled with a drug and capable of releasing the drug over the course of days or weeks. Most hydrogels are formed through a chemical reaction or with a photochemical reaction making it difficult to place the hydrogel inside a patient without a surgical procedure. In this presentation, we will describe our efforts to develop a set of materials and reaction conditions to develop a hydrogel that is injectable and solidifies within a few minutes of injection. When mixed with buffer at 37°C the biocompatible polymers polyethylene glycol diacrylate and ethoxylated trimethylolpropane tri-3-mercaptopropionate (ETTMP) undergo a Michael addition reaction to form a solid crosslinked hydrogel. We show evidence that the purity of the ETTMP is critical to the reaction success and we developed a method to purify ETTMP suitable for making the hydrogel. Finally, we loaded the hydrogel with a model drug, methylene blue, and followed the release of dye over the course of about nine days.

394826 Polymer Film Coatings on Erythrocytes for Extended Release of Chemotherapy

Abstract:

This is an investigation of a novel method of using erythrocytes as carriers for the delivery of chemotherapeutics. Intravenously administered nanoparticle therapies are not optimally effective because nanoparticles have short circulation lives before being cleared from the bloodstream. By trapping chemotherapeutic nanoparticles in films coatings, we seek to prolong the release of the drug. A means of extending drug release and circulation time would alter the pharmacokinetics of the drug, potentially improving the efficacy of treatment with less frequent dose administration

Modular Tissue Engineering: Chitosan-GAG Fibers Formed From an Ionic Complex

Alexander J. Gagliardi and Howard W.T. Matthew
Wayne State University

Modular tissue engineering is a promising approach to current regenerative medicine issues. Modules containing cells can be used as a building block for a larger structure, eventually correlating to an engineered tissue. In general, any module must be able to induce an environment favorable for cell growth while retaining mechanical properties suitable for a specific application. Chitosan-polyanion ionic complex membranes can be used to encapsulate and culture a variety of cell types. Hollow fibers of these materials also have the potential to be used in a variety of tissue engineering applications in vitro and in vivo. The chitosan provides structure, is biocompatible, and can be modified for specific mechanical properties. Glycosaminoglycans (GAGs) provide an environment favorable for tissue engineering applications because of their wide range of biological activity that can be tailored to specific tissue engineered systems. In this study we investigated and developed procedures for continuous generation of chitosan-polyanion hollow fibers with cells or other materials incorporated during fiber formation.

A polyanion solution, consisting of 1.5 wt% medium-viscosity carboxymethylcellulose and 4.0 wt% chondroitin 4-sulfate dissolved in sorbitol-HEPES buffer, was layered over a solution of 0.6 wt% chitosan and 3.8 wt% sorbitol in 0.06 % acetic acid was introduced horizontally into a solution of chitosan in sorbitol via a needle and syringe pump setup. Fibers were drawn from the ionic complex membrane formed at the chitosan-polyanion solution interface. Fibers were pulled coaxially away from the needle and through the chitosan solution using tweezers to initiate extraction. These fibers were collected by pulling out of the chitosan solution and were either rolled onto a cylindrical rod or pulled vertically until failure. The ability to encapsulate materials within the fibers was examined by suspending these materials in the polyanion solution prior to drawing. Materials investigated included suspensions of fixed hepatocytes, mesenchymal stem cells, and microcarriers.

The fibers drawn had diameters ranging from 0.2-1.0 mm after being extracted by hand from a liquid-liquid interface of 7.74mm^2. These fibers achieved a maximum length of ~10cm when lifted vertically. If rolled onto a cylindrical rod, the fiber continued to form until a solution was depleted. The polyanion solution was introduced into the chitosan solution at a rate of 1.0 $^{ml}/_{min}$ Fibers generated were observed to be hollow by the evidence of both microcarriers and cells enveloped within the complex fiber. Fibers with cell suspensions placed in culture fractured and disintegrated within 24 hours and is thus the frontier of our ongoing work.

These fibers are promising candidates for modular tissue engineering applications. These fibers can also be generated with materials of interest inside the fiber walls. Fibers generated with cells and matrix components can be assembled into 2D or 3D structures for engineered tissue assembly and analysis. Our ongoing work entails making these fibers strong enough to withstand static culture conditions with viable cells and matrix components inside. Once a suitable fiber formulation has been identified, characterizing cell growth, organization, and function are important qualities to study for future progress. A system for continuous, controlled rate fiber extraction is also being implemented to determine a correlation between fiber properties and extraction parameters.

Characterization of *E. coli* Encapsulated in mPEG-PDLLA Polymersomes

Jacob Barlow[1], Kevin Gozzi[2]

[1]Deparment of Chemical Engineering, Northeastern University, Boston, MA
[2]Department of Biology, Northeastern University, Boston, MA

Micro-environment fabrication is a way of controlling growth and the interactions between your system and its surroundings. We developed a polymer based micro-environment that will allow us to control the growth and products of bacteria. To selectively control transportation in and out of the system, amphiphilic polymers are used to create a shell mimicking the composition of liposomes. Spherical vessels composed of polymer bilayers with a hydrophilic core and a hydrophobic shell are known as polymersomes.

Self-assembly methods of polymersomes will not allow efficient encapsulation of bacteria so a microfluidic system created at Harvard University was employed [1]. The purpose of using this microfluidic system is to allow for stringent control over polymersome size and loading to generate polymersomes of uniform size and content quantity. In this system, two tapered capillaries are placed facing each other within square tubing to allow different flows through and around the inner capillaries. The polymer is placed in an organic solvent that flows around the upstream capillary while the drug phase is injected into the polymer phase via the capillary. Water flows in the opposite direction around the downstream capillary to generate a shear force capable of pinching off the drug-containing polymer droplets.

This microfluidic device allows monodisperse polymersomes encapsulating bacteria to be formed and also visualized with a fast-motion camera hooked up to a microscope. In order to track cellular growth, *E. coli* constitutively expressing GFP were initially used. A fluorescent microscope was then used to visualize and quantify the growth of bacteria within these polymersomes across many time points and growth conditions. After this proof of concept, these micro-environments are now being optimized for various applications.

These micro-environments allow for bacterial outputs to be incorporated in a system without said bacteria overtaking that system. Possible applications range from probiotic care to culturing unculturable bacteria.

[1] Ho Cheung Shum, Jin-Woong Kim, and David A. Weitz. 2008. *Microfluidic Fabrication of Monodisperse Biocompatible and Biodegradable Polymersomes with Controlled Permeability*. Journal of the American Chemical Society 130 (29), 9543-9549.

398705 Polymer Therapeutics for Protein Drug Delivery System

Abstract:

A slow and controlled drug delivery system is vital to reducing the overall dosage amount and any toxic symptoms associated with the administration of drugs. The purpose of this research is to quantify and optimize the release of protein that has been incorporated within a Poly (vinyl alcohol) (PVA) membrane chemically cross-linked with Glutaraldehyde (GA) hydrogel to serve as a protein drug delivery system. A PVA-GA hydrogel membrane has a three-dimensional structure that allows the protein to permeate through the matrix. The open mesh created by GA results in advantageously high diffusivity that can be controlled by variations in the degree of crosslinking of the polymer matrix. As a part of a development of a drug release system, this study investigates the extent of protein release from the polymeric network of the hydrogel into buffer solution of varied pH. The studied parameters include: the concentration of cross-linking agent (glutaraldehyde) of 0.75% and 1% (v/v), the concentration of protein (0.4 mg/mL and 0.7 mg/ml) and its overall effect on the swelling of the hydrogels, protein diffusion rate, and the release characteristics of the protein. Ovalbumin (OVAL) and Chicken Egg White Lysozyme (CEWL), 45kDa and 14.3kDa, were used as model proteins. Equilibrium Solution Content (ESC), Fourier Transform Infrared Spectroscopy (FTIR) and Thermal Gravimetric Analysis (TGA) characterized the protein-loaded membranes. The ESC experiments show that once protein is incorporated into the hydrogel matrix, the additional hydroxyl groups from the protein permitted additional interaction with the buffer solution and caused an increase in swelling and the mesh size. A higher concentration of protein at 0.7mg/ml resulted in a varied mesh size value. This data suggests that the pores were overloaded and resulted in a dynamic environment, which the protein moved in and out of the pores. Consequently, the 0.7mg/ml resulted in the highest release percentage of the diffusion experiments at 10% higher than the lower concentration of 0.4mg/ml. Also, the protein release demonstrates that from an initial pH 7.4, OVAL and CEWL prefer to diffuse into a buffer solution that is more alkaline of pH 8 than an acidic environment of pH 6.5. There was a 20% decline in the release within an acidic environment. The dissolution rate constant, k_H and diffusion coefficient, D, confirm that the higher pH is a preferred environment as reported. The FTIR show that the protein within the hydrogel cannot be visualized or detected once chemical crosslinking has occurred. However, the TGA results indicate that this is a temporary phenomenon because thermal properties of the hydrogel are not affected by the incorporation of protein but rather chemical crosslinking does affect the thermal properties. The results indicate that the protein was successfully incorporated and the protein does diffuse from the membrane over time. Furthermore, the interaction between the hydrogel and protein is temporary and confirms that the membrane system is ideal for protein drug delivery. Although drug release systems have been studied, our objective is to investigate the mechanism of release and determine the efficiency as it correlates to membrane structure and environment. In our future studies, proteins of various sizes will be used to elucidate the protein release from the hydrogel network.

Optimization of an ultra-fast degrading polymer coating for temporary reinforcement of cortical electrodes during implantation.

Jared Forman[1], Vinod Damodaran[1], Meng-Chen Lo[2], Hilton Kaplan[1], Sagar Singh[2], Shuwu Wang[1], David Shreiber[2], Jeffrey Zahn[2], Joachim Kohn[1]

[1] New Jersey Center of Biomaterials, [2] Department of Biomedical Engineering, Rutgers University, Piscataway, NJ.

Abstract:

INTRODUCTION: Cortical electrodes are important forms of brain-machine interfaces. Characteristically they are large (on the order of 300 µm in diameter) and rigid. We believe that these two aspects contribute to gliosis (scarring), decreasing signal-to-noise ratio and so limiting functional longevity to months at best. In order to reduce gliosis, we propose probes that are: 1) an order of magnitude narrower (~30 µm wide); and 2) highly flexible instead of rigid. To facilitate tissue insertion of these soft microprobes, we have developed a biocompatible, ultra-fast degrading, tyrosine-derived polymer probe-coating, with sufficient mechanical strength to aid proper tissue insertion before rapidly degrading within hours. We report here on the optimization of this polymer.

METHODS: A novel Micromolding in Capillaries (MIMIC) technology was developed to coat 30x5 μm^2 and 80x5 μm^2 non-functionalized electrodes, to uniform overall cross-sectional dimensions of both 100x100 μm^2 and 150x150 μm^2 for each probe size. Optimization of the coating parameters was initially performed on probes fabricated from SU-8. This involved testing various polymer concentrations (5%, 6.7% and 10%), numbers and temperatures of coating layers, and solvents (including anhydrous 1,4-dioxane, acetonitrile, THF, methylene chloride, and DMF). Scanning Electron Microscopy (SEM) was used to characterize the uniformity of the coating, void reduction, and ability to fully mold the capillary features. The optimized process was then implemented for coating probes made from Parylene, a better candidate for *in vivo* testing and electrical functionalization; and polymer erosion kinetics were analyzed using ultraviolet spectroscopy.

RESULTS: It was found that using a 6.7% w/v polymer solution in anhydrous 1,4-dioxane at 75 °C resulted in uniform bubble free coating. Polymer erosion kinetics of both SU-8 and Parylene cortical probes coated in this way, showed complete erosion of the polymer coating within 2 hours under simulated physiological conditions. Our group has also validated this performance in an *in vitro* brain tissue phantom model, and an *in vivo* rat model, where these coated probes exhibited sufficient mechanical strength for insertion.

CONCLUSION: In order to avoid gliosis, cortical electrodes can be made an order of magnitude narrower, and highly flexible instead of rigid, as compared to those in common use. These can be coated to temporarily be sufficiently rigid to insert into *in vitro* and *in vivo* brain models, after which the coating rapidly degrades within hours to leave a soft, flexible, implanted microelectrode.

Development of Polymersomes for Delivery of Therapeutics to the Brain via Intravenous Injection

Development of an intravenously injectable therapeutic is desirable for treatment of the lysosomal storage disease GM1 gangliosidosis. Current studies in a feline model suggest that treatment of patients may require multiple invasive craniotomies and brain injections, with a further disadvantage of therapeutic gradients radiating from the injection sites. A minimally invasive intravenous injection could allow the therapeutic to be more uniformly distributed via the capillary network in the brain. For this delivery method, polymersomes, therapeutic-loaded polymer vesicles, will need to cross the blood-brain barrier (BBB). The BBB is a highly selective, semi-permeable membrane, and based on previous *in vitro* studies, polymersomes will need a diameter of less than 200 nm to transcytose from the blood into the brain [1]. Especially for pharmaceuticals used for rare diseases such as GM1 gangliosidosis, it is important that the formulations are engineered to be stable and have a long shelf life. Thus, therapeutics and carriers need to maintain their properties during processing for long-term storage, which includes cryogenic freezing and lyophilization (freeze-drying).

Polyethylene glycol-b-poly(lactic acid) copolymer with a hydrophilic fraction of 20% was allowed to self-assemble into polymersomes in distilled water. Polymersomes were unable to maintain their size during lyophilization, so the effect of a lyoprotectant on size and structural maintenance was studied. Lyoprotectants are molecules that are hypothesized to protect particle structure and aid in size maintenance during lyophilization due to the presence of hydroxyl groups in their structures. Two lyoprotectants, inulin and mannitol, were tested at concentrations of 2, 5, and 8 wt%/v. Dynamic light scattering was used to measure particle size distributions (PSD) for the polymersomes. Solutions of polymersomes were also filtered through 0.80 and 0.45 μm membranes to remove large polymersomes or aggregates formed during self-assembly. Studies were conducted over two and a half hour time periods with measurements taken every thirty minutes. The self-assembly process was terminated by cryogenically freezing the polymersomes in liquid nitrogen. Lyophilization of the polymersomes was completed using a vacuum apparatus. Polymersomes were then reconstituted from their lyophilized state and diameters measured again to compare to the pre-lyophilization PSD.

Measurement data post-0.45 μm membrane was collected to study size distributions of polymersomes around the 200 nm range, as polymersomes less than 200 nm in diameter could be therapeutically deliverable to the brain, as shown in *in vitro* studies [1,2]. Control polymersomes were those formed without a lyoprotectant, and three sets of measurements were conducted for the control and for polymersomes incorporating either mannitol or inulin at 2, 5, or 8 wt%/v. The average diameters of polymersomes formed with inulin were not statistically different from average diameters of control polymersomes. Polymersomes formed with mannitol were statistically smaller than control polymersomes at the 120 minute time point for 2 and 8 wt%/v mannitol. There were also four instances in which inulin-filled polymersomes were statistically larger than mannitol-filled polymersomes: with 8wt%/v inulin at 90 minutes, 2 and 8 wt%/v inulin at 120 minutes, and 2 wt%/v inulin at 150 minutes. After lyophilization, reconstitution studies were performed to evaluate the size maintenance of the polymersomes. Size maintenance was quantified by dividing post-reconstitution peak diameter by pre-lyophilization peak diameter to provide a normalization value. For control polymersomes, PSD was not well maintained, shown by the normalization value of 4.63 which indicates a much larger post-reconstitution diameter. When either lyoprotectant was incorporated, PSD was much better maintained and resulted in normalization values closer to 1, with a range of 0.54 (2 wt%/v mannitol) to 1.22 (2 wt%/v inulin).

From the studies, it was determined that mannitol decreased the size of polymersomes compared to the control at 2 and 8 wt%/v mannitol and 120 minutes. When using inulin, no statistical differences in size were found versus the control. For reconstitution studies, both lyoprotectants led to enhanced maintenance of polymersome size throughout lyophilization, though inulin was most effective with a normalization value of 0.99 for 8 wt%/v. Current results will inform future studies aimed at creating the optimal polymersome for delivery of therapeutics across the BBB.

References: [1] Costantino, L. Future Medicinal Chemistry, 2010, *2*(11), 1681–1701. Chapter 2.
[2] Gan, C. Biomaterials. 2010; 31:7748-7757.

Nanostructured Nickel-Cobalt Oxide Spinel Films as a Bi-Functional Electrocatalyst in Alkaline Media

Julian A. Vigil, Timothy N. Lambert, Danae J. Davis, Suzanne E. White

Efforts to improve the efficiency of metal-air batteries, fuel cells, and electrolyzers have produced many studies in electrocatalysis. Specifically, high performing electrocatalysts for the oxygen reduction reaction (ORR) and/or the oxygen evolution reaction (OER) in these devices are desired. These reactions exhibit intrinsically slow kinetics and high charge transfer resistances, which impair overall device efficiency. While effective catalysts have been developed and are commercially available, most are precious metal-based and are not economically viable for widespread application. In addition, such commercial catalysts are typically effective at performing either the ORR (e.g. Pt and Pt/C catalysts), or the OER (e.g. Ir and Ir/C catalysts), but not both. An ideal catalyst would be cost effective, easily prepared, stable, and exhibit bi-functional behavior, in that it can perform the ORR and the OER.

We have prepared nickel-cobalt oxide ($Ni_xCo_{3-x}O_4$) spinel nanostructured films through a simple electro-deposition and annealing process. The $Ni_xCo_{3-x}O_4$ catalysts exhibit exemplary performance for both the ORR and OER. Along with an improvement over the parent cobalt oxide (Co_3O_4) catalyst, the $Ni_xCo_{3-x}O_4$ films outperform commercial catalysts for both the ORR (e.g. vs. 20% Pt/C), and the OER (e.g. vs. 20% Ir/C). This presentation will demonstrate the performance of these $Ni_xCo_{3-x}O_4$ catalysts, examine the effect of Ni-doping, and illustrate that these catalysts offers stability greater than or equal to commercial benchmark catalysts. Furthermore, cobalt and nickel are currently about 1,539 and 2,694 times cheaper than platinum, respectively, offering promise as a widespread next-generation renewal energy technology.

The work conducted at Sandia National Laboratories was supported by Sandia National Laboratories: Sandia National Laboratories is a multi-program laboratory managed and operated by Sandia Corporation, a wholly owned subsidiary of Lockheed Martin Corporation, for the U.S. Department of Energy's National Nuclear Security Administration under contract DE-AC04-94AL85000.

Galactomannans as Silicon Binders for Lithium-ion Batteries

Electrochemical energy can be stored in devices, such as the lithium-ion battery, which boasts high energy density and storage efficiency. Improvements are still needed to increase power, capacity, and cycling life. One approach to improving lithium-ion batteries incorporates silicon into the anode because of its high theoretical capacity of ~4200 $mAhg^{-1}$. Silicon suffers from large volume expansions during lithiation, which cause electrode delamination and ultimately electronic isolation from extended cycling, resulting in a shortened battery life. New methods must be explored to increase battery life in order to commercially incorporate silicon as an anode material. Herein, we use galactomannans as silicon anode binders to improve battery cyclability. Galactomannans are environmentally-benign biopolymers that do not require the harsh solvents currently used in carbon anode production. The electrodes are prepared from a slurry of galactomannans, carbon, and silicon and cast onto copper foil. Results for 5, 10, and 15 wt.% guar demonstrated that specific capacity increased with weight percent. However, the 5 wt.% capacity decreased significantly at fast charge and discharge rates. As the weight percent of binder was increased, the columbic efficiency and cycle life decreased. The binder allows for a porous film to be cast with a large surface area as shown in SEM images. Galactomannans are a promising environmentally-benign biopolymer that can be used as binders to increase energy density.

High Performance Nickel Oxide Anodes for Li-ion batteries

Abstract:

Since the market emergence of Li-ion secondary batteries in the 1990s, graphite has been the anode of choice. However, consumer demand for portable electronics with enhanced capabilities and longer times between recharging (i.e. laptops, tablets and smartphones), and the transition of fully-electric and hybrid-electric vehicles to lithium-ion batteries has made it clear that the energy density of future generations of lithium-ion batteries will need to be dramatically improved in order to satisfy weight and power requirements. Therefore, materials with a higher capacity than graphite must be examined. Metal oxide anode materials, such as nickel oxide, have higher theoretical capacities than graphite, but do not possess good capacity retention.

The underlying hypothesis of this work is that the poor capacity retention is caused by low electrical conductivity that is caused by the emergence of Li_2O during the charging cycle of NiO ($NiO + 2Li^+ \rightarrow Ni + Li_2O$). Therefore, in this work, the effect of increasing the electrical conductivity by introducing a non-intercalating carbon to the NiO electrode on the capacity and reversibility of nickel oxide anodes for lithium ion batteries was investigated. Four different nanostructures of nickel oxide were prepared and varying percentages of carbon were added to form the anode material. The results showed that increasing conductivity resulted in higher capacity and capacity retention. However, surprisingly, it was also found that the improved capacity retention was not due to improved structural reversibility.

398601 Studying Voltage Stability of Polymer Electrolyte on Planar Metal Electrodes

Abstract:

Studying Voltage Stability of Polymer Electrolyte on Planar Metal Electrodes

Alexander Rausch, Daniel Hallinan Jr.

Abstract

Lithium batteries offer high energy density, high capacity, and long cycle life. However, these batteries use dangerous organic solvents in the electrolytes that can lead to malfunction, fires, and explosions. Polymer electrolytes are being used to replace these dangerous organic solvents as safer alternatives. In order for polymer electrolytes to become commercially feasible, they must be stable at high voltages. This project aims to determine the voltage stability of solid polymer electrolyte with different metal current collectors for use in solid-state lithium batteries. In particular, it focuses on poly(ethylene oxide) (PEO) doped with lithium bis-triflouromethanesulfonimide salt (LiTFSI) as the electrolyte at a molar ratio of 0.085 mol(Li^+)/mol(EO). Copper, aluminum, and gold are used as the various working electrodes, and lithium is used as the counter electrode. Impedance spectroscopy is used to verify that the half-cells are well assembled by comparing the conductivity to literature values. Voltammetry is used to characterize the voltage stability of the electrolyte in the different half-cell assemblies. The results of these experiments will help determine battery materials that allow for better voltage stability in solid state batteries.

Polymer-based Manufacturing of a Microscale Fuel Cell

Adam S. Hollinger[1], Daniel G. Doleiden[1], Jason L. Williams[2], Gary F. Smith[2]

[1] Department of Mechanical Engineering, [2] Department of Plastics Engineering Technology
Penn State Erie, The Behrend College, 5101 Jordan Road, Erie, PA 16563, USA

Section: 2014 Undergraduate Student Poster Competition

With the rapid advancement of consumer mobile technology, an accompanying demand for energy-dense portable power sources has given rise to increased development of alternatives to traditional rechargeable battery technologies (*i.e.,* lithium-ion) [1]. Microscale direct methanol fuel cells (DMFCs) demonstrate potential as such an alternative, enabling expeditious recharging via fuel solution cartridges and higher energy density than lithium-ion rechargeable batteries [1].

For the purposes of laboratory testing of electrode performance and flow chamber geometry, a multi-layer stacked design is typically used, relying on relatively heavy clamping force to prevent fuel solution leakage and to promote proper electrical contact between electrodes and their respective current collectors [2]. We present ongoing work with Fused Deposition Modeling (FDM or 3D Printing) as an alternative to traditional stacked construction. FDM allows the simplification of test cell components, facilitates the efficient manufacturing of complex design features and geometry, and enables multiple components to be manufactured simultaneously, reducing production time.

In addition, a UNIX-based simulation of fuel flow and diffusion through the flow chamber is presented as a means to predict the effects of varying fuel and electrolyte molarity on DMFC performance. These simulations are benchmarked against experimental data to improve simulation accuracy via modification of the UNIX program.

[1] E. Kjeang, N. Djilali, D. Sinton, Journal of Power Sources, 186 (2009) 353-369.
[2] A.S. Hollinger, P.J.A. Kenis, Journal of Power Sources, 240 (2013) 486-493.

Simulation of a Fuel Cell

***Siyoung Kim*[1]**

[1]KAIST Chemical & Biomolecular Engineering,
Korea Advanced Institute of Science and Technology

Proton Exchange Membrane Fuel Cell(or Proton Electrolyte Membrane Fuel Cell, PEMFC) is one of the energy solutions that does not generate greenhouse gases and any pollutes. Although many scientists have developed a lot of simulations to control and to expect the behavior of fuel cells, only few of them concentrates on a dynamic behavior. If a fuel cell is used as a stationary power source, one should focus on a steady state behavior however if a fuel cell is found in an automobile application, simulation of a dynamic behavior is highly required.

To simulate a dynamic behavior, I assumed catalyst layer was thin enough to be considered as a surface and all reaction happened in the surface of a catalyst. Since the reaction in anode is much faster than that in cathode, I only considered the cathode and I assumed the limiting factor of the cathode was oxygen gases. To describe the dynamic behavior, I proposed a new concept called 'reaction flux' and compared this with a 'diffusion flux'. Not only I got a concentration profile of oxygen gases but also I observed a typical IV polarization curve of a fuel cell. Two different diffusivities were considered in my simulation and this represented how hard it was to penetrate a thin ionomer layer. Finite difference method was used to solve a partial differential equation system.

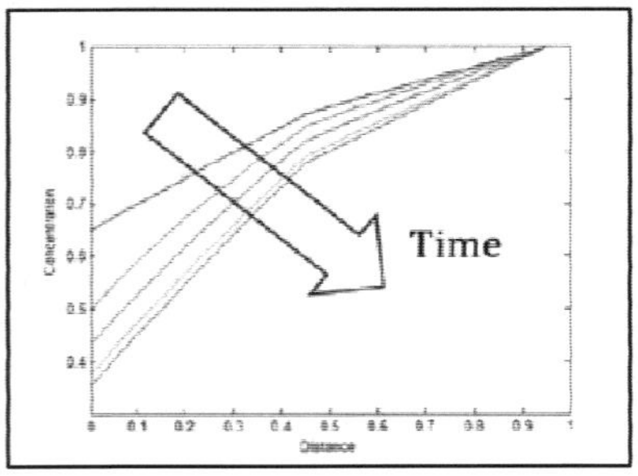

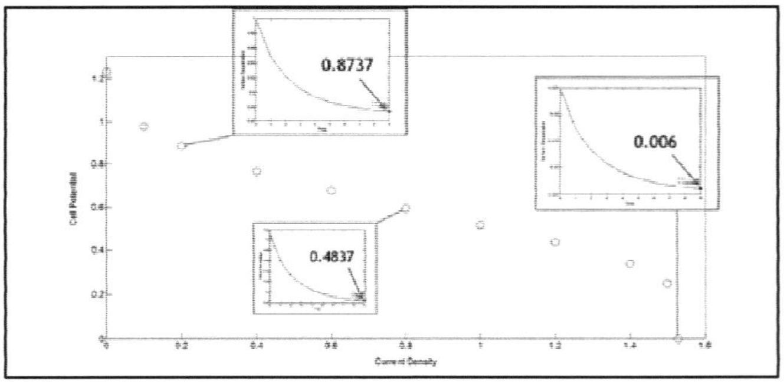

398700 Thermal Energy Storage for the Mitigation of Thermal Cycling in Solid Oxide Fuel Cells

Abstract:

Abstract: There are two main problems inhibiting the progression of (Solid Oxide Fuel Cells) SOFC's as a competitive residential-scale power generation option: materials cost and materials durability due to thermal cycling. This research addresses the latter through a combination of thermal modeling and experimental studies. The proposed solution incorporates a phase change material (PCM) based thermal energy storage system that is used to store thermal energy produced during regular operation, and discharge this energy to the SOFC stack to maintain its temperature while the stack output is varied (or completely shut off). Over a full academic year, the work included a comprehensive techno-economic literature review of SOFC's, testing and modeling heat transfer in thermal energy storage systems, and testing and modeling heat transfer coefficients of various insulating materials (to verify vendor specified heat transfer coefficients) using Matlab. The results show that thermal cycling of SOFC's can be largely mitigated by the use of a phase change material (PCM) based thermal energy storage system.

Advanced Germanium Nanoparticle Composite Anodes using Single Wall Carbon Nanotube (SWCNT) Conductive Additives

M. J. Dzara,[a] M. W. Forney,[b] A. L. Doucett,[a] M. J. Ganter,[b] J. W. Staub,[b] R. D. Ridgley,[c] and B. J. Landi[a,b]

[a] Chemical Engineering, Rochester Institute of Technology, Rochester, NY 14623, USA.
[b] NanoPower Research Laboratories, Rochester Institute of Technology, Rochester, NY 14623, USA.
[c] United States Government, Washington, DC United States

Towards High Energy Density

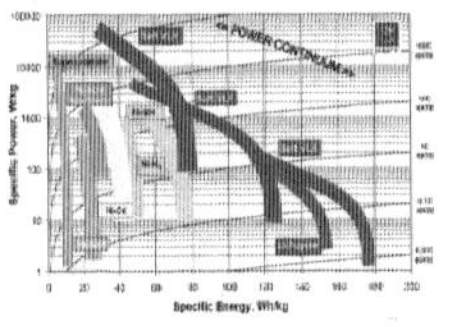

High energy density batteries with good rate performance are needed to meet application demands

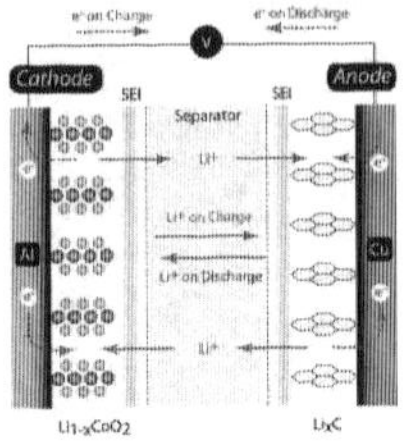

Lithium ion batteries operate through a reversible electrochemical reaction that results in the flow of lithium ions between the anode and cathode

Electrode and Cell Fabrication

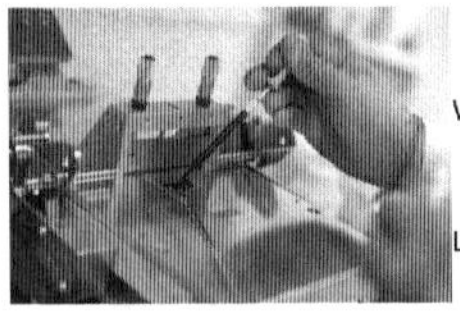

Electrodes are cast from slurries onto conductive foil in blade coating process

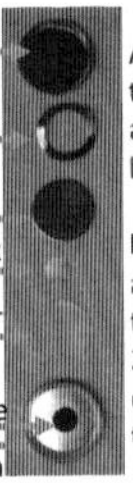

Assembly takes place in an inert glove-box

Finished cells are connected to an Arbin BT 2000 battery cycler for testing

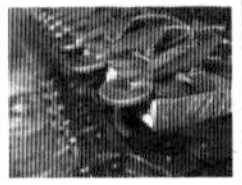

Materials Selection

Germanium Nanoparticles (Ge-NP)

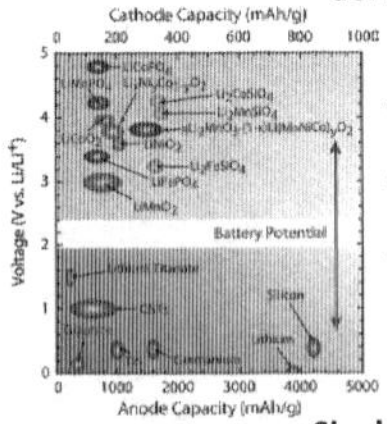

- Ge-NP: Higher theoretical capacity than all but Li and Si
- Li^+ diffusivity of Ge is ~300x that of Si
- Ge-NP purchased from SS-Nano at APS 80-120 nm

Single Wall Carbon Nanotubes (SWCNT)

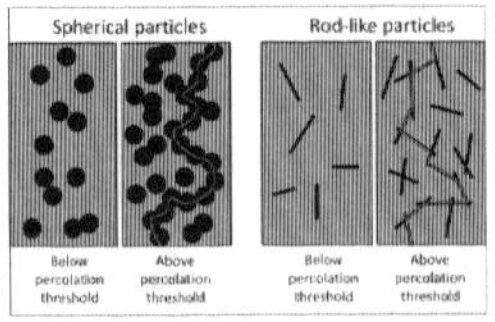

- SWCNT have high aspect ratio, forming percolation threshold at lower mass loadings
- Allows for higher percent active, more energy density
- SWCNT are synthesized and purified in-house

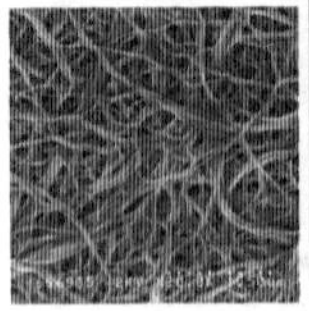

Ge-NP:SWCNT vs. Conventional

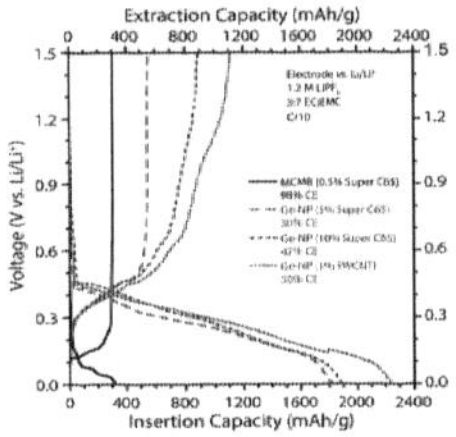

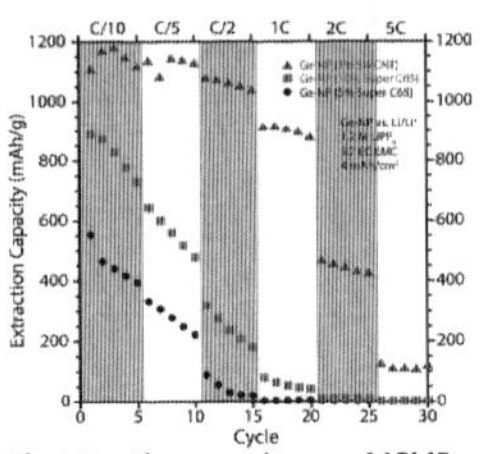

- Ge-NP electrodes show an ~3x improvement in extraction capacity over MCMB anodes
- Higher carbon black loadings are required for Ge-NP to reach higher capacities
- Using SWCNT instead of carbon black yields better rate performance than conventional at similar conductive additive loading

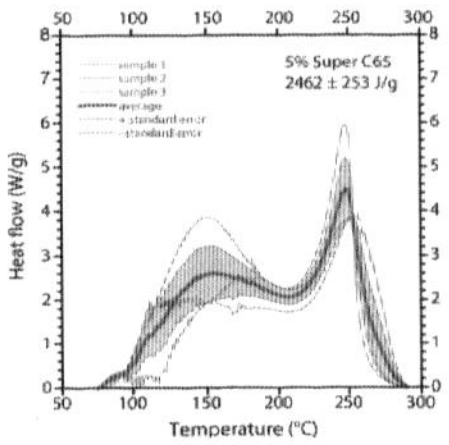

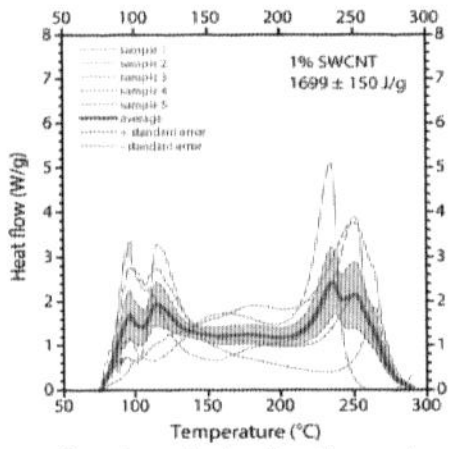

- Differential scanning calorimetry (DSC) can be used to characterize the thermal stability of electrodes
- Switching from carbon black to SWCNT results in a 30% reduction in exothermic energy release during DSC

Electrode Characterization

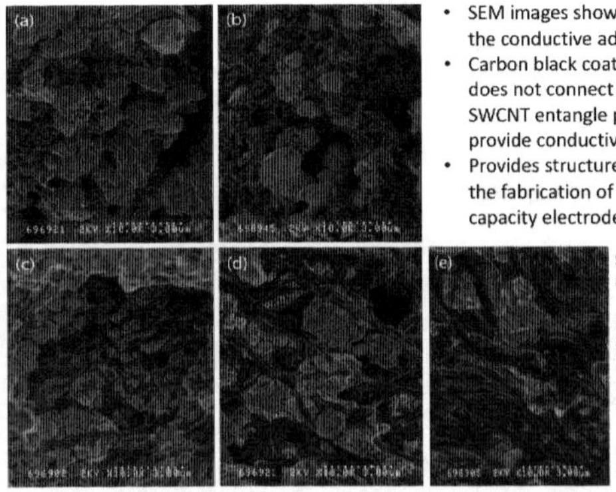

- SEM images show electrodes with the conductive additive in orange
- Carbon black coats particles but does not connect them, while SWCNT entangle particles and provide conductive pathways
- Provides structure that allows for the fabrication of high areal capacity electrodes

SEM images of Ge-NP anode with:

(a) 5% Super C-65,
(b) 10% Super C-65,
(c) 1% SWCNT,
(d) 2% SWCNT, and
(e) 3% SWCNT.

Conductive additive is highlighted in orange

Electrochemical Testing

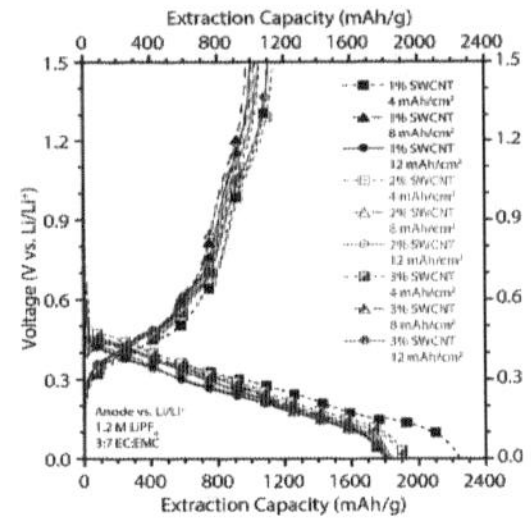

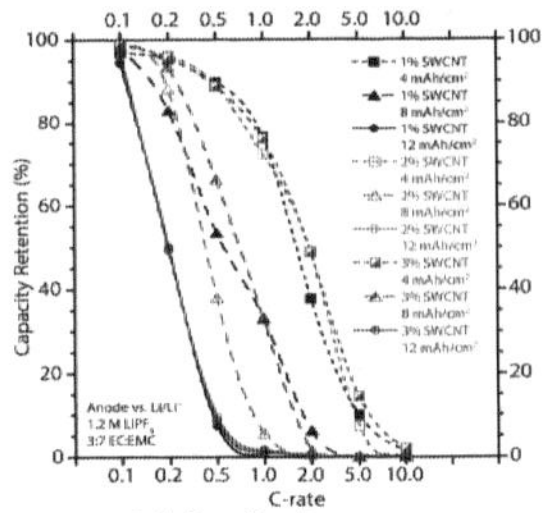

- First cycle extraction capacity is 1000-1100 mAh/g for all cells
- Rate performance decreases with increased areal capacity

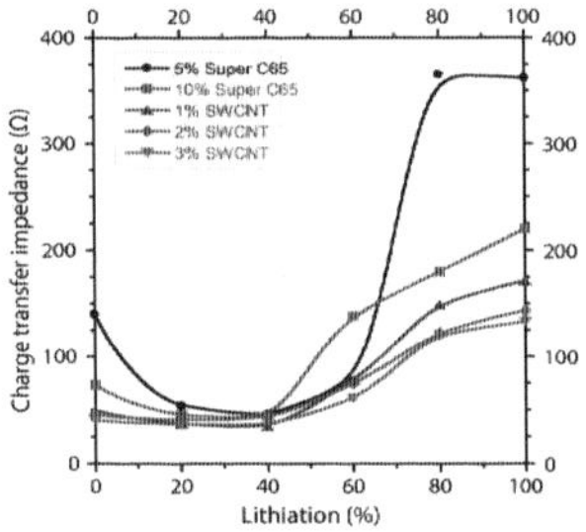

- Electrochemical impedance spectroscopy was used to find charge transfer (CT) impedance as a function of lithiation
- At high percentages of lithiation, SWCNT compounds have significantly lower CT impedance
- There is little difference in CT impedance from 2% to 3% SWCNT, while 1% SWCNT has a slightly higher CT impedance

Energy Density Modeling

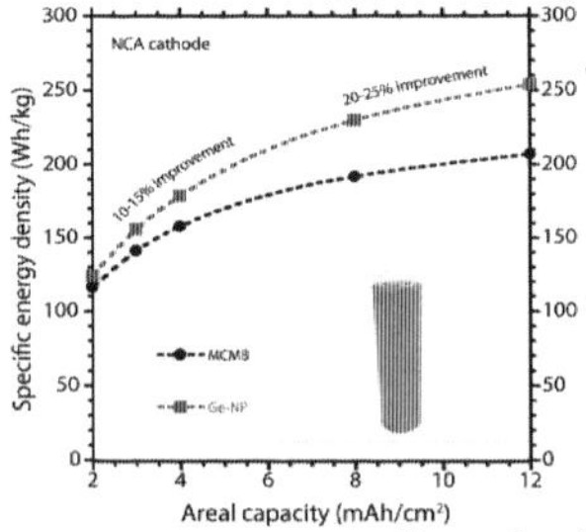

- Ge-NP:SWCNT anodes and conventional MCMB anodes were modeled using an 18650 form factor to compare energy density in a full battery
- Ge-NP:SWCNT anodes would result in a full battery with up to 25% higher energy density than conventional materials

Conclusions

- Ge-NP:SWCNT anodes are predicted to result in a full battery with up to 25% higher energy density than conventional Li^+ batteries
- By replacing conventional conductive additives with SWCNT, Ge-NP:SWCNT anodes have a 30% decrease in exothermic energy release, retain 3x the capacity at a C/2 rate, and have significantly less charge transfer impedance
- Ge-NP:SWCNT anodes present an opportunity to develop full batteries with high energy density, good rate performance, and good thermal stability

Acknowledgements: The authors gratefully acknowledge financial support from the U.S. Government, including a grant from the Intelligence Community Postdoctoral Research Fellowship Program through funding from the Office of the Director of National Intelligence. This material is based upon work funded in whole or in part by the U.S. Government and any opinions, findings, conclusions, or recommendations expressed in this material are those of the authors and do not necessarily reflect the views of the U.S. Government.

Reference: "Advanced Germanium Nanoparticle Composite Anodes using Single Wall Carbon Nanotube Conductive Additives" M. W. Forney, M. J. Dzara, A. L. Doucett, M. J. Ganter, J. W. Staub, R. D. Ridgley, and B. J. Landi, *J. Mater. Chem. A*, 2014, 2, 14528-14535. http://dx.doi.org/10.1039/C4TA02011A

398784 Effect of PEM Equivalent Weights and Carbon Electrodes on the Performance of All-Vanadium Redox Flow Batteries

Abstract:

Vanadium redox flow batteries (VRB) are a potentially enabling technology for intermittent, renewable energy sources such as wind and solar power and offer a means for large-scale energy storage. Much of the emphasis of recent research on VRBs has focused on improving the energy density of the device as this will result in significant reductions in the ultimate system cost. Towards that end, we have explored a series of experimental PEM-based materials and carbon electrodes. We present a detailed analysis of battery performance including charge-discharge cycling to determine the voltage, charge, and energy efficiencies as well as polarization curve analysis that identified losses due to kinetic activation polarization, ohmic polarization due to DC resistance, and mass transport limitations. The results are compared directly to commercially available Nafion 212® and suggest markedly improved performance and power densities in excess of 500 mW/cm^2.

Abstract

The mixture of a non-flammable, ionic electrolyte has been analyzed and tested with the varied addition of an organic carbonate additive. This improved the cycling performance of a lithium-ion battery half-cell when using a lithium iron phosphate ($LiFePo_4$) powdered cathode and lithium metal anode. In this study I investigated the electrochemical response of a lithium-ion half cell when varying concentrations of N-methyl-N-propylpiperidinium bis(trifluoromethanesulfonyl)imide, (Pp-13TFSI) and 1-ethyl-3-methylimidazolium bis(trifluoromethylsulfonyl)imide, (EMITFSI). To assist in ionic mass transfer between the electrodes, varied amounts of ethylene carbonate/di-ethylene carbonate, (EC/DEC) organic additives were mixed with both ionic solutions. The charge and discharge capacity tests indicated that the EMITFSI based electrolyte performed best when using a 0.125 M concentration. The ionic liquid was tested to perform optimally when total solution volume consisted of 10 to 20% EC/DEC additive. This battery configuration resulted in high a high discharge capacity and enhanced cycling ability. In a 0.125 M solution of EMITFSI with 10% EC/DEC additive the specific discharge capacity reached 109 mAh/g during the first cycle, and diminished by less that 10% over the additional three cycles. The Pp-13TFSI solution did not cycle as well and discharge results were around 80% less than that of EMITFSI. Therefore, the non-flammable EMITFSI-based solution can be used effectively and safely with lithium-ion battery based applications

398894 The Electrolyte Equilibrium in Ion Exchange Polymers

Abstract:

Recent developments in a number of technical areas, such as in grid-scale batteries, desalination and other water purification technologies and so on have revealed gaps in our understanding of the interaction of polymer electrolyte membranes with concentrated electrolyte solutions. In such systems, the electrolyte concentration exceeds a threshold below which co-ions are excluded from the membrane. Significant quantities of 'molecular' acid, salt and so on are then imbibed into membrane, dramatically altering its properties. A step toward a quantitative explanation of the resulting phenomena is a description of the driving forces favoring and opposing the uptake of solution components into the membrane. This balance is between osmotic pressure and activity of external solution components as a driving force for uptake balanced against the chemical potential of the pore fluid in the polymer membrane. The chemical potential will be defined by terms related to the Donnan potential opposing acid or salt uptake, individual ion properties such as the enthalpy of solvation and the 'swelling pressure' of the membrane. The latter is related to the mechanical properties of the polymer. Here we provide a first description of some of these properties, emphasizing the sorption equilibria and the mechanical properties of the membrane, for a membrane-acid system typical of that used in vanadium redox flow batteries.

398849 Characterization of Microbial Fuel Cell Materials

Abstract:

Microbial fuel cells (MFC) are an ever-growing clean energy technology. Like hydrogen fuel cells, they harmonize the relationship between system design parameters and operating parameters; however biological aspects must be accounted for as well. MFCs function to recover energy from organic waste through the transfer of electrons from bacteria to an anodic surface (ex. carbon felt electrode). Optimizing the transfer of electrons from bacteria to the anodic surface is the ultimate goal. Characterization of the carbon felt electrode was carried out with the ferri/ferrocyanide redox couple using cyclic voltammetry in an attempt to minimize the overpotential of the reduction and oxidation, thus demonstrating the electrode material's activity. High frequency resistance was also targeted for study. Finally, cathode catalysts (ex. platinum v. non-precious metal catalyst) will be considered. The MFC design is single-chambered and was constructed to minimize contact resistance.

398957 Acetic Acid Tolerance in Bio-Hybrid Fuel Cells

Abstract:

Glucose has been shown to oxidize into ethanol, and bio-hybrid fuel cells take advantage of this by using ethanol with limited preparation in a direct ethanol fuel cell (DEFC). Although this technique is promising, a variety of variables have been shown to influence fuel cell performance. One of these variables is acetic acid production by ethanol oxidation at the fuel cell's cathode. Using amperommetry and linear sweep volatammetry (LSV), we were able to monitor the fuel cell's performance with a variety of acetic acid concentrations. We concluded that although low pH can aid the performance of the proton exchange membrane (PEM) it degrades the performance of the metal catalyst and can be catastrophic to fuel cell power output.

398964 Microbial Fuel Cells: Effect of Electron Mediation on Alcohol Production in Baker's & White Wine Yeast

Abstract:

Given the high energy efficiency of some known organisms and their wide range of acceptable 'food' molecules, interesting possibilities for clean power generation have arisen. A number of genera (such as yeast and some types of bacteria) are capable of using macroscopic electrodes in place of oxygen as electron acceptors. These electrodes can facilitate the use of a Microbial Fuel Cell (MFC) to generate current. MFCs hold much promise for clean energy and waste remediation, but they do have limitations. One of these is the rate of electron transfer at the cell membranes of the organisms being used. The other may be the effect of electron transfer on the production of ethanol as a by-product.

In order to investigate the efficiency of electron transfer and to examine the viability of using a dye as an electron mediator, as well as examine the effect on alcohol production, Methylene Blue (MB), were added to an electrolyte solution containing a yeast suspension. The effect on electron transfer and alcohol production was observed through current vs time and refractive index vs time curves. Knowing the effect of electron mediators on the mechanisms by which electron transfer occurs and alcohol production may enable different possibilities for utilization of electronic processes within cells of the genera being tested.

Experiments have been run with baker's and white wine yeast. Using a potentiostat, potential-time and current-time data was collected across a 10kµÙ resistor. This data was then used to calculate the current output of each species in its MFC. Trials were run using baker's yeast, and white wine yeast while changing mediators and solutions in order to determine the effect of the electron mediator on current generation and alcohol production. It was determined that the optimal current and alcohol production was observed with wine yeast. Graphite rods are being tested as electrode materials and the current production is being compared to carbon felt electrodes.

394472 Solar Fuel Production Via Thermochemical Dissociation of H2O/CO2 Via Ferrite Based Redox Reactions

Abstract:

Solar fuels (produced with the help of H_2O, CO_2, and solar energy as the raw materials) replacing fossil fuels is one of the most promising options to overcome the problems related to the continuously rising of oil prices, dependency of the world's energy economy on the abundance of fossil fuels, rapid depletion of the easily accessible oil reserves, and environmental problems caused by the CO_2 induced greenhouse effect. One of the ways of producing solar fuels is by metal oxide based two-step solar thermochemical cycles. In these cycles, the first step consists of the endothermic reduction of a metal oxide at elevated temperatures releasing O_2. The second step corresponds to the exothermic re-oxidation of the reduced metal oxide at lower temperatures by H_2O, CO_2, or a mixture of the two producing H_2, CO, or syngas.

In this study, mixed ferrite based spinel redox materials were synthesized using sol-gel method. As-synthesized gels of mixed ferrites were dried and calcined upto different temperatures in air. Obtained calcined powders were characterized using various analytical techniques such as powder X-ray diffraction, BET surface area analysis, scanning (SEM) & transmission electron microscopy (TEM), and Inductively coupled plasma spectrometer (ICP). Synthesized mixed ferrites were further examined for their solar fuel production ability by performing multiple H_2O and CO_2 splitting cycles via thermal reduction/oxidation reactions. The obtained results indicate that the sol-gel derived mixed ferrites are capable towards producing high and constant levels of solar fuel in multiple thermochemical cycles. The results related to synthesis, characterization and solar fuel production experiments will be presented in detail.

394824 Biofuels: Conversion of Microalgae to Diesel

Abstract:

Microalgae have the potential to become a major global renewable fuels source, beneficially utilizing sunlight, CO2 and nutrients to rapidly grow long chain lipids which can be extracted from the algae biomass. Typically, the extracted algae oil contains about 12wt% oxygen which can be removed via hydrodeoxygenation to produce straight-chain hydrocarbons, mimicking green diesel. The conventional catalyst for hydrotreatment in petroleum refining is the sulfided NiMo catalyst, however, crude oil contains an insignificant amount of oxygen (<0.5%), so its effectiveness for algae oil has not been established. In this project, precious metal-based catalysts are being evaluated in a microreactor for hydrotreatment of algae oil extracted from Nannochloropsis Salina. A performance study will be conducted to elucidate the effects of various operating conditions, such as temperature, pressure, liquid hourly-space velocity (LHSV), weight hourly-space velocity (WHSV) and H2/oil ratio on hydrocarbon yield, product selectivity and residual oxygen content.

396965 Butanol Production Improvement By Metabolically Engineered Clostridium Tyrobutyricum with Proteomics Analysis

Abstract:

Some metabolically engineered clostridia have the potential to economically produce biobutanol, a sustainable and environmentally friendly fuel with similar energy intensity to petroleum gas. The goals of this study were to develop a fundamental understanding of carbon redistribution to butanol and achieve higher butanol production using engineered *C. tyrobutyricum* ATCC 25755. The 2-L free-cell fermentations were performed in stirred tank bioreactors at pH 6.0 and 37°C using butyrate producing strains (control) and a high butanol producing ACKKO-*adhE2* mutant (butanol titer of 14-16 g/L). The global proteomics profiling of these strains were investigated and compared. It is found that, as compared to wild type, the ACKKO-*adhE2* had decreased expression of thiolase (*thl*), 3-hydroxybutyryl-CoA dehydrogenase (*hbd*) and crotonase (*crt*) and high expression of heterologous bifunctional acetaldehyde/alcohol dehydrogenase (*adhE2*). These findings indicated that we could further improve butanol production by refluxing carbon to butyryl-CoA.

398194 Effect of Water on Whole Cell-Catalyzed Transesterification of Bean Oil Via Rhizopus Oryzae for Biodiesel Production

Abstract:

Biodiesel produced by the methanolysis of plant oils has received a lot of research attention due to its potential as a renewable and environmentally-friendly energy resource. The use of a whole-cell biocatalyst eliminates the need for purification that is necessary for most modern industrial processes and appears to be a more efficient and promising technique in the process of producing biofuel through the transesterfication of triglycerides. The effect of water on biodiesel production was investigated with the use of bean oil and *Rhizopus oryzae* whole cell with polyurethane foaming plastic as a whole-cell biocatalyst. The optimal water content was found to be 10% after 60 hours. The results presented could be beneficial for developing an efficient and cost-effective biodiesel production technique.

398683 Biodiesel from Alligator Fat: A Comparison Between Supercritical and Conventional Transesterification Conditions

Abstract:

As the world expands and the population continues to rise, it is becoming more important to diversify where the world gets its energy from. Biodiesel as an alternative fuel source has been a steadily growing industry for years and offers a lot of promise. The traditional method for producing biodiesel is to mix some sort of vegetable or animal fat with a low molecular weight alcohol, as well as a NaOH catalyst, and react them in a batch reactor. After the reaction is complete the biodiesel, excess alcohol, filter cake, water, and glycerol must be separated. The alcohol can be reused, but the caustic catalyst cake must be disposed of properly and the conversion is only around 60%. An alternative method to this traditional process is doing the reaction via a supercritical process.

Research has been conducted using an autoclave to do a supercritical batch process with just raw alligator fat and methanol. The fat and methanol are mixed in the autoclave and it is put in a muffle furnace to react. Alligator fat is used as feedstock due to its availability in south Louisiana, but pork fat and beef fat have been used as well and it is thought that the process can translate well to any sort of triglyceride feed stock. Rendered oil can also be used in this process, which has a conversion of up to 95%. This process is advantageous because no solid waste is formed and no catalysts are required. Another process has been developed uses a supercritical flow reactor to produce biodiesel using an oil-methanol mixture. Oil is rendered from alligator fat and put in a mixer with methanol. An HPLC pump sends the emulsion through a small metal tube that is kept in a muffle furnace. The pressure is maintained by a high pressure valve on the end of the metal tube. The excess methanol can be flashed off and recycled. This process has shown promise and is currently being researched.

398926 Understanding the Formation and Thermal Dissociation of Methane Hydrates

Abstract:

The continuously increasing demand for fuel has called upon the utilization of unconventional sources of natural gas. Locked inside the earth's crust in permafrost and sub-ocean floors in the form of methane hydrates, is about 1000 times more natural gas than what is present on earth today. Methane hydrates are solid, non-stoichiometric crystalline cage-like structures. They are formed when water molecules under high pressures and low temperatures entrap individual gas molecules. Conditions ideal for natural methane hydrate formation are created. An in-situ thermal simulation is used for the dissociation of hydrates. It is seen that there is a high efficiency, between 75-92% in extracting the methane when there is higher thermal input (about 50 W) for a 50% saturation of methane hydrates.

Carbon dioxide gas delivery to thin-film aqueous systems via hollow fiber membranes

Lauren Merriman, Alex Moix, Robert Beitle, Jamie Hestekin*

University of Arkansas, Ralph E. Martin Department of Chemical Engineering, Fayetteville, Arkansas

* jhesteki@uark.edu

Uneven distribution of energy throughout the world has caused political and economic tension, bringing renewable energy research to the forefront of scientific focus. The United States Department of Energy estimates that if biofuel produced from algae replaced all of the petroleum in the United States it would require only 0.42% of the country's total land area for production, 1/7 of the total land area used by corn crops in 2000 [1]. With promise for algae as a possible feedstock for renewable energy, this research focused on reducing the production cost and greenhouse gas emissions of algal production. Carbon dioxide addition to algal growth systems has shown to increase biomass production with the possibility of using industrial flue gas to reduce carbon dioxide emission to the atmosphere. This research used hollow fiber membranes as a carbon dioxide delivery mechanism into thin-film aqueous systems as an experimental representation of their performance in open algae growth systems such as algae raceways. Comparison with other gas delivery mechanisms, such as bubbling via a diffuser and open tube, shows hollow fiber membranes are far superior in terms of mass transfer efficiency. In under 25 minutes at a liquid depth of about 4 centimeters, the hollow fiber membranes delivered 60 times more carbon dioxide to the system than an open tube as well as 6 times more than a porous diffuser. This research has revealed that a large surface area, small pore size, and certain surface properties are key factors to hollow fiber superiority in carbon dioxide delivery into thin-film aqueous systems.

References

1. Dyer, Gwynne (17 June 2008). "A replacement for oil". The Chatham Daily News. Retrieved 24 January 2014.

398927 Incorporation of High Pressure Clc into IGCC Systems for Carbon Capture

Monday, November 17, 2014
Galleria Exhibit Hall (Hilton Atlanta)

Oscar Nordness[1], Zhiquan Zhou[2] and George M. Bollas[1], (1)Department of Chemical & Biomolecular Engineering, University of Connecticut, Storrs, CT, (2)Department of Chemical, Materials & Biomolecular Engineering, University of Connecticut, Storrs, CT

Incorporation of High Pressure CLC into IGCC Systems for Carbon Capture

Oscar Nordness*, Zhiquan Zhou, George M. Bollas
Department of Chemical & Biomolecular Engineering, University of Connecticut, Storrs, 191 Auditorium Road, Unit 3222, Storrs, CT, 06269-3222, USA. Email:oscar.nordness@uconn.edu

Coal is nowadays considered to be the world's "fastest growing" fossil fuel [1] and currently provides 29.7% of the global energy supply, primarily in the form of electricity [1]. Coupled with this energy however, is the production of considerable greenhouse gas emissions, mainly in the form of CO_2. Currently coal combustion produces over 44% of global CO_2 emissions [1]. Integrated gasification combined cycle (IGCC) plants offer an attractive option for electricity production from coal gasification, while allowing for mercury removal and CO_2 capture at much lower additional costs. At the same time, chemical-looping combustion (CLC) offers the highest estimated efficiency for CO_2 capture [2]. Raising the operating pressure of an IGCC increases the thermal efficiency of the process, while decreasing the energy required for CO_2 compression [3]. However, when IGCC is combined with CLC, the effect of the high-pressure on the synthesis gas-CLC is not well understood, as it depends on the selection of oxygen carrier and other process conditions [4]. In this presentation the effect of elevated pressure on CLC of CH_4 and syngas was studied using copper and nickel oxygen carriers in an alternating-flow fixed bed reactor (Figure 1). Experiments were conducted at the conditions shown in Table 1 at pressures of 1, 5, and 10 bar. These conditions were selected based on standard operations conditions, but at an extended time scale in order to achieve reduction and oxidation at elevated pressure. Figures 2 and 3 show experimental results of the reduction of NiO and CuO with CH_4 at different system pressures (1, 5 and 10 bar) and a bed temperature of 800 °C. Figures 4 and 5 present the experimental results of the NiO-syngas and CuO-syngas systems at 1 and 5 bar. It was found that the reduction of CuO with both CH_4 and syngas occurred with slower reaction kinetics than that of NiO across all pressures, however CuO achieved higher CO_2 selectivity. The elevation of pressure with consistent gas flowrates resulted in increased residence time of the reduction and oxidation gases within the reactor. This allowed for a longer period of reaction for CH_4, which significantly increases the CO_2 selectivity. Elevated pressure contributed to increased solid carbon formation for reduction of CuO with CH_4 and syngas as well as for NiO with syngas. The formation of solid carbon was found to decrease at elevated pressures in the case of NiO reduction with CH_4. Although increased CO_2 selectivity with elevated pressure shows promise for IGCC integration and carbon sequestration, the added carbon formation presents new challenges within the IGCC + CLC system.

Table 1: Experimental Conditions

Oxygen Carrier	20 wt.% NiO/Al_2O_3, 36 wt.% CuO/SiO_2
Load	2.2 grams
Reactor Dimensions	I.D. 9.6 mm, height 482 mm
Total Flow	100 sccm
Temperature	800 °C
Pressure	1, 5, 10 bar
Oxidation	8 min, 20 vol.% O_2/Ar
Reduction (CH_4)	3 min, 10 vol.% CH_4/Ar
Reduction (syngas)	60 min, 5 vol.% CO, 5 vol.% H_2 /Ar

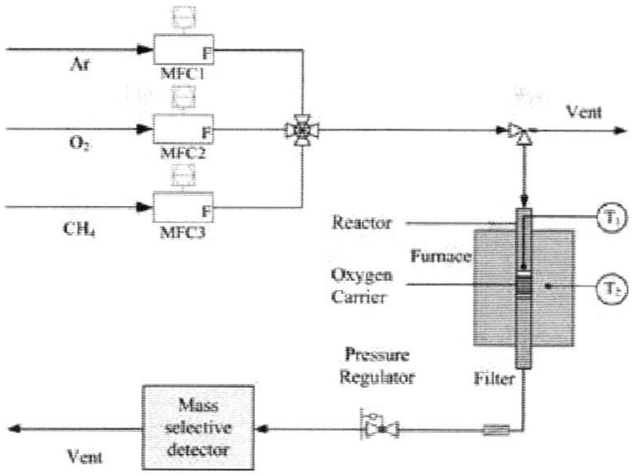

Figure 1: Fixed-bed chemical-looping setup used in this work.

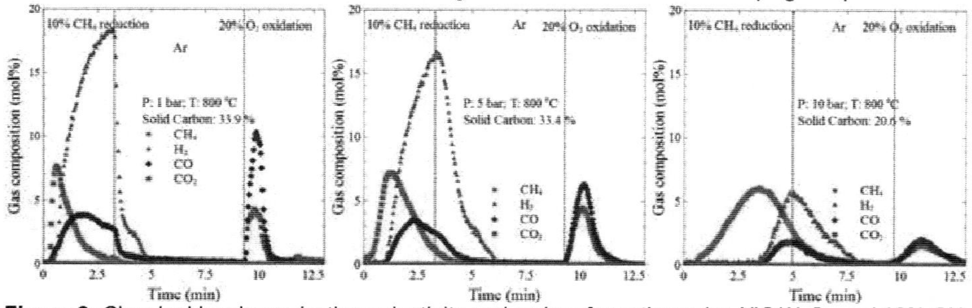

Figure 2: Chemical-looping reduction selectivity and carbon formation using NiO/Al_2O_3 and 10% CH_4/Ar in fixed-bed reactor at (a) 1 bar and 800 °C; (b) 5 bar and 800 °C and (c) 10 bar and 800 °C.

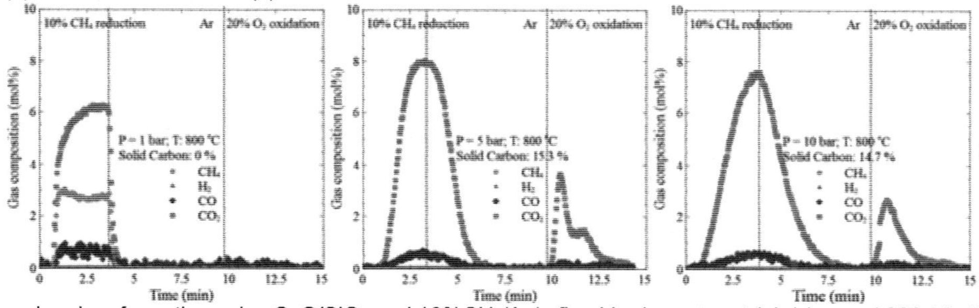

Figure 3: Chemical-looping reduction selectivity and carbon formation using CuO/SiO_2 and 10%CH_4/Ar in fixed-bed reactor at (a) 1 bar and 800 °C; (b) 5 bar and 800 °C and (c) 10 bar and 800 °C.

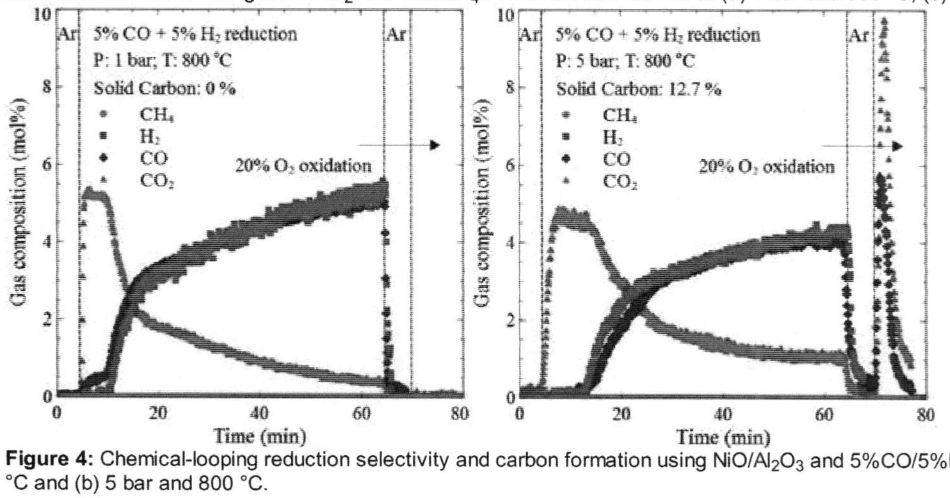

Figure 4: Chemical-looping reduction selectivity and carbon formation using NiO/Al_2O_3 and 5%CO/5%H_2/Ar in fixed-bed reactor at (a) 1 bar and 800 °C and (b) 5 bar and 800 °C.

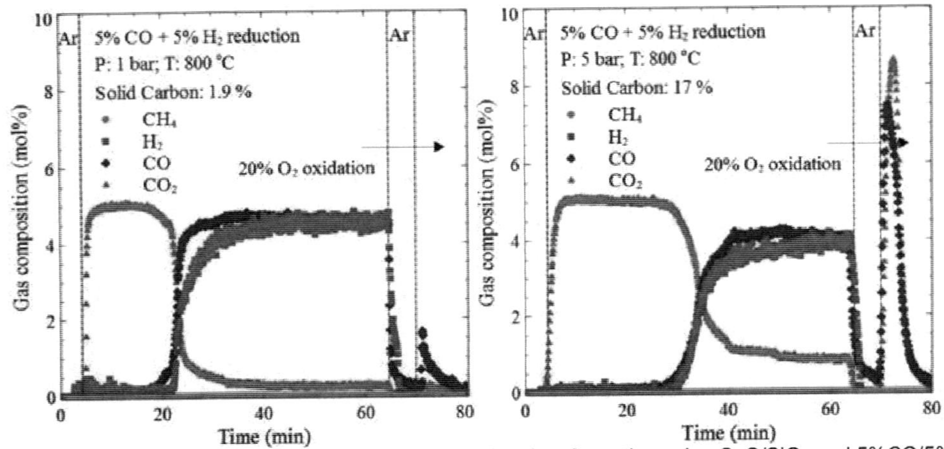

Figure 5: Chemical-looping reduction selectivity and carbon formation using CuO/SiO_2 and $5\%CO/5\%H_2/Ar$ in fixed-bed reactor at (a) 1 bar and 800 °C and (b) 5 bar and 800 °C.

Acknowledgement
This material is based upon work supported by the National Science Foundation under Grant No. 1054718. Support by W.R. Grace & Co. by providing the Al_2O_3/SiO_2 matrices is gratefully acknowledged.

References

[1] "Coal," *Center for Climate and Energy Solutions*, 2014. [Online]. Available: http://www.easybib.com/cite/view.
[2] L.-S. Fan, *Chemical Looping Systems for Fossil Energy Conversions*. American Institute of Chemical Engineers, Wiley & Sons, 2010.
[3] *IPCC Special Report on Carbon Dioxide Capture and Storage*. New York: Cambridge University Press, 2005.
[4] J. Adanez, A. Abad, F. Garcia-Labiano, P. Gayan, and L. F. de Diego, "Progress in Chemical-Looping Combustion and Reforming technologies," *Prog. Energy Combust. Sci.*, vol. 38, no. 2, pp. 215–282, Apr. 2012.

The Kalman Filter Applied to Glucose Levels in Critically Ill Patients

Weiss, Samantha[1,2]; Knab, Tim[1]; Pritchard-Bell, Ari[1]; Parker, Robert[1] PhD; Clermont, Gilles[1,3,4] M.D.

[1]Department of Chemical and Petroleum Engineering, Swanson School of Engineering, University of Pittsburgh, PA

[2]Department of Mathematics and Statistics, Smith College, MA

[3]: The CRISMA (Clinical Research, Investigation, and Systems Modeling of Acute Illness) Laboratory, Department of Critical Care Medicine, University of Pittsburgh, Pittsburgh, PA, USA

[4]: The McGowan Institute for Regenerative Medicine, University of Pittsburgh Medical Center, Pittsburgh, PA

Many critically ill patients suffer from fluctuating blood glucose levels, which leads to hypoglycemia and/or hyperglycemia and worsens health outcomes for these patients. Targeted Glycemic Control (TGC) may improve outcome for patients in the intensive care unit (ICU) [1]. A critical challenge in controlling blood glucose is the reliability of subcutaneous (subQ) sensor measurements. Continuous Glucose Monitors (CGM) yield noisy and potentially biased measurements. 24 patients underwent simultaneous measurement using two Dexcom Platinum G4 CGMs, and the raw signals were processed using Kalman filters (unmodified, adaptive, and dual rate) to fuse the CGM data into a single, smoothed glucose measurement. The Kalman filters were tuned using sensor-specific estimates of the standard deviation of the sensor noise. Sensor noise estimates were calculated by (i) smoothing the subQ sensed glucose data with a moving average filter (#-point horizon); identifying maxima and minima in sensed glucose concentration; fitting a least-squares regression line between the successive extrema; and regressing the error between subQ sensor readings and the line. From these residuals, standard deviation estimates for sensor noise were calculated. The Kalman filters were employed retrospectively in this study, looking at estimating glucose concentrations for data with known calibration end-points, it is possible to employ the Kalman filters in a more prospective manner by estimating the patient- and sensor-specific noise parameters over a brief calibration window after sensors are placed and then simply applying the filter to later measurements, allowing this algorithm to be usable in real time. Kalman filter effectiveness was analyzed using the residuals between the filtered glucose measurements and more accurate, but less frequent, fingerstick measurements. The average residual from the filtered signal to the fingersticks was 1.697 mg/dL indicating improved accuracy of the filtered signal over the raw signal. The average residual between the filtered signal and the raw signal is -.4905 mg/dL with an average standard deviation of 14.2 indicating that the filtered signal is often making substantial corrections to the raw signal. The Kalman filters improve signal accuracy and could be used to drive a TGC algorithm.

1. Van Den Berghe, Greet, et al: Intensive Insulin Therapy in Critically Ill Patients. New England Journal of Medicine 2001: Vol 345, No. 19.

Electrochemically Active Composite Nanofibers by Electrospinning

Ziev Basson[a], Bharat Gattu[b], Prashanth Jampani Hanumantha[b], Prashant N. Kumta[b]

[a]Department of Chemical Engineering, Bucknell University
[b]Department of Chemical and Petroleum Engineering, University of Pittsburgh

With global energy demands on the rise, much research is ongoing into the field of batteries as energy storage devices. Whether it is a handheld device or larger electric vehicles, storage needs are currently met with the lithium-ion battery (LIB) technology. LIB technology is limited in energy density and there is a need to explore new materials' chemistries with greater theoretical specific energies to complete the transition away from fossil fuels. This work investigated several nanofibrous materials as potential electrodes in lithium-sulfur (Li-S) and sodium-ion batteries (SIB). Using electrospinning technique, polymerized nanofibrous mats containing sulfur (for Li-S) and tin (for SIB) were created. Electrospun fibers are preferred due to their high area/mass ratio and the ease of creating conductive composites. Following characterization of the mats by SEM and XRD, the materials were tested for battery cycling stability, areal capacity and rate capability in a two-electrode coin cell.

396157 Central Africa Energy: Utilizing NASA Earth Observations to Explore Flared Gas As an Energy Source Alternative to Biomass in Central Africa

Abstract:

Much of Central Africa's economy is centered on oil production. Oil deposits lie below vast amounts of compressed natural gas. The latter is often flared off during oil extraction due to a lack of the infrastructure needed to utilize it for productive energy generation. Though gas flaring is discouraged by many due to its contributions to greenhouse emissions, it represents a waste process and is rarely tracked or recorded in this region. In contrast to this energy waste, roughly 80% of Africa's population lacks access to electricity and in turn uses biomass such as wood for heat and light. In addition to the dangers incurred from collecting and using biomass, the practice commonly leads to ecological change through the acquisition of wood from forests surrounding urban areas.

The objective of this project was to gain insight on domestic energy usage in Central Africa, specifically Angola, Gabon, and the Republic of Congo. This was done through an analysis of deforestation, an estimation of gas flared, and a suitability study for the infrastructure needed to realize the natural gas resources. The energy from potential natural gas production was compared to the energy equivalent of the biomass being harvested. A site suitability study for natural gas pipeline routes from flare sites to populous locations was conducted to assess the feasibility of utilizing natural gas for domestic energy needs. Analyses and results were shared with project partners, as well as this project's open source approach to assessing the energy sector. Ultimately, Africa's growth demands energy for its people, and natural gas is already being produced by the flourishing petroleum industry in numerous African countries. By utilizing this gas, Africa could reduce flaring, recuperate the financial and environmental loss that flaring accounts for, and unlock a plentiful domestic energy source for its people.

Fuels, Petrochemicals and Energy

The Study of and Co-Relations to the Global Fuel and Biochemical Research and Development.

Atlantic International University,
Chemical and Chemistry Engineer
Undergraduate presentation
AIChE
Atlanta Summit 2014

ABSTRACT

The Study of Engineering within the engineering industry fields of Fuels, Petrochemicals, and Energy Engineering; he Co-Relations to the Global Fuel and Biochemical Research and Development. A post presentation will be presented.

Ethenia A. Scott
Student
Atlantic International University
Undergraduate Student Poster Session:
Fuels, Petrochemicals and Energy
Abstract
Title:
The Study of Engineering the industry Fuels, Petrochemicals, and Energy Engineering; the Co-Relations to the Global Fuel and Biochemical Research and Development.

Chemical Engineering is to cope with complex, real –world problems, chemical engineers develop strong core expertise in engineering thermodynamics, transport processes, and chemical kinetics, creating a powerful and widely applicable combination of molecular knowledge and engineering problem solving.

I am a current undergraduate student at Atlantic International University; our University is a humanitarian and adult learning focused University. Our University has been recognized world wide from UNICEF to Global events in South America and International Continents outside of the United States. Our University holds Awards and Accolades of recognition for their educational effects and humanitarian involvement. Our professors at Atlantic International University are Professors at some of the most Prestige's University throughout the world. My first Chemical Engineering study was at King Fahd University of Petroleum and Minerals (KFUPM)

Chemicals derived from petroleum or natural gas - **petrochemicals** - are an essential part of the chemical industry today.

"Igniting" our Fuel and Global Energy.

Motivate Formulate Enhance
Atlantic International University AIChE Presentation Ethenia A. Scott
Atlanta, Georgia Summit 2014

Keywords

Fuel
Energy
Petrochemicals
Environmental Regulations
Crude Oil
Natural Gas
Safety
Facility Security
Jobs
Labor
Taxes
Congressional
Environmental Testimony
Agency
Comments
Crude by Rail
Chemical Safety
Chemical Safety Issues
Everyday Use

N-Doped Graphene-Based Bulk Materials for Energy Storage Applications

Ashley Ware, David Baah, Jonathan Mbah

Tuskegee University Department of Chemical Engineering, Tuskegee, AL 36088

Abstract: Graphene is a two-dimensional network in which sp^2-hybrized carbon atoms packed into hexagonal structure and arranged into two different triangular sub-lattices (A and B). Chemically modified graphene nanoplalets have shown great promise in select applications due to their enhanced electrical properties. Chemical doping is an effective method to tune the electronic properties of graphene materials. In this work, simple chemical methods (hydrothermal and acid catalyzed condensation reaction) are used to produce nitrogen-doped graphene aerogel. The N-doped graphene aerogel possess redox active pseudocapacitance creating an addition to the already existing double layer capacitance resulting in a hybrid capacitor with increased energy density.

396755 Increasing Hydrogen Production Using Nuclear Integrated High Temperature Electrolysis for a Coal to Liquids Plant

Abstract:

The possibility of producing synthetic fuels using coal gasification with low carbon emissions is dependent on the availability of hydrogen gas to be mixed with syngas. Finding a new process for producing hydrogen gas to replace steam methane reforming, a CO2 emitting process, centers around the integration of nuclear process heat with chemical processes or electrolysis. High temperature steam electrolysis offers an opportunity for conventional light water reactors to play a role in hydrogen production even with their lower relative temperatures. The coal gasification process offers an alternative feed, high pressure steam from the gasifier, which could also be utilized for electrolysis. The electricity required to power the coal to liquids plant demands too much electricity from the nuclear reactor system, resulting in lower hydrogen production.

397172 Theoretical Studies on Functionalized Graphene-Metal Cluster Interactions

Abstract:

Functionalized graphene – metal nano-cage hybrid platforms are used in recent years for trace level identification of explosives and environment contaminating materials using Raman fingerprints. In the present project, nature of interactions of functionalized graphene systems with model metal (Au_4 and Ag_4) clusters are investigated using state of the art density functional theories (DFT) including B3LYP functional and LANL2DZ basis set of atoms. Functionalized (5, 5)-graphene sheets are initially modeled with full geometry optimization and gold and silver clusters (Au_4 and Ag_4) are attached to the functional sites (with full geometry optimizations). The affinities of such clusters towards these systems are computed in terms of interaction energies. Vibrational analysis is carried out to confirm the nature of the computed geometries. Pyridine is attached to these model systems and Raman spectrum is computed for these complexes to monitor the nature of Pyridine vibrations. Isolated pyridine has very weak Raman intensity (RI). The chemical effect due to such attachments increases the RI of important Raman modes several times. The results show that such models pave way to generate the further increase of RIs of pyridine through resonance/surface further enhanced Raman spectroscopy and could prove to be an alternate way of detecting such molecules.

Conference Abstract

Of recent concern in offshore drilling is sub-sea oil pipeline flow. As companies are forced to drill further away from shore, crude is exposed to cold temperatures and high pressures for a longer period of time. Without additives, paraffins crystallize and form volume spanning networks in the pipe, restricting oil flow. Much research has gone into additives to prohibit gelation but very little has gone into understanding the formation of the gels themselves. Tests in this experiment analyzed the effects of solvent chain length on the clouding and gelation of monodisperse and polydispersed systems of C_{28}, C_{32}, and C_{36}. Solvents tested include C_{10}, C_{12}, and C_{16}. Findings show that although C_{10} and C_{12} systems show similar gelation trends while C_{16} systems follow a separate trend.

Electrodeposition of Palladium from Extracted Nuclear Waste

Eriugen Gjoka[1*], Sujan Shrestha[1], Elizabeth J. Biddinger[1+]
[1]*Department of Chemical Engineering, City College of New York, New York, NY 10031*
**Presenting author*
[+]*Corresponding author: ebiddinger@che.ccny.cuny.edu*

There are more than 57,000 tons of spent fuel being stored at U.S. nuclear power plants today[1]. The spent fuel contains a mixture of radioactive fuel, and radioactive and nonradioactive fission products. Nuclear fuel reprocessing, as an essential part of nuclear fuel cycle, is of great significance from the point of view of both nuclear safety and sustainable development of nuclear energy. A goal for long-term handling is to separate fission products by half-life, recycle the unreacted fuel back to the reactor, and then sequester the much smaller volume, long half-life products in solid form. In the Green Chemistry and Energy Laboratory at the City College of New York, we are assessing the utility of ionic liquids (ILs) in the electrodeposition of palladium, a non-radioactive noble metal part of spent fuel fission products. ILs are salts with melting points below 100°C, can be used as non-flammable, minimally volatile, selective extractants for the separation of fission products from the aqueous system resulting from the acid digestion of spent fuel rods[2] and also act as both the electrolyte and the solvent in elecrodeposition. Ionic liquids [Bmim][NTf_2], [Bmim][DCA] and [Bmpyrr][DCA] will be investigated in the poster. The presentation will focus on how the IL selection and electrochemical conditions impact the electrodeposition efficiency of palladium. The effect of cation and ion, metal substrate and electrochemical conditions on the experiments will be examined. The criteria used for analyzing the electrochemical windows and redox behavior of palladium salts in ILs will be evaluated. The experiments with palladium are imperative because palladium has the highest concentration of noble metals in fission products and has non-radioactive isotopes available that can be used as replica. The work with palladium is also crucial to the overall separation of the product nuclear waste stream because it can be further used to model the recovery of other fission products in ILs.

---

[1] Fact sheet on licensing Yucca Mountain, U.S. NRC, 2012, http://www.nrc.gov/reading-rm/doc-collections/fact-sheets/fs-yucca-license-review.html. Date accessed 11/26/12

[2] J. F. Wishart, *Energy & Environmental Science*, 2, 956 (2009)

First principles quantum chemical modeling of radium in barite for fracking wastewater remediation

Wagner, Gina; Saravanan, Karthikeyan; Keith, John

gswagner11@my.trine.edu; kas389@pitt.edu; jakeith@pitt.edu

Chemical and Petroleum Engineering Department

University of Pittsburgh

Wastewater from natural gas extraction by hydraulic fracturing (fracking) can contain potentially hazardous naturally occurring radioactive materials (NORMs). A potential remediation method is to use mineral particulates to absorb the NORMs in water so that they can then be safely disposed, but the thermodynamics and kinetics of NORM absorption processes are not well understood. As a starting point to elucidate and understand these mechanisms, we carried out first principles quantum mechanics studies to predict how radium binds within barite, a mineral under consideration for wastewater remediation. We find that thermodynamics of radium binding within barite are unfavorable at substitution concentrations of 25%, but they are favorable at substitution concentrations of 12.5%. Calculated data using state of the art Kohn-Sham density functional theory with and without van der Waals dispersion are reported.

397289 Exploring Properties of Conceptual Models for Hydraulic Fracture Network Growth

Abstract:

Three conceptual models for complex fracture network growth due to hydraulic fracturing (simple geometry, PFSSL, and MMS) were explored using discrete fracture network simulation. Simple geometry models, variants of which are commonly used in industry, neglect the effects of natural fractures present in the reservoir pre-stimulation. PFSSL (primary fracture with shear stimulated leakoff) and MMS (mixed-mechanism stimulation) are alternative models which account for interactions between natural and newly formed hydraulic fractures. The investigation centered on finding physical consistency with the parameters of fracture network length, stress and heterogeneity effects, and pressure trends. It is found that simulations of the simple geometry model yield results that are not physically realistic, while results of PFSSL and MMS simulations show better correlation with experimental data, including fracture length and geometries predicted by microseismic studies. This study uses CFRAC, a leading 2D simulator of fracture network growth that efficiently couples fluid flow and geomechanical deformation in low-permeability reservoirs.

Effect of Salt on the Rheological Properties of Guar- Based Fracturing Fluid

Tatiana M. Colon[a], Swastika S. Bithi[b] and Siva Vanapalli[b]

[a] Chemical Engineering, Polytechnic University of Puerto Rico
[b] Chemical Engineering, Texas Tech University

Abstract: Guar gum is a pseudoplastic polymer, which is commonly used as viscofying agent in guar-based fracturing fluid. In this research project, we study the effect of guar concentration and alkaline metals salts (e. g. KCl, NaCl, and $CaCl_2$) on the rheological properties of different guar solution at different shear rates. The shear rate is a key factor to evaluate the performance and application of hydraulic fracturing fluid. We chose the viscosity of solution as the measuring parameter for the rheological properties. We experimentally measure the apparent viscosity of guar solutions using a rheometer in parallel plate geometry. Viscosities of two dilute concentrations of guar solutions (0.24 and 0.48 wt.%) are measured over a range shear rates (0.1- 1000 S^{-1}) at a constant temperature of 25°C. The viscosity exhibits shear thinning behavior, *i. e.* decreases significantly when the shear rate increases at a constant temperature. On other hand, the viscosity increases with increasing guar concentration at low shear rate. Viscosities are also measured for guar solutions in three different alkali metal salts solutions. The viscosity of guar in salt solutions is lower than that of in aqueous solutions. The reduction in viscosity facilitates the flow of residual polymer after providing quick recovery of the proppant. This study provides approaches to obtain the effects of salinity on the properties of fracturing fluids. The viscosity test conducted in this study would indicate the optimal salt concentration at which the viscosity decreases as the shear rate increases.

Key words: Guar gum, pseudoplastic, fracturing fluid, hydraulic fracture, rheological, proppant.

398582 Characterization of Anaerobic Digesters during FOG Co-Digestion Shock Loading

Abstract:

Biogas (a mixture of methane, carbon dioxide and trace amounts of other gases) is produced as a byproduct of anaerobic digestion in wastewater treatment plants. While the methane in biogas can be converted into electrical energy, most often it is flared due to non-economical quantities being produced. The addition of Fats, Oils, and Greases (FOGs) can increase the production of biogas, and thus methane, to levels that make it economical to convert to electricity. However, adding excessive amounts of FOG to an anaerobic digester, referred to as shock loading, can negatively affect the anaerobic bacteria and lead to a decrease in cumulative biogas production. In this study, the effects of shock loading anaerobic digesters were tested for six different FOG sources. The anaerobic digesters, set up as batch rectors, were used to test various loading rates for each FOG source. The FOG sources tested include Food Manufacturing Waste, Bakery Waste, Vitamin Waste, Yogurt Waste, Cherry Waste, and Mayonnaise Waste. Biogas and methane production were measured daily. In addition, total solids (TS) and volatile solids (VS) were analyzed and used to assess the performance of the anaerobic digesters. Of the six FOG sources tested, four did not demonstrate any signs of shock loading when a maximum of $1.50*10^2$ mg VS of FOG was added to $2.25*10^3$ mg VS of inoculated sludge. An optimal loading rate of 75.8 mg VS FOG per $1.65*10^3$ mg VS of inoculated sludge was determined for one of the two FOG sources that exhibited signs of shock loading.

Role of Aromatics in Synthetic Jet Fuels

Moiz Bohra, Muaz Selam, Hafis Mohamed, Wajdi Ahmed, Nimir O. Elbashir

Texas A&M University at Qatar, PO Box 23874, Education City, Doha, Qatar

This paper describes our investigation into the effects of adding aromatic compounds to synthetic aviation fuels on the performance of these fuels. The work was done as an undergraduate research (UREP) project and was an extension to our group's previous studies which were aimed at characterizing blends of conventional and synthetic aviation fuels [1] and identifying optimum paraffinic building blocks for SPK [2]. Aromatics play a key role in aviation fuels as they enhance the density and elastomer swelling properties of these fuels. However, they are absent from synthetic gas-to-liquids (GTL) derived aviation fuels such as synthetic paraffinic kerosene (SPK) and thus must be added in from another source. In the first part of this study, standard analytical tests were conducted on blends of SPK and aromatic additives (styrene and Shell A150) in different concentrations from 0 to 25 % to measure certain fuel properties (i.e. density, flash point, freezing point, heat content and viscosity) using relevant ASTM procedures. The measured properties were then compared with the desired values for certification according to the ASTM D-1655 and D-7566 standards for conventional and hybrid (i.e. mixtures of oil-derived and synthetic fuels obtained from either natural gas or coal) jet fuels, respectively. In particular, ASTM D-7566 specifies the use of the 50-50 synthetic fuel blend with jet A-1; our group's efforts are aimed at increasing the workable synthetic ratio in the fuel. The experimental work was conducted in the Fuel Characterization Lab at Texas A&M University at Qatar. The data generated from these experiments was used to identify correlations between blend compositions and properties.

In the second part of this study, small amounts (i.e. 1 vol %) of specialty aromatic compounds (such as benzaldehyde) were added to the blends from the first part of the study, specifically to enhance the elastomer swelling characteristics of these blends. This is a crucial property of aviation fuels as it ensures a tight seal in fuel tanks, thus preventing potentially catastrophic leakages. The added aromatic compounds were identified based on a Hansen solubility parameter analysis, and it was found that their addition favorably increased both the density and elastomer swelling characteristics of the SPK. Notably, three of the blends tested outperformed the elastomer swelling characteristics of conventional Jet A-1 by about a factor of 2. With respect to freezing point, viscosity, heat content, and flash point, the performance of all of the blends tested were in agreement with ASTM standards. These experimental results demonstrate the favorability our approach to synthetic fuel design i.e. the focused introduction of additives to improve the performance of synthetic fuels to standards at par with or, in some cases, surpassing those of conventional jet fuel.

References:

[1] I. Al-Nuaimi, "A Path to the Formulation of New Generations of Synthetic Jet Fuel Derived From Natural Gas," Master's thesis, Texas A&M University at Qatar, 2013.

[2] M. Bohra, A. Rahman, E. Elmalik, J. Blank and N. Elbashir, "Development of Visualization Models to Guide the Design of Future Generations of Synthetic Fuels from Natural Gas," in *Qatar Foundation Annual Research Forum* , Doha, 2012.

Reactor Design and Analysis of a Simulated Moving Bed Reactor for Chemical-Looping Combustion

Clarke Palmer, Lu Han, George M. Bollas

Department of Chemical & Biomolecular Engineering, University of Connecticut

http://pdsol.engr.uconn.edu

Objective

- Address the existing challenges facing chemical-looping combustion and propose a novel process design for higher conversions and efficiencies.

Chemical-Looping Combustion (CLC)

- Energy production from combustion with low-cost CO_2 separation
- **Reduction**
  - Hydrocarbon fuel (i.e., natural gas) oxidized by metal oxide oxygen carrier (i.e., NiO), yielding CO_2 and H_2O
- **Oxidation**
  - Air regenerates reduced oxygen carrier (i.e., Ni), yielding N_2 and O_2
- Current reactor designs for CLC are shown in **Figure 1**

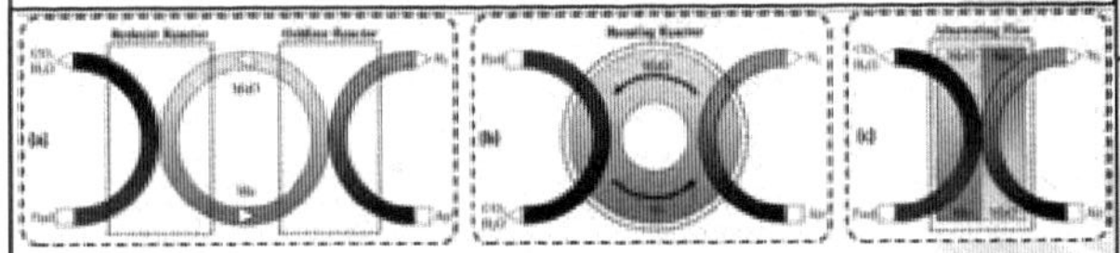

Figure 1: *Reactor designs for CLC: (a) circulating fluidized bed; (b) rotating reactor; and (c) alternating flow over a fixed-bed.*

Processes and Challenges

Fixed-bed: Plug flow, packed oxygen carrier (OC) bed (**Figure 2a**)

- Flexible particle size
- High fuel conversion
- Solid carbon build-up
- Incomplete OC conversion

Moving bed: Countercurrent flow of fuel and fluidized OC solids

- Higher CO_2 selectivity and equilibrium restrictions overcome [1]
- Product contamination
- Large OC particles
- High abrasion factor

Simulated moving bed: Simultaneous switching of inlet/outlet ports along fixed-bed to simulate countercurrent solid flow (**Figure 2b**)

- Higher conversion and product selectivity [2]
- High temperature valve system

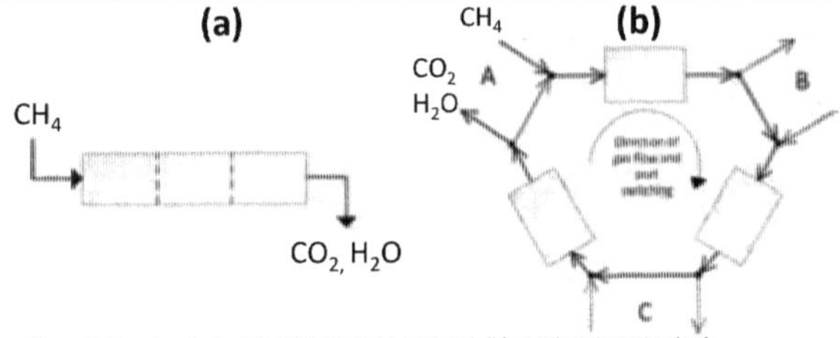

Figure 2: Reactor designs for CLC: (a) fixed-bed and (b) simulated moving bed.

Model

- Process simulation and optimization
- Reaction network, model equations and performance evaluation equations shown in **Tables 1 and 2** [3-5].

Table 1: CH_4-NiO reaction network.

Oxygen carrier reduction reactions	
Partial CH_4 oxidation	$CH_4 + 2NiO \rightarrow 2Ni + CO_2 + 2H_2$
H_2 oxidation	$H_2 + NiO \rightarrow Ni + H_2O$
CO oxidation	$CO + NiO \rightarrow Ni + CO_2$
Partial CH_4 oxidation	$CH_4 + NiO \rightarrow Ni + 2H_2 + CO$
Reactions catalyzed by Ni	
Steam reforming	$CH_4 + H_2O \leftrightarrow 3H_2 + CO$
Water gas shift	$CO + H_2O \leftrightarrow H_2 + CO_2$
Dry reforming	$CH_4 + CO_2 \leftrightarrow 2CO + 2H_2$
Methane decomposition	$CH_4 \leftrightarrow 2H_2 + C$
Carbon gasification by steam	$C + H_2O \leftrightarrow CO + H_2$
Carbon gasification by CO_2	$C + CO_2 \leftrightarrow 2CO$

Table 2: Mathematical model for fixed-bed chemical-looping combustion.

Mass balance:	$\frac{\partial(\varepsilon_b C_j)}{\partial t} + \frac{\partial(uC_j)}{\partial z} = \varepsilon_b \frac{\partial}{\partial z}\left(D_{ax,j}\frac{\partial C_j}{\partial z}\right) + \rho_b \sum R_j$
Energy balance:	$\frac{\partial(\varepsilon_b C p_g C_g T)}{\partial t} + \frac{\partial(C_g C p_g u T)}{\partial z} = \varepsilon_b \frac{\partial}{\partial z}\left(\lambda_{ax}\frac{\partial T}{\partial z}\right) + \rho_b \sum(-\Delta H_j)(R_j)$
Momentum balance:	$\frac{dP}{dz} = -\left(\frac{1-\varepsilon_b}{\varepsilon_b^3}\right)\left(\frac{\rho u_0^2}{d_p}\right)\left(\frac{150}{Re_p} + 1.75\right)$
Boundary conditions:	$\varepsilon_b D_{ax,j}\frac{\partial C_j}{\partial z}\Big\|_{z=0} = u(C_j - C_{j,in}); \varepsilon_b \lambda_{ax}\frac{\partial T}{\partial z}\Big\|_{z=0} = uC_g Cp_g(T - T_{in})$; $\frac{\partial C_j}{\partial z}\Big\|_{z=H} = \frac{\partial T}{\partial z}\Big\|_{z=H} = 0; P\|_{z=H} = P_{out}$
Initial conditions:	$T\|_{t=0} = T^o; C_j\|_{t=0} = C_j^o$

Performance Evaluation:		
	Methane Conversion	$X_{CH_4} = 1 - \frac{F_{CH_4,out}}{F_{CH_4,in}}$
	CO_2 Selectivity	$Y_{CO_2} = \frac{F_{CO_2,out}}{F_{CH_4,in}}$
	Selectivity Towards Solid Carbon	$S_C(mol\ C_{(s)}/mol\ CH_4) = \frac{\int_0^H C_{(s)}(z)dz}{\int_0^{t_{red}} F_{CH_4,in}dt}$

Design

- Oxygen carrier, fuel composition, temperature, flow rate, adiabatic conditions, and reactor diameter and total length identical for fixed-bed and simulated moving bed simulations [6]
- Overall dimensions optimized for fixed-bed model
- Simulated moving bed model is instantaneously switched to the next configuration after certain time durations (**Figure 2b**)

Configuration	Duration (seconds)
A	70
B	10
C	120

Results

- CLC performance in each reactor configuration:

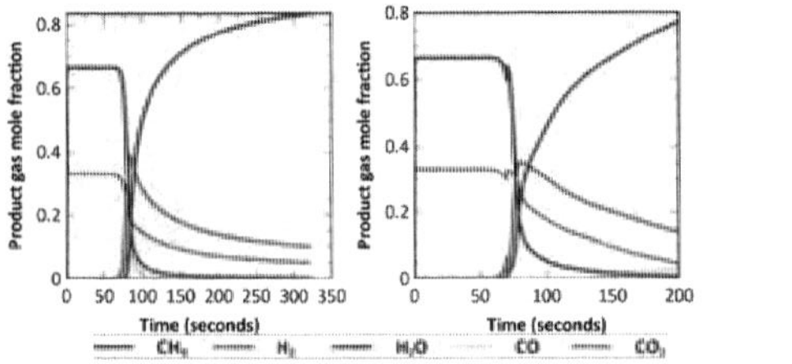

Figure 3: Mole fraction of gas species exiting the reactor vs. time for fixed-bed (left) and simulated moving bed (right) processes.

- Analysis of CLC efficiency:

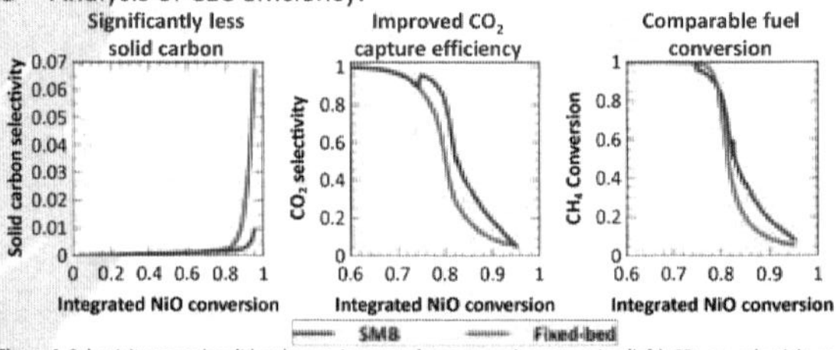

Figure 4: Selectivity towards solid carbon vs. integrated oxygen carrier conversion (left), CO_2 gas selectivity vs. integrated oxygen carrier conversion (center) and CH_4 conversion vs. integrated oxygen carrier conversion (right) for fixed-bed and simulated moving bed (SMB) processes.

Conclusions and Future Work

- SMB process shows increased selectivity to CO_2 with decreased selectivity to solid carbon compared to fixed-bed process
- Further investigation of SMB reactor design, switching schemes and switching times

Undergraduate Research Experience

Research has allowed me to explore my own creativity and insight on a novel process and understand why so much time and thought is invested into these research projects. It has provided an important link between classwork and hands on engineering. While classwork teaches concepts and demonstrates past methods, research allows engineers to apply these concepts and design novel ideas.

References

[1] L. Fan, F. Li, S. Ramkumar, *Particuology, 6.3* (2008) 131–142. [2] Ray, A. K., & Carrt, R. W. Aris, R. (1994). *Chem. Eng. Sci., 48(4)*, 469-480. [3] Han, L., Zhou, Z., & Bollas, G. M. (2013). *Chem. Eng. Sci., 104*, 233–249. [4] Iliuta, I., Tahoces, R., & Patience, G. (2010). *AIChE J., 56*, 1063–1079. [5] Zhou, Z., Han, L., & Bollas, G. M. (2013). *Chem. Eng. J., 233*, 331–348. [6] Han, L., Zhou, Z., & Bollas, G. M. (2014). *Int. J. Greenhouse Gas Control*, (In Review).

Acknowledgements: This material is based upon work supported by the National Science Foundation under Grant No. 1054718 and the UConn Prototype Fund.

Center for Clean Energy Engineering

UCONN

Performance of an IC Engine Fueled by Landfill Gas and Syngas

Student: Leah Curry
Advisor: Marco J Castaldi

According to the Environmental Protection Agency's Landfill Methane Outreach Program, municipal solid waste landfills are the third largest contributor of human-related methane emissions in the United States. Methane, a potent greenhouse gas can be prevented from going into the atmosphere by either flaring or by capturing it and converting it to energy. The second option, known as Landfill Gas to Energy (LFGTE), is preferable because it has the added benefit of offsetting our dependence on non-renewable fuels. Generation of electricity can be carried out with a number of technologies, internal combustion engines and turbines being the most common.

The gas from landfills, approximately a 50:50 ratio of CH_4 and CO_2 is a low quality fuel and presents some problems when it comes to emissions and efficiency. A possible method to improve the performance of the engine is to add syngas to the fuel stream. The most cost effective method to do this is to authothermally reform a portion of the landfill gas in-situ. Experiments show that the ratio of H_2 to CO produced is a function of reactor temperature and can range from 1.2 to 4.7. Furthermore, it has been observed that the ratio of H_2 and CO in the syngas impacts the performance of the engine more than the total amount of syngas that is added to the landfill gas.

This research used a small spark ignition, four-stroke, single cylinder internal combustion engine connected to a generator to replicate real field engines on a laboratory scale. The engine had been modified to run on gaseous fuel. For the experiments, simulated landfill gas was used to fuel the engine with varying additions of syngas and emissions were monitored. For a 5% syngas addition at 0.4 kW load, CO emissions decreased by 7% and unburned hydrocarbon emissions by 66% compared to fuel of only CH_4 and CO_2. Engine efficiency increased by 10%.

398955 Flash Point Estimation of Ternary Alcohol Mixtures

Abstract:

Author: Katie De Hoedt

Email: Katie-dehoedt@uiowa.edu

ABSTRACT

The objective of the research is to obtain flammability limits, limiting oxygen concentrations, flash points, and minimum ignition energies for various materials to aid in establishing safety parameters for release valves. A flash point tester will be utilized, along with two unique pieces of equipment: a flammability chamber, and a minimum ignition energy tester (MIE). The flash points of flammable liquid mixtures will be recorded to develop models that will predict the flash point of alcohol mixtures of three or more components. The flammability chamber will be used to determine the flammability limits and limiting oxygen concentrations needed to ignite methane. The MIE will be used to test the minimum ignition energy of methane and propane gas mixtures. The focus of this presentation will be the flash point calculation method that was developed to determine the flash point of a mixture of three or more alcohols. No previous research has developed an effective method for these calculations. Developing a method for estimating the flash point of alcohol mixtures will effectively enhance chemical process safety.

398983 Special Topics in Multiphase Mixing - Cloud Height

Abstract:

The goal of the study was to determine the effect of particle size and density on the cloud height of particles. A Lightnin 8.5 centimeter diameter stainless steel model A310® impeller was used to suspend the particles. Aluminum, urea formaldehyde, and polyoxymethylene (acetal) spheres were used as testing material with specific gravity values of 2.7, 1.2, and 1.4 respectively. A literature review was conducted in order to form a basis for experiments. Assessments could be drawn from previous literature findings. A Plexiglas, flat bottom tank 28.3 centimeters in diameter was used. The tank was filled with water to a height of 56 centimeters for each test. The tank contained 3 centimeter baffles. The impeller height off the bottom of the tank was 9.5 centimeters, about one third the tank diameter. The cloud height was measured at different impeller speeds (measured in RPM) ranging from nearly just suspended speed (NJS) to 2355 RPM, the maximum speed of the equipment. The particle concentrations used within the tank were 1%, 2.5%, and 5% by weight. Particle mixtures as well as pure materials were tested. It was determined that cloud height is dependent on speed, particle size, and particle density. Concentration also has a small effect. A notable observation is certain mixtures produce homogenous clouds, such as a mixture of aluminum and acetal. Other mixtures produce a stratified cloud, in which one particle forms a distinctly higher cloud height. This is seen in a mixture of acetal and urea. It is hypothesized that large differences in particle size produce this stratified cloud.

Thermodynamics Curriculum Redesign

Abstract

Author: Hayley Neebe

Redesign Project Team: Patrick J Pinhero, Professor, Chemical Engineering (Lead)
Carmen Beck, Instructional Designer, ET@MO
Mary A. Myers, Associate Teaching Professor, Chemical Engineering

The current Thermodynamics I course was adapted into a modular structure with a focus on mastery learning, because the current curriculum is overly theoretical. This creates difficulty for the students with visualizing the concepts and their application. The modular structure evolved by identifying key course concepts and developing sequential, measurable Learning Objectives, clearly communicating them to the students from the start. These were analyzed and used to design and develop resources to help students achieve mastery learning and aid in visualizing concepts and applications, making the course both easier to learn and easier to teach. The course redesign utilizes the Learning Objectives, online resources, and interactive activities to add greater flexibility for when and how students engage with the course material. The course redesign is formatted to keep students on pace and to promote student learning as well as innovative thinking. The curriculum was reorganized to immerse the students into the difficult core concepts earlier in the course using realistic case studies. Lessons encourage the use of computer software such as Excel and MATLAB in order to teach more advanced problem solving techniques. In addition, new tools were designed and are being developed to illustrate concepts such as process visualization, heat capacity, energy and entropy, and cyclic processes. These applications are being integrated to the current Thermodynamics I class, with results recorded and approaches manipulated, to furnish students with a better academic framework in order to learn thermodynamics. This is accomplished through providing students with greater interaction, instantaneous feedback, and multiple opportunities to recall and repeat exercises so that skills are mastered. Swift feedback amongst professors, research assistants, and students is integral to the course redesign; it quickly ameliorates misunderstandings and provides assessment data for continuous improvement. Thermodynamics is generally considered to be one of the most difficult courses in Chemical Engineering. It is one of the first STEM courses to redesign pedagogy at the University of Missouri, potentially being applied to other classes, with hopes of being a marketable "Learn Thermodynamics" teaching module.

395911 Incorporation of Plant Architectures within Microfluidic Devices

Abstract:

One of the most significant challenges in microfluidics and tissue engineering is the efficient delivery of fluids and solutes throughout the system. Many natural organisms have already developed a solution to this problem, in the form of active or passive circulatory systems. This research focuses on developing a procedure for creating a PDMS microfluidic device with a pre-existing system: venation in plant leaves taken directly from nature. Such a device has great potential as synthetic vascular scaffold for tissue engineering purposes, as the geometry allows effective transport, but is still relatively planar, making it more adaptable to microfabrication processes. In this work, the surface topography and vein cross-section from both the leaves and molds were characterized, along with the fluid flow within the channels. An effective method of isolating a leaf's vasculature was additionally found and evaluated. Although there are some limitations in alignment, resulting in an altered cross-sectional morphology and fluid flow, the individual portions of the mold effectively capture the dimensions and patterning of the vasculature.

398480 A Unified Cloud-Based Solution to Large-Scale Multidimensional Student to Project Assignment

Abstract:

As a part of the overall engineering curriculum, senior capstone projects provide the necessary opportunity for engineering graduates to apply their technical skills towards solving contemporary, open-ended, and challenging engineering problems. At Rowan University, both juniors and seniors engage in such hands-on projects across the last four terms of their degree. Each term, students are free to select projects that interest them, potentially working on four different projects over their Junior and Senior years. Projects are pitched to the students at the beginning of each term to encourage student interest and provide a project overview. Based on the pitch and personal interest, students submit a list of their preferred projects. In order to address the need for an efficient and accurate student-project assignment in the face of an increasing number of projects and students, a cloud-based platform was designed and implemented. This solution used Google Forms to generate a list of projects with details for students to review and submit their project preferences. In turn, this data was used to generate project and student objects. Next, these objects were processed by an algorithm that assigned students to projects using multi-level criteria, including project funding, student seniority and experience, and professor requests. To create a more robust platform, more detailed selection criteria was added that was too complicated to account for by hand. In addition, several frequently used tasks with varying degrees of complexity were automated. The algorithm utilized the recently released Google-App Script's ability to generate and manipulate data objects while having full access to Google's variety of cloud services. During the Fall 2014 term, the automated approach's results were compared side-by-side to the manual sorting performed by each of the four engineering departments with varying results. This poster describes the structure of the automation algorithm in detail and justifies design decisions. Due to the wide spread use of Google Apps and Google Forms in an academic setting, this poster can serve as a template for the development of similar cloud-based platforms by other institutions. The project described here details the utility and unification that can be provided by the use of modern cloud-based services to academia as a whole.

396551 Wavelet-Based De-Noising of Cellular Images

Abstract:

Wavelet analysis is a prevalent tool for image compression (JPEG2000 uses wavelets to compress images), data storage and de-noising, yet its application in biophysics has been largely unexplored. One challenge in the measurement and analysis of image based parameters for biological specimens is the removal of undesired noise from the image, which creates ambiguity and inaccuracy when segmenting via thresholding. The accuracy of measurements of these image based parameters could be improved significantly by incorporating wavelets as a de-noising tool.

After image acquisition, a discrete wavelet decomposition (with a variety of wavelet families) was used to separate the underlying frequencies in the cell image, thus isolating the high frequencies, characteristic of noise. The high frequencies were then attenuated and the image is reconstructed to provide a de-noised image. This method was used to de-noise fixed phase contrast and fixed fluorescent images of cells and nuclei. This alone was able to improve segmentation, allowing more accurate measurement of imaged-based parameters. In combination with Spatial Filtering, the segmentation was even greater improved and was comparable to methods currently used in image processing.

The results from wavelet de-noising can be benchmarked against existing de-noising methods such as Gaussian blur and anisotropic diffusion to determine what has optimal performance in terms of information retrieval (in particular, image based parameters) from cell images. Some basic cell information includes cell parameters such as the cell area, perimeter, boundary curvature, length-to-width ratio or average protrusion length. With better resolution, these cell parameters can reveal important, and more accurate, details about cell migration, drug resistance and proliferation that are otherwise unavailable due to image noise. This improved cell tracking and segmentation can be a powerful tool to advance medical studies in topics such as cancer development, guided nerve regeneration and wound healing.

Title: Expression for Aqueous Electrolyte Solution Viscosity based on Andrade Equation

Author name: Matt Kovalski

Author affiliation: Texas Tech University, Department of Chemical Engineering, Chen Group

Author E-mail: matt.kovalski@ttu.edu

Author bio: Matt Kovalski is a junior undergraduate chemical engineering student with the Department of Chemical Engineering at Texas Tech University. He studies the viscosity of electrolyte solutions from a computational perspective under Dr. Chau-Chyun Chen.

Abstract: Viscosity is considered to be one of the most important physical properties within many applications. In terms of chemical species, electrolytes are considered to be an important, and often difficult, class of chemicals to model the physical properties of. As such, modeling the viscosity of electrolyte species can be considered to be an important, yet often difficult task. In order to be most relevant to industrial conditions, modeling of this property for this class of chemical species should be performed with multiple components, and also high temperatures and solute concentrations in mind.

Methods exist to calculate the viscosity of electrolyte solutions, such as the Jones-Dole or Andrade expression, however; interpretations of these methods sometimes have qualities that make them undesirable or inapplicable for the calculation of viscosity some applications. These issues can include a lack of testing of the expression with a diverse range of electrolyte species, tedious and complex calculations rendering the expression undesirable, and lack of theoretical origin of the expression. Here we propose an expression based on the Andrade expression for liquid viscosity to calculate the viscosity of aqueous solutions. The expression is based on theoretical observations and does not require tedious calculations. Furthermore this expression keeps in mind the relevance of industrial conditions by incorporating a mixing rule for multiple electrolytes, temperature effects, high concentrations of solutes, and testing with a broad range of electrolytes species.

398390 A New Approach to Vapor-Liquid Equilibrium Property Prediction from Cubic Equations of State

Abstract:

Vapor pressure prediction is important in a broad range of industrial applications, but general methods applying equations of state require iterative algorithms. In this work, a novel method for vapor pressure calculations was applied to the Soave-Redlich-Kwong (SRK) cubic equation of state. This allowed for an exact solution for the vapor pressure of any substance with a dimensionless expression combining reduced temperature and a function of the acentric factor. Several useful variables in vapor-liquid equilibrium – including reduced phase densities, compressibility, and the ratio of reduced pressure to reduced temperature – could be expressed as functions of this single dimensionless variable. Exact results could be derived for an inverse problem beginning with reduced phase densities as independent variables. When the standard independent variables of temperature and acentric factor were combined with empirical relationships for reduced phase densities, vapor pressure was approximated within 0.05 percent deviation from the exact results for acentric factors from 0.05 to 0.97 and reduced temperatures from 0.16 to 0.98.

Rizan Baig
Texas A&M University at Qatar

Study of the Thermal Decomposition of Dicumyl Peroxide Under Runaway Conditions Using Adiabatic Calorimetry

Abstract

In a thermal runaway, exothermic reactions experience uncontrolled self-heating due to the failure of the cooling mechanism of the vessel containing them. The temperature increases at extremely high rates, boiling the contents of the vessel and increasing the reaction rate. This causes the pressure to increase to a point higher than what the vessel can withstand. The vessel fails that causes a release of flammable and toxic chemicals. To avoid this, the maximum specific gas generation rate during a runaway reaction needs to be determined and according to that rate, proper modifications to the vessel need to be made to avoid loss of containment.

Many runaway reactions involve the decomposition of organic peroxides as they are thermally unstable. Consequently, their decomposition properties need to be studied. To serve this purpose, calorimetry is used. In a calorimeter, small samples are used where heat loss is minimal (adiabatic conditions). This simulates large scale conditions where heat transfer in large vessels is small as these vessels have a larger volume than surface area.

In this project, Dicumyl peroxide (DCP) was dissolved in 2,2,4-trimethyl-1,3-pentanediol diisobutyrate and cumene and its runaway profiles (temperature and pressure during the runaway reaction) were constructed using the PHI-Tec I and II calorimeters. Both solvents are commonly used in industry. Three different concentrations of DCP (20, 30 and 40%) in each solution were tested at two different fill levels (55% and 70%). Two different cell arrangements, closed cell and open cell were also tested. In the open cell arrangement, two different initial pressures were selected, 1 and 5 bar, to study the effect of pressurizing the vessel on the runaway profile.

It was found that the maximum gas generation was considerably higher when DCP was dissolved in 2,2,4-trimethyl-1,3-pentanediol diisobutyrate. The gas generation rate was higher when the initial pressure of the open cell was also higher. As the fill level was increased, for the same concentration, the gas generation decreased.

All of these results will be utilized by INERIS in France and HSL in the UK as a basis for their design of larger scale experiments.

394417 Improving Plant Economy By Preventing Heat Loss in Edible Oil Refinery

Abstract:

An exhaustive study of the processes involved in the refining of Ground Nut oil was undertaken. It was found that for complete processing of 10 MT of crude oil, the plant under study took 21.25 h which was too high. Subsequently, based on the process variables readings throughout the process, the point of maximum heat loss was determined and calculations were carried out to determine the effect of insulation over that point. Prior to calculating, storage tank's modeling and simulation were carried out in order to determine its liquid fill percentage (84.1%) and wetted surface area (24.24 m^2). The findings show that 82% reduction in the total heat loss of 80.57 MJ (at quiescent air conditions) is achieved when a Glass/Wool insulation of 25 mm thickness is employed over the bleached-oil storage tank. Further, doubling the insulation thickness prevents the heat loss by an additional 8%. The resultant decrement in heat loss significantly reduces the workload over deodorizer and also brings down the production time. The effect of wind over heat loss was also studied and a graph was plotted between wind speed and the corresponding heat loss values. Moreover, it was found that employing of insulation over the storage tank generates a profit of US$ 23598 per annum for the refinery.

Developing Amphiphilic Diblock Co-Polymers for the Delivery of Cytosolically Active Immunostimulants

Kathryn Bumila, Max Jacobson, Sema Sevimli, John T. Wilson

A significant challenge facing synthetic vaccine development is inefficient delivery of antigens and immunostimulatory adjuvant molecules to the cytosol of antigen presenting cells. In order to overcome this challenge, we are developing pH-responsive diblock copolymers that can be utilized as nano-carriers for these adjuvants. We postulate that a successful carrier must form sub-100 nm particles under physiological conditions as well as have endosomal escape capabilities to promote delivery to cytosolic pathogen recognition receptors. A library of amphiphilic diblock copolymers synthesized with a poly(ethylene glycol) (PEG) first block and a pH-responsive and hydrophobic 2nd block composed of dimethylaminoethyl methacrylate (DMAEMA) and either butyl methacrylate (BMA) or hexyl methacrylate (HMA). The second blocks were designed to have compositions of 20%, 30%, and 40% BMA or HMA, and the effect of block composition and alkyl chain length on micellar nanoparticle assembly and intracellular delivery were evaluated. To determine the ability of polymer chains to form micellar nanoparticles, dynamic light scattering (DLS) was performed over a range of pH 5.8-7.4, representing acidification during endo/lysosomal trafficking. Polymers containing 30% HMA, 40% HMA, and 40% BMA successfully formed 13-18 nm diameter micelles at pH 7.4 and transitioned to unimeric polymer chains with decreasing pH. Next, a red blood hemolysis assay was performed across the same pH range to evaluate the pH-dependent membrane destabilizing activity of polymers. Results showed that polymers composed of 40% BMA displayed the greatest membrane disruptive characteristics at pH 5.8 and pH 6.2, however, at other pH values less than pH 7.4 polymers containing 20% HMA and 30% HMA were significantly more hemolytic than their BMA counterparts. Finally, a cell viability assay was carried out to determine the cytotoxicity of polymeric carriers. The results showed an average of 80% or higher cell viability at concentrations ranging from 0.01-10 µg/ml, with the 40% HMA polymer demonstrating the least cytotoxicity. In conclusion, initial results suggest that PEG-*bl*-DMAEMA-*co*-BMA and PEG-*bl*-DMAEMA-*co*-HMA are promising cytosolically active nano-carriers and that the endosomolytic activity of such carriers can be tuned through both composition and the alkyl chain length of hydrophobic monomers. Ongoing evaluations are focused on utilizing DMAEMA residues to electrostatically complex nucleic acid adjuvants and evaluating drug delivery efficacy in cell culture.

Zhengyuan (Jung) Fang

fangz@rose-hulman.edu

Chemical Engineering Department of Rose-Hulman Institute of Technology

864-494-6519

Research Title: Spin-Coated, Photo-Polymerized Temperature and Humidity Sensitive Polymer Thin Films

Abstract:

We demonstrate a framework for constructing humidity and temperature-responsive thin films containing poly (N-isopropylacrylamide) (PNIPAAm) with applications in environmental sensing and drug delivery. The films were fabricated via Layer-by-Layer (LbL) assembly and photo-polymerization. Humidity hysteresis, a difference in degree of swelling by a polymer in response to increasing and decreasing the relative humidity of the ambient environment, was investigated. The hydrogen bonded polymer multilayers exhibit significant swelling-deswelling hysteresis, swelling substantially less from a dry state (relative humidity ≈ 0%) than from a hydrated state (relative humidity ≈ 95%) as measured by in-situ reflectometry. Our results indicate that 5min UV curing time during the photo-polymerization leads to smaller maximum swelling ratios than that cured for the 30min. The films cured for 5min also exhibit less hysteresis at higher temperatures than that at room temperature due to the collapse of a portion of film networks leading to an increase in rigidity.

Alkaline Fuel Cell Performance with a Polymerized Ionic Liquid as the Anion Exchange Membrane

Rishon R. Benjamin[1], Jacob R. Nykaza[1], Yossef A. Elabd[2]
[1] Drexel University, Department of Chemical & Biological Engineering, Philadelphia, PA 19104, U.S.A.
[2] Texas A & M University, Artie McFerrin Department of Chemical Engineering, College Station, TX 77843, U.S.A.

Proton Exchange Membrane Fuel Cell (PEMFC) versus Alkaline Fuel Cell (AFC)

PEMFC	AFC
Converts chemical fuel directly to electrical energy	Converts chemical fuel directly to electrical energy
Requires robust polymers with high proton conductivity	Requires robust polymers with high hydroxide conductivity
Requires extensive noble metal catalysts (Pt)	Can be used with inexpensive non-noble metal catalysts (e.g., Ni)

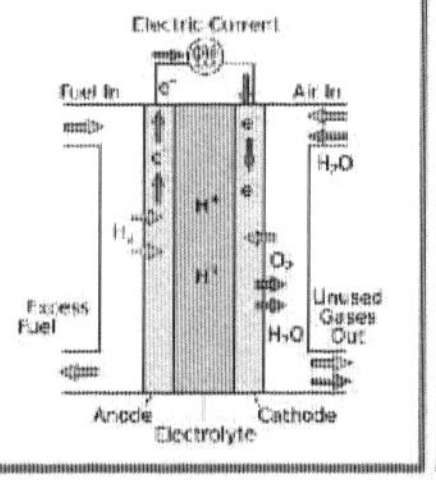

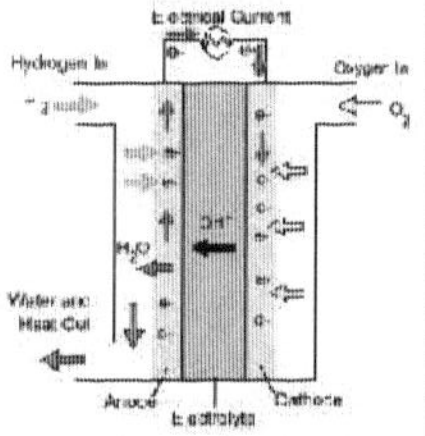

Polymerized Ionic Liquid (PIL) Block Copolymers

Polymerized ionic liquids (PILs), a new class of materials, are of interest for energy conversion and energy storage devices including fuel cells, batteries, super capacitors, and solar cells.

PIL block copolymers as anion exchange membranes (AEMs) in alkaline fuel cells are of particular interest due to the following unique properties:

- **High conductivity and chemical stability**
- **Low volatility and non-flammability**
- **High mechanical and thermal stability**
- **Tunable nanostructured morphology and domain size**
- **Variable macroscopic properties due to nanostructured morphology**

Objectives

- Synthesize PIL block copolymer AEMs
- Fabricate Membrane Electrode Assemblies (MEAs) with new membranes
- Measure fuel cell performance of new MEAs as a function of :
  - ***Electrode composition***
  - ***Back pressure***
  - *Fuel purity*
  - *Membrane thickness*
  - ***PIL content in block copolymer***
  - ***Heat press technique***

MEA Fabrication and Fuel Cell Conditions

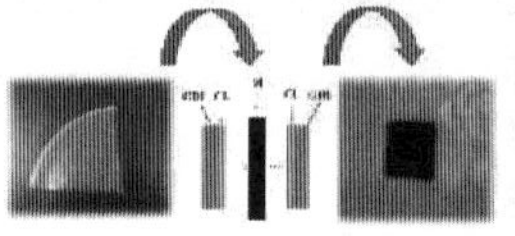

GDL: Microporous carbon-fiber treated with PTFE
CL: 33wt% polymer solution, 13wt% Pt in a 3:1 IPA:H_2O solution sonicated for 5 mins.
M: Solution cast using acetonitrile as the solvent. Annealed at 120 °C for 48 h and ion-exchanged as needed.

Unless otherwise mentioned, fuel cell tests were conducted with:

Anode/Cathode Temp: 60 °C
Cell Temp: 60 °C
Anode/Cathode back pressure: 25 psig
Anode Fuel Flow Rate: 0.42 L/min
Cathode Fuel Flow Rate: 1.00 L/min

The cell was allowed to equilibrate for 4 hours before collection.
The cathode and anode fuels were fully humidified prior to injection into the fuel cell.

Polymer Synthesis: Reversible Addition Fragmentation chain Transfer (RAFT) Polymerization

Synthesis of IL Monomer

Synthesis of Neutral PIL Block Copolymer

Functionalizing PIL Block Copolymer

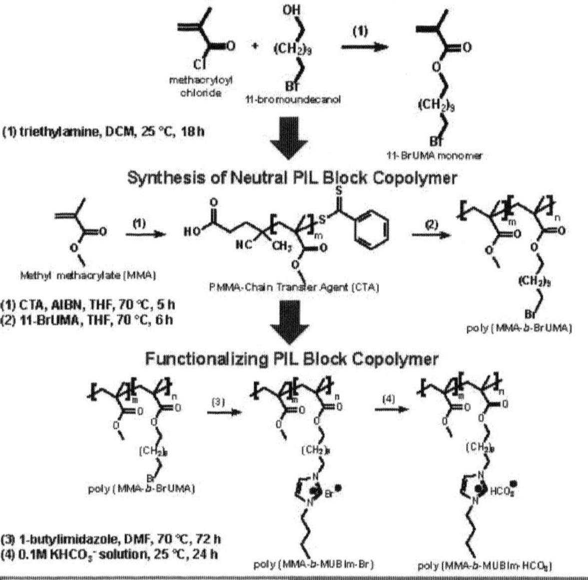

(1) triethylamine, DCM, 25 °C, 18h

(1) CTA, AIBN, THF, 70 °C, 5 h
(2) 11-BrUMA, THF, 70 °C, 6h

(3) 1-butylimidazole, DMF, 70 °C, 72 h
(4) 0.1M $KHCO_3^-$ solution, 25 °C, 24 h

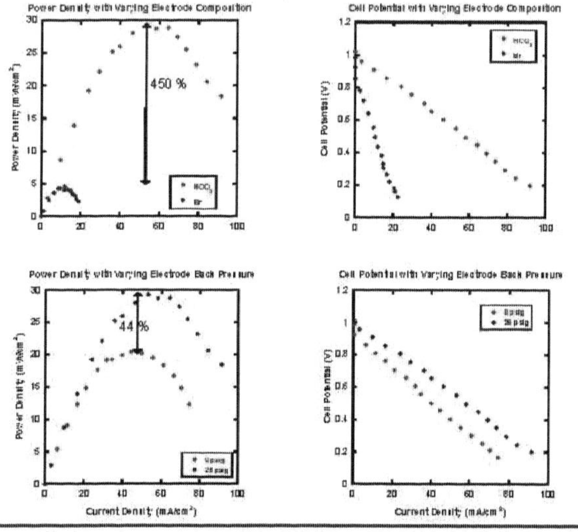

Alkaline Fuel Cell

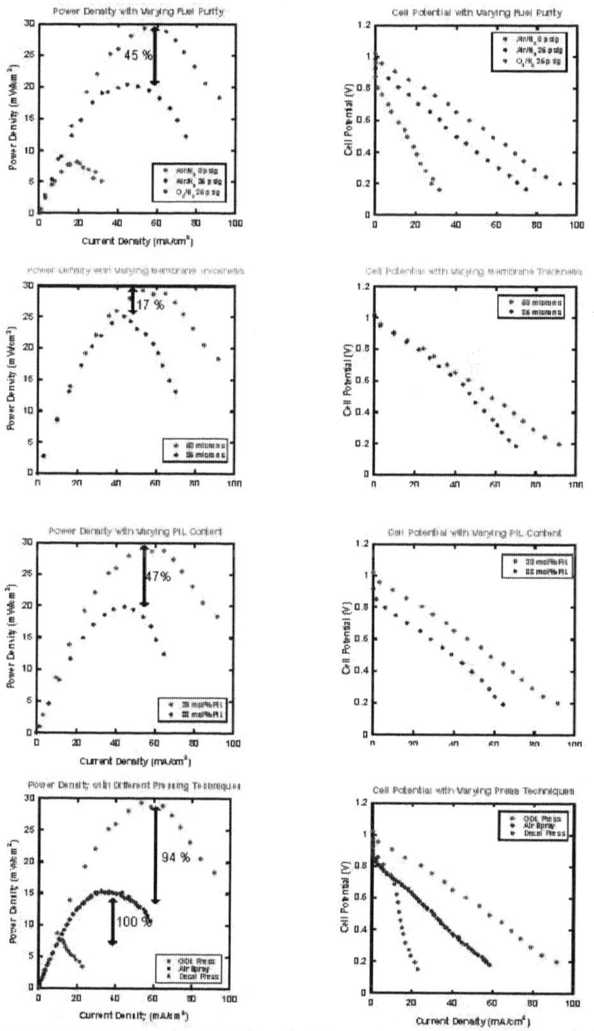

Conclusions

- ✓ Bicarbonate versus bromide form of polymer in electrodes improved AFC performance by 450 %
- ✓ Increasing back pressure from 0 to 25 psig increased AFC performance by 45 %
- ✓ Pure oxygen at the cathode resulted in a 45 % increase in performance over air
- ✓ Increasing AEM thickness from 35 to 50 microns increased AFC performance by 17 %
- ✓ Decreasing PIL content from 35 to 20 mol % increased AFC performance by 47 %
- ✓ The GDL press MEA fabrication technique resulted in the highest AFC performance

Funding

(NTI 1001-087)
(W911NF07-1-0452)

397482 A Novel Synthesis for Tough Polymer Coatings

Abstract:

A new method of preparing tough, chemically-resistant polymer films is presented. Surface-initiated ring-opening metathesis polymerizations (SI-ROMP) were used to grow crosslinked poly(dicyclopentadiene) films on gold coated substrates. Film growth was increased twenty-fold by utilizing vapor-phase monomer, which is believed to grow in a thin solvent film at the metal surface. Thickness and roughness were both found to be dependent on the solvent used during rinse stages, further supporting this hypothesis. These highly crosslinked films exhibit the superior structural characteristics of bulk poly(dicyclopentadiene) according to infrared spectroscopy, forming an excellent protective coating that resists both mechanical scratching and aqueous penetration.

398392 Ruthenium-Poly(Vinyl Pyridine) (RuPVP) Metallopolymer for Catalyzing Self-Oscillating Gels

Abstract:

Stimuli responsive polymer gels, in which a single stimulus (e.g. temperature, pH, etc.) causes a change in volume, have been a subject of intense interest for applications such as drug delivery and biological sensors. A periodic change in external stimulus is required for a gel to exhibit periodic oscillations. However, many biological systems maintain periodic oscillations under constant environmental conditions, converting chemical energy into mechanical work. Materials capable of emulating this biological behavior represent exciting opportunities for extending responsive behavior through energy harvesting and autonomous function. Autonomous oscillations can be achieved by the oscillating Belousov-Zhabotinsky (BZ) reaction within gels containing the BZ catalyst. When a gel containing a catalyst metal, such as ruthenium, is placed in a solution containing the BZ reactants (minus the Ru), the catalyst within the gel undergoes oscillation in its redox state. Due to the difference in the hydrophilicity of the polymer network at the Ru^{2+} and Ru^{3+} states, the gel displays swell-deswell oscillations. Currently used catalysts are either cost prohibitive or overly difficult to synthesize. To alleviate this problem, a facile, relatively inexpensive synthesis of ruthenium catalyst complex was attempted following previously reported procedures in the coordination polymer literature. Using readily available precursors, *cis*-Dichlorobis(2,2'bipyridine) ruthenium(II) and poly(4-vinylpyridine), RuPVP metallopolymer that successfully triggers BZ reactions was prepared. The catalyst was successfully immobilized in poly(N-isopropylacrylamide) to make hydrogels that capture chemical oscillations in the form of waves when placed in a BZ reaction solution.

398439 Synthesis and Electrochemical Characterization of Azulene Polymers for Use in Electric Double-Layer Supercapacitors

Abstract:

Although lithium-ion battery technology has been improving in recent years, there is still concern for eco-friendly production, overcharging, and end-of-life disposal. Electrochemical double-layer (EDL) supercapacitors have a much smaller energy density than traditional batteries, but when used complementary to a battery can offer up to 100 times the power density of the battery alone. EDL supercapacitors offer nanoscopic charge separation, are less expensive to manufacture, have greater cyclability, and are more power dense and eco-friendly than traditional batteries. In this study, the electrical properties of polymeric and co-polymeric azulene and trimethylazulene films prepared by electrochemical oxidation were studied for implications in EDL supercapacitors. Electrochemical quartz crystal microbalance (EQCM) was utilized to calculate the mass of polymer electrochemically deposited onto an electrode, allowing for the determination of a thin film's specific capacitance. Contrary to literature precedent, trimethylazulene was unable to be electropolymerized into a conducting film, while azulene was, with fast charge/discharge rates and good cyclability.

Abstract

Polymer thermoelectric (TE) devices are desirable because of their convenient properties for mobile applications and their ability to be produced at low-cost using well-established solution based techniques. Radical polymers are an emerging class of conducting polymers that have been shown to have electrical conductivities on par with π-conjugated polymers and are readily synthesized by reversible addition-fragmentation chain transfer (RAFT). Poly(2,2,6,6-tetramethylpiperidinyloxy methacrylate) (PTMA) is a well-studied radical polymer that has demonstrated moderate TE properties. The relatively high glass transition temperature of PTMA (T_g ~170 °C), however, causes significant fracturing when deposited as the thick films (~40 μm) required for TE devices which significantly limits device performance and lifetime. As such, it is desirable to synthesize a radical polymer with a lower glass transition temperature that could produce continuous films. Specifically, we propose to remove the methyl group on the polymer backbone which promotes steric hindrance and significantly increases the glass transition temperature of the polymer. Here, 2,2,6,6-tetramethylpiperidinyl acrylate (TMP) is synthesized at room temperature, polymerized via free-radical polymerization (PTMP), and then has the pendant groups functionalized to stable radical sites to form poly(2,2,6,6-tetramethylpiperidinyloxy acrylate) (PTA). The success of the synthesis of the monomer (TMP) and polymer (PTMP) can be characterization via ^{1}H nuclear magnetic resonance (NMR) spectroscopy. Finally, we can demonstrate the lower glass transition temperature of PTA via differential scanning calorimetry (DSC) and its effects on the TE performance.

SIDE CHAIN STRUCTURE MODIFICATION FOR IMPROVING THE PROPERTIES OF

Π-CONJUGATED SEMICONDUCTING POLYMERS

YUNDI JIANG,[1] BOYI FU,[1] JOSE BALTAZAR,[1] ASHWIN RAVI SANKAR,[1] PING-HSUN CHU,[1] SIYUAN ZHANG,[2] DAVID M. COLLARD,[2] AND ELSA REICHMANIS[1,2,3*]
[1]*SCHOOL OF CHEMICAL & BIOMOLECULAR ENGINEERING, GEORGIA INSTITUTE OF TECHNOLOGY*
[2]*SCHOOL OF CHEMISTRY & BIOCHEMISTRY, GEORGIA INSTITUTE OF TECHNOLOGY*
[3]*SCHOOL OF MATERIALS SCIENCE AND ENGINEERING, GEORGIA INSTITUTE OF TECHNOLOGY*

Continuous efforts have been devoted to improve the performance of optoelectronic devices such as organic field-effect transistors (OFETs) and organic photovoltaic cells (OPVs). Device performance significantly depends on the charge transport and transfer properties of the organic semiconducting polymers that serve as the active layer of the devices. To investigate the potential strategies to enhance the properties of π conjugated semiconducting polymers, our group designed three polymers having the same backbone structure, but different side chains: one with a linear side chain, **pTBTD-OD**, and two with branched side chains at different branching sites, **pTBTD-5DH** and **pTBTD-2DT**. The polymers were synthesized and characterized in terms of their electrochemical, photophysical, thermal, and charge carrier transport properties. Characterization results indicated that, compared to **pTBTD-OD** which possesses a linear side chain and **pTBTD-2DT** whose side-chain has a branching site close to the polymer main chain, **pTBTD-5DH** with a branching site relatively remote from the backbone has significantly superior properties including better solubility and charge carrier transport. The improved solubility of **pTBTD-5DH** allows for further solution processing, and the superior charge carrier transport enhances the performance of optoelectronic devices. This result emphasizes an important correlation between side chain molecular structure and the ultimate properties of π conjugated semiconducting polymers in device configurations. Future research will explore the impact of alternative polymer processing methods on the high-performance semiconducting polymer, **pTBTD-5DH**, to further improve its properties when applied to optoelectronic devices.

398610 Improving Photopolymerization Rates and Conversion of Common Monomers through Formation of Coordinated Ionic Liquids

Abstract:

Coordinated ionic liquid monomers were synthesized via dissolution of lithium bistriflimide (a bulky salt) in small, polar organic monomers including: acrylamide, acrylic acid, methacrylate, methacrylic acid, methyl acrylate and HEMA. The coordinated ionic liquids were photopolymerized and Fourier transform infrared spectroscopy (FTIR) was used to dynamically monitor monomer conversion and photopolymerization kinetics. It was found that certain stoiochiometries of monomer:salt (e.g. 3:1) greatly improved conversion and kinetics in comparison to the neat monomer. This presentation will discuss our methods, results and potential applications for polymers produced from coordinated ionic liquid monomers.

398661 Anti-CD123 Antibody Labelled Polymer Micelles Target Leukemia Cells

Abstract:

Self-assembled polymer micelles have been shown to be effective vehicles to transport hydrophobic anticancer drugs. Micelles composed of poly(styrene-*alt*-maleic anhydride)-*b*-poly(styrene) (PSMA-b-PS) have been shown to display aqueous stability, drug loading, and functionalization. Interleukin-3 receptor alpha chain (CD123) is specifically expressed in many hematological malignancies, including acute myeloid leukemia (AML). Therefore, anti-CD123 antibody was conjugated to (PSMA-b-PS) micelles in order to fabricate nanoparticles that target cells with CD123 expression. Two methods were used and evaluated in linking the antibody to micelles. Conjugates were formed via carbodiimide crosslinking reagents or Copper(I)-Catalyzed Azide-Alkyne Cycloaddition, a type of "click chemistry". The antibody concentration was quantified with fluoraldehyde (OPA) assay, and conjugation efficiencies of 20.1% and 16.4% were found for carbodiimide and click reactions respectively. Antibodies retained immunoreactivity post conjugation. Micelles were fluorescently labeled post conjugation for *in vitro* studies. Interactions between antibody-conjugated micelles and a CD123 expressing leukemia cell (MV-4-11) were assessed with flow cytometry. It was found that anti-CD123 conjugated micelles effectively bound to MV-4-11 cells with a 5-fold increase in cellular fluorescence compared to untargeted micelles. Our results suggest that anti-CD123 micelles serve as a useful targeted vehicle for delivery of AML therapies.

Block Copolymers of Bio-based Materials via. Reversible Addition Fragmentation Chain Transfer ~~(RAFT)~~ P~~p~~olymerization

Due to the diminishing supply and rapidly increasing prices of petroleum in today's society, numerous solutions are being sought to replace petroleum based products~~-~~, *i.e.* polymers and block copolymers~~or to increase the products' longevity~~. One option our group is studying ~~such potential solution~~ is the~~o~~ creation of~~e a~~ a ~~biorenewable~~ biobased ~~option for~~ substitute to the butadiene present in commercially available ~~block copolymers that is to replace butadiene in~~ styrene-butadiene-styrene (SBS) block copolymers. ~~that are used in many products and applications in industry, including the doping of asphalt to make it last longer.~~ We make use of controlled radical polymerization (Reversible-Addition Fragmentation Chain Transfer Polymerization) of chemically modified ~~Polymerizing modified~~ soybean oil ~~via RAFT polymerization is economical and safe compared to using butadiene~~ and styrene to synthesize the block copolymers. This~~ese~~ technique is able to create diblock and triblock copolymers, resulting in polymers that are predominantly non-crosslinked linear or ~~with~~ lightly branched chains. These materials behave as elastomers/rubbers at room temperature but are susceptible to common processing techniques at elevated temperatures, making them suitable for a wide range of applications.~~. The RAFT polymerization method is much easier to perform and allows much more versatility than the most common polymerization technique today, anionic polymerization. When copolymerized with polystyrene to create thermoplastic elastomers, also via RAFT, the resulting polymers have unique properties in comparison to its petroleum based counterpart. Due to the high functionality of acrylated epoxidized soybean oil (AESO), which is made through the acrylation of epoxidized soybean oil (ESO), the polymers created from this material, called poly-acrylated epoxidized soybean oil (PAESO), are extremely branched, enhancing the interaction between the molecules and the phenolic groups in the asphalt binder. The binder is much more elastic and structured with polystyrene-PAESO block copolymers and provides an environmentally friendly, economical substitute to styrene-butadiene-styrene in asphalt modification.~~ In this contribution, we will discuss the modification of soybean oil to form a suitable monomer for polymerization and the Reversible-Addition Fragmentation Chain Transfer Polymerization mechanism used to create our polymers. The benefits of our polymers when compared to their petroleum based counterparts will also be touched upon in this presentation.

398782 Dart Mass Spectrometry of Non Polar Polymers and Development of a Novel Data Analysis Program (MIRA-MER)

Abstract:

Silicone solutions are typically a heterogeneous mixture of silicone species. Identifying the molecular length distribution and chemistry can be important to the compatibility of the polymer for its intended use and its properties, such as viscosity. The aim of this project was to ascertain the chemical composition of silicones using a self-written code to analyze data from a newer technique, DART Mass Spectrometry. DART Mass Spectrometry was a helpful characterization tool but a data analysis tool was required to process the data in a useful manner.

396465 Characterizing Protein Corona Formation on Polystyrene Nanoparticles Using Single-Molecule Analysis

Abstract:

It is well known that nanoparticles in biological media form a protein coating, called a protein corona. The corona has been hypothesized to be one of the most important factors in determining a nanoparticle's fate and effect on cells. Here we have used fluorescence correlation spectroscopy (FCS), a non-invasive, single-molecule analysis technique, to characterize the adsorption of fluorescently labelled human transferrin and bovine serum albumin onto 50nm diameter polystyrene (PS) beads. The size of the corona was found to increase with protein concentration, with evidence of bilayer corona formation shown in both transferrin and albumin. The time span of corona formation was 120 minutes in transferrin and 180 minutes in albumin. Albumin coated PS beads showed significantly improved salt stability characteristics, an important result for medical potential. Results from this study will inform future studies of protein corona formation on nanoceria particles, specifically for drug delivery across the blood-brain barrier via transferrin-receptor mediated endocytosis.

396602 Synthesis of Polypeptides for Hydrogel Scaffolds

Abstract:

Biocompatible polypeptides are the main structural component of hydrogels, which can be used for cell studies and biocompatible materials such as wound filler. Two of these synthetic polymers, poly(propargyl-L-glutamate), PPLG, and poly(allyl-L-glutamate), PALG, are of particular interest due to the accessibility of their pendant alkyne and alkene groups, respectively, for click chemistry. The attachment of various functionalities to the polymer backbone allows for mimicry of the extracellular matrix. A general scheme for the traditional organic synthesis of these polymers has been developed and shows promise for extended study, further modification of properties, and a variety of applications.

Mechanistic Understanding of Distearyl Thiodipropionate as an Antioxidant in the Cross-Linking of Polyethylene

Charles Winslow, Fiaz Mohammed, Mark Conley, Harris Eldridge, Pamela Pollet, Charles Liotta, Charles Eckert

School of Chemical and Biomolecular Engineering, Georgia Institute of Technology, 311 Ferst Drive, Atlanta, GA

Cross-linked polyethylene (XLPE) is used as insulation in high and medium voltage electric cables. It is primarily synthesized through the free- radical induced crosslinking of polyethylene (PE) using organic peroxides as the radical initiator. One such initiator which is commonly used is dicumyl peroxide (DCP). However, during the processing of the PE, a small percentage of DCP undergoes homolytic cleavage into its cumyloxy radicals which subsequently causes premature cross-linking, or scorch. Scorching affects the ability of the PE to be processed and also reduces the physical and chemical properties of the final end product. To prevent scorching, different antioxidants and stabilizing additives are used to inhibit the free radicals produced during the processing stage. This ensures that uniform cross-linking will occur and the XLPE will have enhanced inductive properties. A frequently used additive is the secondary antioxidant distearyl thiodipropionate (DSTDP). However, despite its use in conjunction with other primary antioxidants, a detailed investigation of PE blends using only DCP and DSTDP has not been proposed in literature to the best of our knowledge. This work therefore explores the underlying chemistry of the thermal decomposition of DCP in the presence of DSTDP to determine how this antioxidant prevents scorch.

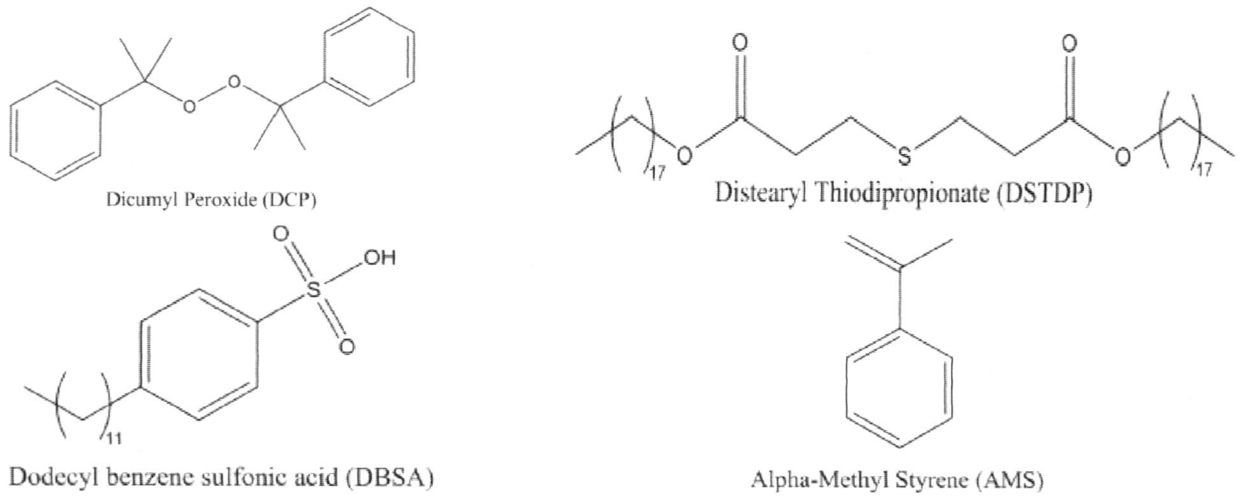

Figure 1: Main Species Used in Studies

The work was primarily performed in a liquid system using dodecane as a solvent to simulate polyethylene. A temperature range from 70 °C to 140°C was used for kinetic studies and to determine the overall effects of temperature on our system. The temperature most commonly used was 120°C, a temperature where thermal degradation of DCP was still relatively slow but the significant chemistries between DCP and DSTDP were accelerated. At this temperature, experiments were performed in both inert environments and environments open to the air with varying concentrations of DSTDP as well as its oxidized derivatives (DSTDP-sulfoxide and DSTDP-sulfone). A series of acid studies were also performed where low concentrations of dodecyl benzyl sulfonic acid (DBSA) were introduced into the DCP and DSTDP environments under argon. Species identification and quantification were determined using proton NMR, GCMS and HPLC.

By tracking DCP concentrations over the reactions of varying DSTDP concentrations in an environment open to air, it was observed that the rate of DCP decomposition increased with the concentrations of DSTDP until a maximum rate was observed at .25 molar equivalents of DSTDP. Furthermore, the rate of DCP degradation decreased as the oxidation level of DSTDP increased. However, under argon, this rate was comparable to the rate that DCP degrades by itself under a similar environment and temperature; this was also observed with DSTDP's oxidized derivatives. When DBSA was introduced into the inert environments, though, the rate of DCP degradation was comparable to that of the above DSTDP and DCP system that was open to air. Another important observation is the presence of α-methyl styrene (AMS) was observed only in the systems where DSTDP was present in air or the acid was present under argon. .

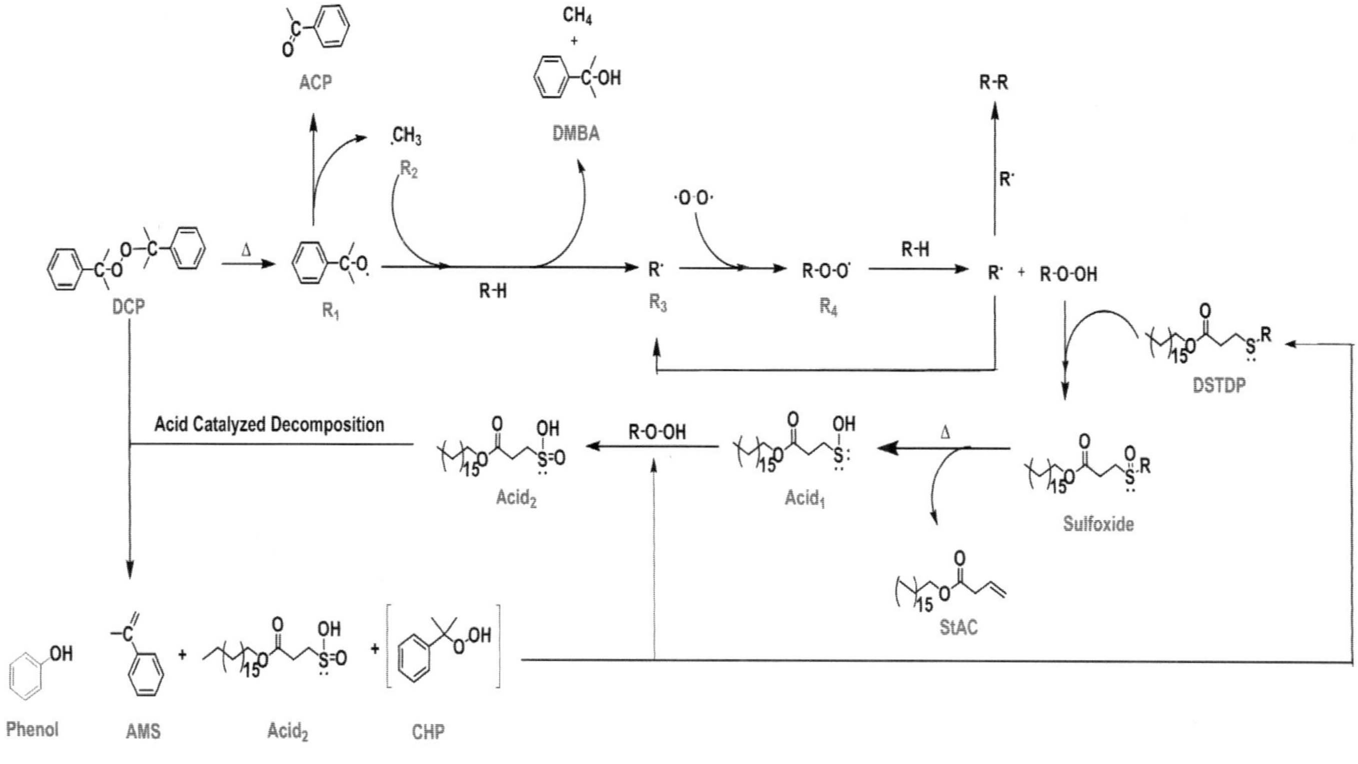

Figure 2: Working Mechanism for Reaction of DSTDP and DCP in Air

These observed effects can be explained through a detailed mechanism, shown in Figure 2, by which alkylhydroperoxides, formed when alkyl radicals reacted with triplet oxygen in the air, oxidize DSTDP into DSTDP-sulfoxide. The in-situ DSTDP-sulfoxide can then go under a retro-michael addition at the processing temperature, forming a hyposulfenic acid and a subsequent acrylate. It is conjectured that the hyposulfenic is then further oxidized into a sulfenic acid, subsequently catalyzing the decomposition of

DCP into cumene hydroperoxide (CHP) and AMS. As a reactive intermediate, CHP contributes to the oxidization of DSTDP into DSTDP-sulfoxide, forming an auto-catalytic cycle as more hyposulfenic acid is formed by the retro-michael of the newly formed DSTDP-sulfoxide. Lastly, the AMS, formed from both acid-catalyzed decomposition of DCP and the dehydration of cumyl alcohol, could then dimerize through both cationic and radical pathways to form a known radical sink.

In conclusion, the study of DCP and DSTDP has led to a working mechanism by which DSTDP is oxidized before undergoing a retro-michael addition and decomposing DCP into AMS and CHP. This AMS formed can quench the cumyloxy radicals by dimerizing and acting as a radical sink, thus preventing premature cross-linking, which improves the processing efficiency of XLPE.

Alexandra Sutlief

Chemically Amplified, Positive Tone Polynorbornene Dielectrics

Permanent dielectrics are used in microelectronics as insulators between electrical components. Polymers can be used as permanent dielectrics because they are more easily processed compared to inorganic materials, and they can also generally be patterned by photolithography. When using photolithography, it is desirable to have a highly light sensitive dielectric so as to minimize the energy required to pattern.

To pattern thick films or films with high aspect ratio structures, a material with good lithographic properties is needed, but the film also must be cross-linked to make it a permanent part of the electronic device. Our lab has developed a functional chemistry that uses chemically amplified photo-patterning and subsequent cross-linking. Chemical amplification involves a light generated catalytic species that starts subsequent chemical transformations. In our case, UV irradiation generates a strong acid to catalyze a reaction that leads to a solubility switch in the film. The film can then be developed in aqueous base to remove the exposed regions. Cross-linking of the film provides improved mechanical and electrical properties.

The polymer used has a polynorbornene backbone with hexa-fluoro alcohol and tert-butyl ester moieties. Films of this polymer are cast with a small amount of photoacid generator (PAG). These films are originally insoluble in base, but when exposed to ultraviolet light, the PAG will generate a strong acid that catalyzes the deprotection of the tert-butyl ester. The deprotection results in the formation of a carboxylic acid, which increases the solubility of the film in aqueous base. Our lab measured the D_{100} to be 8.09 mJ/cm^2 and a contrast of $\geq$14.2, both of which are a large improvement over the current positive tone dielectrics. The dielectric constant was measured to be 2.2 to 2.8, depending on the composition of the film. All of this was done with low PAG loading (3 parts per hundred).

To ensure good mechanical and electrical properties of the permanent dielectric, the film is thermally cured for 2 hours at 250□C after photolithography. Covalent cross-linking of the film occurs when the carboxylic acid (deprotected tert-butyl ester) undergoes a Fischer esterification with the alcohol functionality. Even though both the esterification and original deprotection are acid catalyzed, the deprotection caused by UV irradiation is much faster than the esterification. This means the exposed regions can be deprotected without cross-linking.

The two different functional groups were studied to determine how their presence impacts the mechanical and electrical properties. Different monomer ratios of the tert-butyl ester and hexa-fluoro alcohol have a profound effect on the physical properties. Increasing the hexa-fluoro alcohol content decreased the dielectric constant. This is most likely due to the high carbon-fluorine content, which is known to have a low dielectric constant. The increase in the alcohol group also decreased the modulus and hardness. The carboxylic acid formed will cross-link with other carboxylic acids to form an anhydride (confirmed by infrared spectroscopy). With a greater tert-butyl ester content, the greater amount of carboxylic acid cross-linking and thus a higher modulus and hardness. A polymer with 60% hexa-fluoro alcohol was measured to have a dielectric constant of 2.78, a modulus of 2.95 GPa, and a hardness of 243 MPa. Like the photolithography properties, these values either exceed or are similar to the current positive tone dielectrics.

398570 Pressure Sensors with Microstructured Polydimethylsiloxane Dielectrics Fabricated Via the Breath Figures Method

Abstract:

Flexible pressure sensors with high sensitivity show promise in applications such as touch screens, biomedical diagnostic technologies such as wearable heart rate monitors, and artificial "electronic skin" for robotics or prosthetic devices. For "electronic skin" applications, there exists a need for low-cost and scalable methods for producing pixels to sense both medium (10-100 kPa) and low pressures (<10 kPa). In this work, pressure sensors have been fabricated by incorporating compressible, microstructured polydimethylsiloxane (PDMS) and poly(styrene-butadiene-styrene) (SBS) dielectric layers into capacitors. The "breath figures" (BFs) technique, a simple, self-assembly based method for producing honeycomb-structured polymer films, was used to structure the PDMS and SBS films. Capacitive sensors incorporating microstructured PDMS dielectrics varied widely in structure and quality, but consistently showed high reproducibility, little hysteresis, and high sensitivity at lower pressures with decreased sensitivity at higher pressures. SBS dielectrics, on the other hand, resulted in more desirable linear responses to pressure, but exhibited lower sensitivity and some hysteresis. Future work to increase reliability of PDMS and SBS film formation will lead to the incorporation of these dielectrics into transistor-based pressure sensing devices and, eventually, biomedical applications.

Block Copolymer Interfacial Control: Photo-Patternable Interfaces

The integration of block copolymers (BCP) with lithographic processes shows promise in reducing semiconductor feature size—a necessary step for the next-generation of microelectronic devices. The self-assembly of BCPs into sub-20nm lamellar domains in thin films is critical to their application. This assembly is governed by the interactions at the surfaces adjacent to the BCP. By synthesizing polymeric "topcoats" and surface treatments, these interactions can be manipulated through the polymeric composition. The addition of acid-labile functional groups in these polymers, in conjunction with photo-acid generators (PAG), allows for the surface to undergo a large chemical change when UV-exposed. The local orientation of BCPs can then be controlled via these directly photo-patternable surfaces.

398916 Synthesis of a Block Copolymer Containing and Self-Immolative Block

Abstract:

Block copolymers (BCPs) can self-assemble into periodic nanostructures, which make them attractive for a multitude of applications. For lithographic patterning, the ability to selectively remove one block is particularly beneficial. This process often requires harsh conditions and usually damages the other block, diminishing the fidelity of subsequent pattern transfer steps. To solve this problem, a BCP was synthesized which contained a polymer that undergoes rapid depolymerization when exposed to strong acids. Poly(dihydroxy-tetrahydro naphthalene) (poly(DHTN)) was synthesized under high mechanical stirring from DHTN and dibromomethane in a two-phase reaction mixture. The poly(DHTN) homopolymer was coupled to azide-terminated poly(trimethylsilyl styrene) by the copper-catalyzed azide-alkyne click reaction to form the block copolymer. Selective depolymerization of the poly(DHTN) block was demonstrated in solution when mixed with a strong acid catalyst.

Analyzing the dynamic behavior of amphiphilic peptides on the surface of lipid-coated microbubbles

Pierre Desir, Joseph Badami, and Raymond Tu
Department of Chemical Engineering, The City college of New York

The City College of New York

CUNY Summer Undergraduate Research Program

Introduction

Microbubbles (MBs) are surfactant stabilized spherical cavities containing a certain type of gas surrounded by an aqueous environment. These particles are of a peculiar interest as their outer layer can be functionalized to include various targeting molecules. Moreover, microbubbles are used as contrast agents by enhancing images of the deep tissue of biological structures such as tumors. In addition, they play a therapeutic role by acting as drug carrying agents targeting specific cells in the body. These particles typically range in size from 1 - 10 μm, but smaller microbubbles would be useful for transport through small capillaries, 5-10 μm, or for the penetration of epithelial walls, 100-500 nm. However, the preparation of smaller and smaller microbubbles can prove to be problematic as these structures become less stable as they shrink.

Objectives

- Create stable microbubbles by using amphiphilic α-helical peptides that can partition to the bubble interface
- Investigate the influence of charge distribution on the partitioning of such peptides

Bubble Structure

- Core Material: Perfluorobutane Gas (C_4F_{10})
- Lipid Coating: DPPC and DPPA Lipids

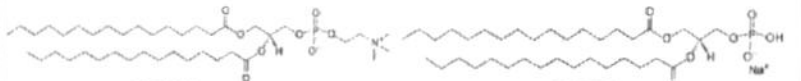

- Stabilizing Agent: Pluronic P85 Block Copolymer Surfactant
- Peptides: K-2.5 and K-6.0

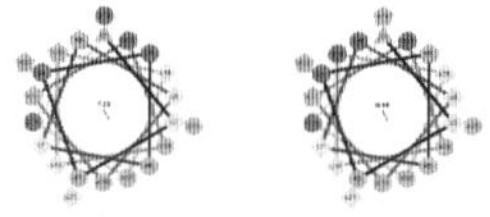

Bubble Synthesis

- The microbubbles were generated by applying Shear Mixing

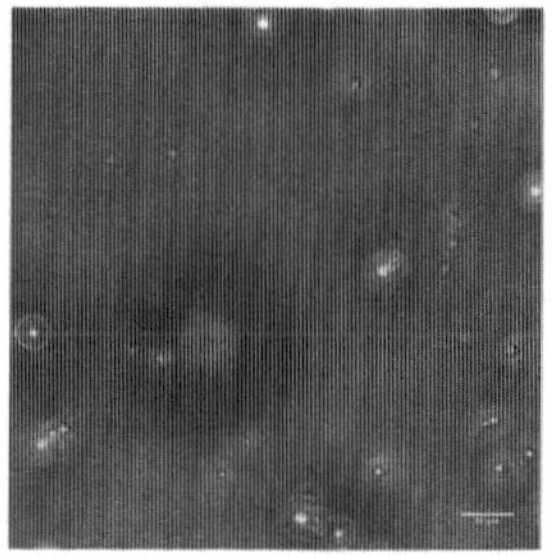

Bright field image of MBs in solution with K-2.5 at 5:1 lipid ratio and 6% initial peptide concentration

Bubble Analysis

- ➢ Bubble Separation:
  - Centrifugation of bubble solution using syringes as flotation columns
- ➢ Isolation of Molecular Constituents:
  1. Overnight freezing of samples
  2. Drying in pressurized chambers using a "Labconco Lyophilizer"

Lyophilizer

Centrifuge

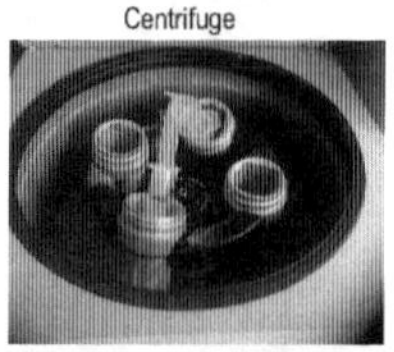

Peptide Calibration Curve

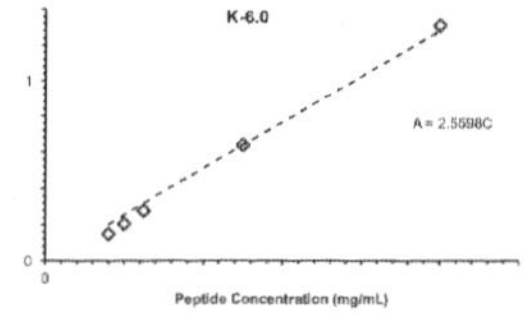

*A: absorbance
*C: concentration (mg/ml)

- Plot of peptide absorbance against concentration was obtained by using a "BIOMATE 3S UV-Visible Spectrophotometer".
- Beer's Law states that the absorbance of a solution varies with its concentration in a linear way: A = εcl where "A" is the absorbance, "ε" is the molar absorptivity coefficient of the sample ($M^{-1}.cm^{-1}$), "c" is the concentration (M^{-1}), and "l" is the pathlength (cm).
- Using a linear fit of Peptide Absorbance vs. Concentration, the amount of peptide in a bubble sample could be determined.

Peptide Partitioning

- In order to determine the effects of charge distribution on peptide partitioning, two sets of microbubbles were prepared with the following lipid mass ratios:
  - 5:1 DPPC:DPPA
  - 5:2 DPPC:DPPA
- DPPC is neutrally charged whereas DPPA is an anionic lipid.
- DPPC directs peptide partitioning mainly through hydrophobic interactions.
- DPPA binds the peptide to the interface via stronger electrostatic attractions between the negatively charged lipid and the positively charged peptide.
- Different lipid mass ratios examine the extent of charge distribution on peptide partitioning.
- Each peptide accounts for 5.5% of the total mass of the bubble system.
- The amount of peptide bound to the molecular constituents of the bubbles and the self-assembled structures is measured by spectroscopy using a "BIOMATE 3S UV-Visible Spectrophotometer".

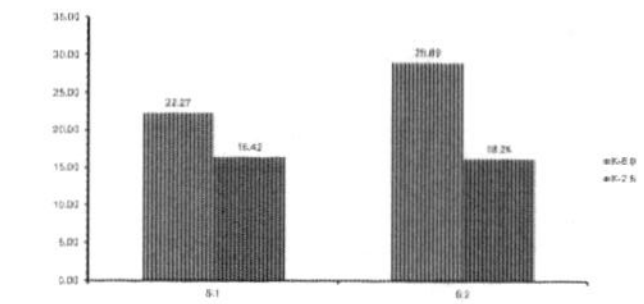

The amount of k-6.0 adhering to the bubble surface increases with the larger relative amount of DPPA used whereas the amount of k-2.5 partitioned varies slightly with the different lipid mass ratios. Also, the charge separation of the peptide influences its interfacial partitioning. Therefore, each peptide's secondary structure was analyzed on the surface of the self-assembled structures in solution via Circular Dichroism (CD).

Secondary Structure

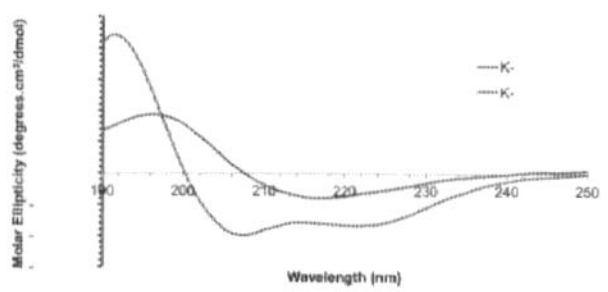

- K-6.0 conserved its α-helical structure on the liposome.
- K-2.5 changed from an α-helical to a β sheet configuration once partitioned to the liposome.
- Both peptides have the same amino acid composition.
- The greater charge separation on K-6.0 helped maintain the helical secondary structure.
- The smaller charge separation on K-2.5 resulted in charge repulsion between the positive charges, which could explain the change in configuration.

Bubble Size Assessment

We investigated the influence of peptide binding on bubble size and stability using a "Partica Laser Scattering Particle Size Distribution Analyzer LA-950V2".

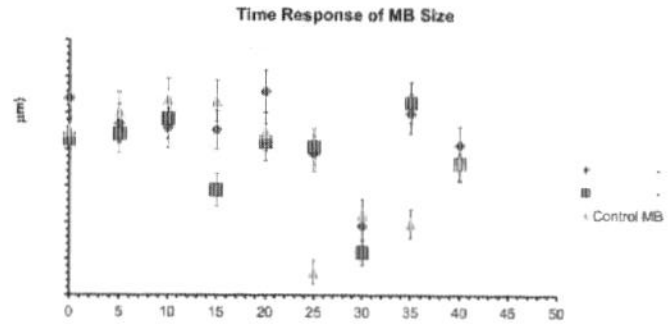

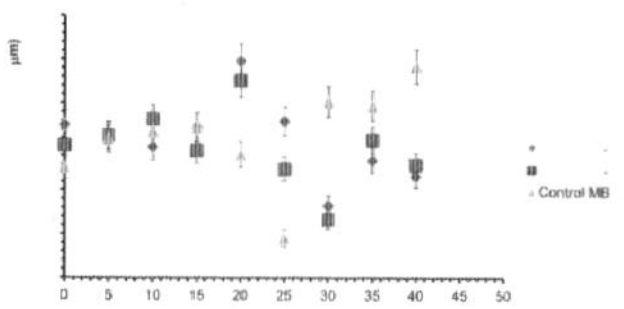

- Two assumptions can be made given the size distribution overtime:
  1. The higher pressure inside the bubble causes the Gas Core to escape into the solution, which reduces the size of the bubble
  2. The bubbles collide and coalesce to form larger systems

Bubble Shrinking

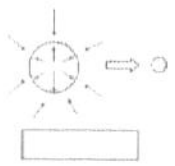

Bubble Coalescence

Conclusions

- The partitioning of K-2.5 and K-6.0 onto the microbubble surface is governed by electrostatic forces coupled with hydrophobic interactions between the peptides and the lipid monolayer.
- Peptide self-assembly on the lipid monolayer is strongly dependent on peptide charge separation.
- Microbubbles have a relatively constant size distribution over a period of 40 minutes, which reflects their stability in solution.
- Both peptides K-2.5 and K-6.0 generate microbbubles with the same average size overtime.

Future Work

- Investigate temperature effects on peptide secondary structure at the bubble interface
- Examine peptide secondary structure at air/water interface both in 2D and on the bubble surface
- Explore bubble shell mechanical properties using AFM

Acknowledgement

Dr. Raymond Tu, Joseph Badami, C-SURP, and NSF-DMR #1006407

396633 Determining the Rate of ROS Generation from Composite Magnetic Nanoparticle Hydrogels in an Alternating Magnetic Field

Abstract:

The surfaces of superparamagnetic iron oxide nanoparticles (IONPs) can generate reactive oxygen species (ROS) in the presence of hydrogen peroxide via the Fenton and Haber-Weiss reactions. The ROS were used in an advanced oxidative process (AOP), where the radicals degrade methylene blue, a commonly used model dye for an environmental pollutant. Previous research showed that when exposed to an alternating magnetic field (AMF), free IONPs in dispersion were able to enhance ROS production; however, as the exposure time and IONP concentration increased, the enhancement decreased. This was attributed to possible AMF-induced nanoparticle aggregation, which reduced the amount of surface area available for reaction. Here, IONPs were synthesized and entrapped in the crosslinked network of a poly(ethylene) glycol (PEG)-based hydrogel. It was hypothesized that immobilized IONPs would not aggregate under the AMF and would thus be able to sustain enhancement of ROS production. It was found that ROS production using immobilized IONPs under an AMF was indeed enhanced to a much higher degree than for free IONPs in dispersion, but the enhancement was still seen to decrease as exposure time increased. This may have been due to a change in oxidation state of the surface iron ions or hydrogen peroxide degradation under high local temperatures. Though a decrease in enhancement was observed, this result was promising in that the enhancement factor did not drop to 1 (no enhancement) even after 30 min of AMF exposure.

397607 Shape Actuation of Dual-Cured Networks

Abstract:

Conventional two-way shape-memory polymers switch between two shapes in response to temperature in the presence of an external load. In this work, we demonstrate a single phase, two-way shape actuator which is thermal-responsive without an external load. Double networks were formed by Thiol-acrylate Michael addition, followed by additional chemical cross-linking using UV irradiation under pre-strained conditions. Prepared samples exhibit contractile forces along the pre-strained direction upon melting of crystalline domains due to entropic elasticity of the first network. Samples reversibly elongate, upon cooling, due to internal stress-induced crystallization along the same direction. The degree of orientation and reversible crystallization was further investigated using a combination of X-ray diffraction and calorimetry. Shape actuation was studied at various pre-strained conditions. Under optimized conditions, reversible actuation of around 15% strain can be reproducibly achieved and cycled many times with nearly uniform actuation.

398394 Thermally Reversible Hydrogels

Abstract:

Responsive hydrogels are a type of soft material made up of lightly cross-linked polymers that are highly hydrophilic and have the ability to respond to stimuli. Thermally reversible hydrogels respond to a change in temperature and have the ability to exhibit either lower or upper critical solution temperatures where phase separation occurs. Self-oscillating hydrogels combine an oscillating chemical reaction with responsive hydrogels. In these gels chemical oscillations produced by the Belousov-Zhabotinsky (BZ) reaction lead to mechanical swell-deswell oscillations. These materials have the potential to be used as soft actuators, microfluidic pumps, or as a form of chemical energy harvesting. A primary objective for our project is to improve on existing self-oscillating gel systems by synthesizing a new self-oscillating hydrogel material that has the ability to be easily patterned. The patterning property that is being explored can then be used to create a large array of synchronized oscillating gel patches, which in turn can be used to produce large cooperative and complex motion. Thermal reversibility makes printing gels more controllable because the polymer solutions exhibit phase separation leading to gelation under certain conditions. A copolymer composed of acrylic acid and N-isopropylacrylamide was synthesized and successfully displayed gelation above 35°C. A copolymer that displayed a gelation below 25°C was also synthesized by a reaction of poly(allylamine) and potassium cyanate. The next step will be to chemically crosslink the gel to make it permanent and incorporate the metal catalyst for the BZ reaction.

Cavitation Rheology of Alginate Gels

Kinsey Naas
Meysam Hashemnejad
Santanu Kundu

Alginate is a polymer extracted from brown seaweed. The extracted alginate ionically crosslinks in contact with calcium ions, resulting in a gel. Polymer gels made from alginate have uses in various fields such as drug delivery, tissue engineering, wound dressing and cell culture[1].

The alginate gels used for cavitation rheology experiments are prepared in the lab. Initially, alginate powder is mixed in 40mL of water. Once the alginate powder is fully dissolved, dibasic calcium phosphate is added and the contents are mixed until the dibasic calcium phosphate is fully dissolved. In a separate beaker, GdL(gluconic acid δ-lactone) is dissolved into 40 mL of water; the GdL and water are then poured into the beaker containing the alginate and again the contents are mixed. The GdL provides an acidic environment for crosslinks to develop resulting in a gel. The mass of alginate and calcium salts are varied to obtain four different concentrations of gel.

Cavitation rheology is used to characterize the mechanical properties of soft gels. Elastic modulus (Young's modulus) is a characteristic of material to resist deformation. Elastic modulus can be measured using the cavitation rheology technique and such information is useful for designing gels for various applications. Cavitation rheology is performed by inserting a needle attached to a syringe into the gel being tested. The syringe is filled with air and the air is pumped into the gel at 5mL/min. As air is compressed in the syringe, the pressure inside the gel increases. Finally, at the critical pressure, P_c, a cavity is formed inside the gel (Figure 1). The value of the critical pressure is directly related to elastic modulus and surface tension as follows[2]:

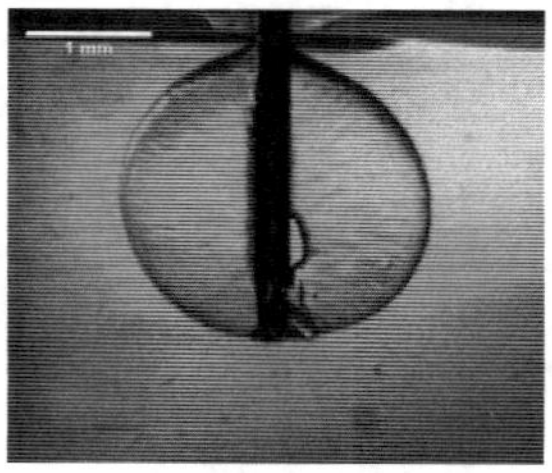

Figure 1: Cavity developed inside an alginate gel.

$$\frac{\square}{r}$$

Where P_c is the critical pressure obtain from experimental data, E is the elastic coefficient, γ is the surface tension of water, and r is the inside radius of the needle. With the elastic modulus, it can be predicted how much deformation a material can withstand during its application. The data collected from the cavitation rheology is being related to the local mechanical properties.

---

1 Lee KY, Mooney DJ. Alginate: Properties and biomedical applications. Prog Polym Sci 2012;37:106-26

2 Zimberlin, J. A. and Crosby, A. J. (2010), Water cavitation of hydrogels. J. Polym. Sci. B Polym. Phys., 48: 1423–1427. doi: 10.1002/polb.21968

Nature Inspired Tetherless Metallic Spiked Surfaces for Tissue Collection

JOHNS HOPKINS WHITING SCHOOL of ENGINEERING

Alex Abramson, Evin Gultepe, and David Gracias.

Johns Hopkins University Department of Chemical and Biomolecular Engineering. Baltimore, MD.

Project Rationale & Goals

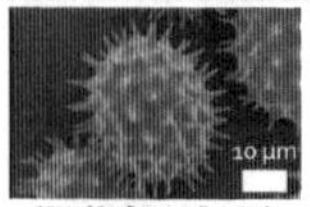

SEM of Sunflower pollen. Each spike is 4 µm long.

Forceps Biopsies require invasive procedures.

- **Hypothesis: Barbed metallic spikes, similar to those found in pollen, can be created by electroplating and selectively etching metals. These spikes can be fabricated onto microdevices and used for tissue biopsy.**
- By varying the amount of metal deposited, the photolithography patterns, and the etching times, we can create differently sized and shaped spikes.
- This new non-invasive form of biopsy can replace the use of forceps and needle biopsy tools.

Gaia. Pollen. 2009. The Beauty of Chaos. Web. 18 October 2014. http://www.gihealth.com/html/education/photo/biopsy.

Spike Creation in Nature

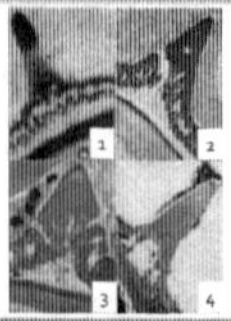

- Tricolpate pollen in sunflowers (*Helianthus Annuus)* develop spikes on the sexine due to the contribution of sporopollenin supplied by tapetum and microspores at sites with locular fluid on the exine (1-3). The sporopollenin polymerizes.
- The rest of the tapetum then is decomposed leaving only the spikes (4).

Harry T. Horner Jr. and Christine B. Pearson. American Journal of Botany, Vol. 65, No. 3 (Mar., 1978), pp. 293-309

Fabrication Model and Process

- The force of the spikes on the tissue can be modeled using the Hertz mechanics equation for contact between a rigid cone and an elastic half space:

$$F = \frac{2E}{\pi(1-\nu^2)} * \frac{d^2}{tan\theta}$$

E= Elastic modulus of tissue ($8.03*10^5$ dynes/cm^2)
ν= Poisson ratio of tissue (0.45)
d= Indentation depth
θ= ½ supplementary tip angle (90°)

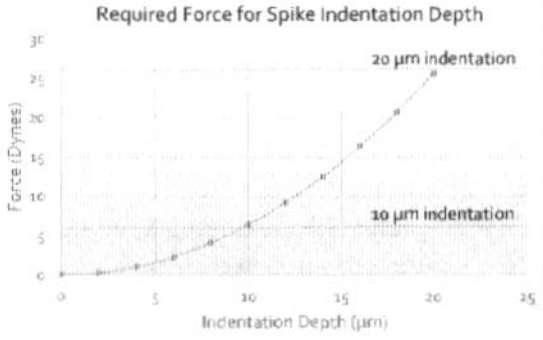

Spike Fabrication Process:

1. Electroplate panels of various metals onto a sheet.
2. Cover the panels in patterned photoresist.
3. Etch the panels to create spikes.

The thick layer of photoresist is left on the wafer to prevent the etchant from seeping through the sides

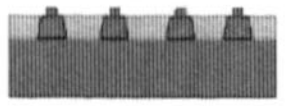
Small circles are added to the panels to create the spike shapes. When the panels are large, then more circles are added.

The spikes must be at least one spike diameter away from the next spike. This prevents the etching patterns from overlapping.

Differently Shaped Spikes

Regular vs Barbed Spikes

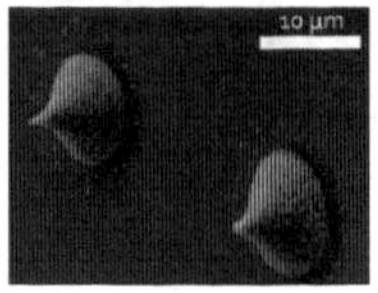

These images, taken under SEM, show the spikes' shapes. If the spikes are etched for longer, then the points become sharper. These spikes are made of Copper.

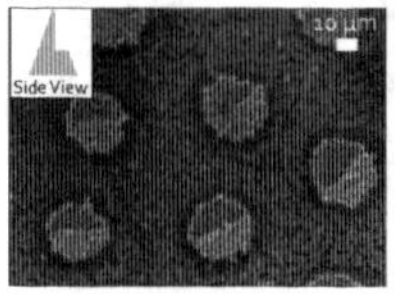

By covering the spikes with another differently shaped layer of photoresist and etching again, we create barbed spikes. These Nickel spikes have a flat shaped barb.

Copper vs Nickel Spikes

Copper spikes are patterned on top of tetherless surgical microgrippers used to take biopsies of the colon. The spikes create a rough surface which helps them collect tissue.

Nickel Spikes are patterned on top of panels used to take tissue biopsies. The spikes scrape the tissue from the wall and collect the samples in crevasses between spikes.

In Vitro Tissue Collection

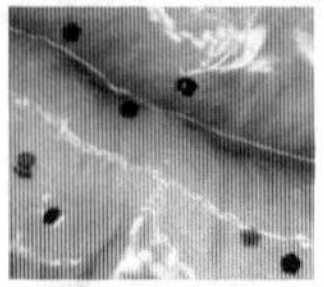

- Hundreds of spiked panels were propelled out of a tube at 15 atm towards a piece of pig intestine. The tissue was rinsed with water, and the remaining panels were collected with a magnet
- All panels had an attachment rate ~11%
- The tissues on the panels were fixed with formaldehyde, dyed with DAPI, and analyzed under an optical microscope to determine the amount of tissue biopsied based on spike profile.

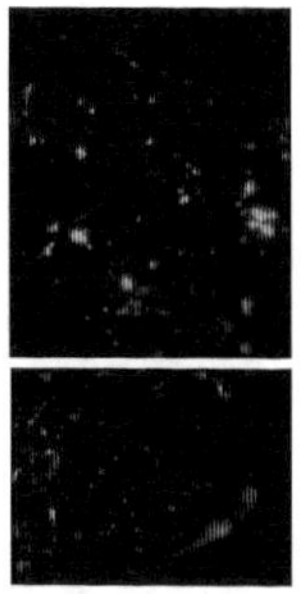

- The pictures to the right shows Small Barbed closely patterned spike panels with DAPI staining.
- The tissue tends to be organized around the spikes, showing the importance of the spikes in the biopsy.
- Most of the spikes on the panels contain some tissue, showing that the compact nature of the spikes does not affect the tissue yield.

Results and Analysis

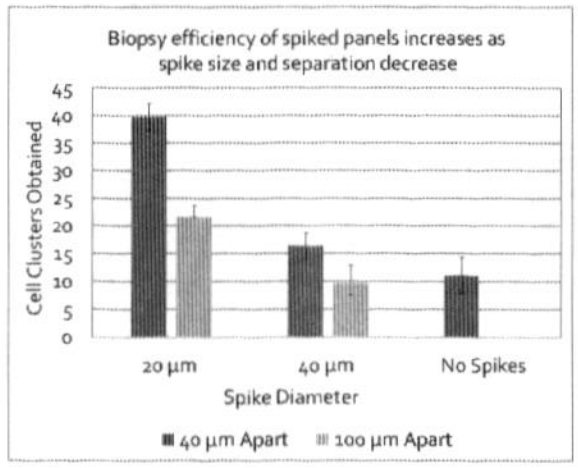

- The closely packed spiked panels with small spikes collected four times more tissue than the panels without spikes (p= 3.6e-8)
- The smaller spikes gathered over two times the tissue of the larger spikes (p= 1.3e-7). This likely happened because the small spikes exerted a greater force per area than the larger spikes allowing for tissue breakthrough.
- When we increased the density of spikes, the tissue collection increased twofold (p=1e-4).

- While the panels with spikes further apart showed some tissue collection, a higher density of tissue could be found on the panels with the spikes closer together.
- The smaller spikes had more cell clusters than the larger spikes.
- Tissue found near the edges of the panels was not counted as tissue collected by the spikes. These cells were collected by the edges of the panels.

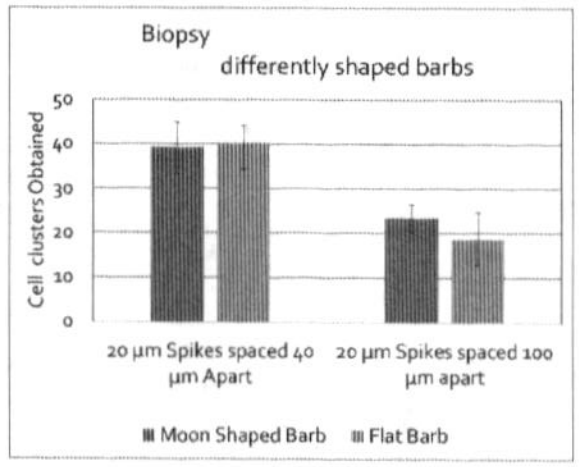

- We tried two differently shaped barbs, flat barbs and crescent moon barbs. The shape of the barb did not effect the spike's tissue collection (p=0.88 [close], p=0.59 [far])
- The barb did not change the tip of the device and therefore did not change the piercing pressure of the spikes.
- The different barb shapes can be seen in the *Applications to Microdevices* section.

Uses in the Field

Microdevices with spikes can potentially be used as both biopsy and scraping devices. We have already added spikes to the following devices:

1. Tetherless surgical microgrippers for colon biopsy (top)
2. Self folding dodecahedra which mimic the shape of pollen (bottom).

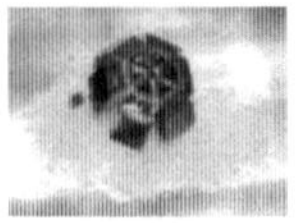

We theorize that these spikes can be added to devices for several other applications as well, including:

1. Attaching to clothing and fabrics
2. Scraping plaque from arteries
3. Increasing drug delivery rate by puncturing tissue

Gracias, David. Microgrippers. 2009. Headlines at Hopkins. Web. 6 March 2014.

Conclusions

- The fabrication of barbed metallic spikes onto microdevices increase the effectiveness of tissue biopsies.
- Panels patterned with spikes 20 um in diameter and 40 um apart collect four times as much tissue compared to smooth panels
- Spikes made of various metals, created in various shapes and sizes can be fabricated onto any microdevice.
- These spikes can be easily collected via a magnet.

Future Directions

- We will perform *in vivo* studies with the spiked panels.
- We will create a method to fabricate hooked spikes in order to allow greater attachment to tissues and fabrics.
- We will test the spikes on plaque, to see if they are effective at removing blockages without damaging the arterial wall.

Acknowledgements

- Thanks to Shivendra Panday for helping me take SEM pictures of the spikes.
- This research was made possible in part by a grant from the Provost Undergraduate Research Award Committee at Johns Hopkins University.

398093 Deformation and Bonding of Cold Sprayed Iron-Based Amorphous Metal Particles

Abstract:

Name: Maryvivian Okwara

Year: Class of 2016

Faculty Mentor: Professor Constance Ziemian (Mechanical Engineering department)

ABSTRACT

Deformation and Bonding of Cold Sprayed Iron-Based Amorphous Metal Particles

To date, various attempts have been made to produce amorphous iron (Fe)-based metallic coatings using conventional deposition methods such as thermal spray techniques. These methods, however, have been shown to compromise the amorphous structure of the coatings and adversely affect the superior properties of the metallic glass, such as corrosion and wear resistance. The overarching goal of this project is to determine the feasibility of depositing fully amorphous coatings using a unique, kinetic energy-based process known as the cold spray. The project specifically tests the hypothesis that cold-sprayed $Fe_{48}Cr_{15}Mo_{14}C_{15}B_6Y_2$ at.% (SAM1651) metallic glass particles will sufficiently deform and bond to a mild steel substrate with the proper optimization of process parameters. This is accomplished by analyzing the deformation and adhesion of individual Fe-based metallic glass particle impacted upon a steel substrate, known as cold spray splat tests. Several splat tests were completed with a variety of gas temperatures (850°C, 900°C, 950°C, 975°C, 1000°C), gas pressures (3MPa, 4MPa), powder feed rates (0.2rpm, 0.25rpm, 1 rpm), and spray raster speeds (1200mm/s, 2400mm/s, 3600mm/s). The resulting specimens were gold-coated and analyzed using scanning electron microscopy (SEM). The deformed particle morphologies were found to vary as a function of primarily gas temperature, ranging from well-deformed to un-deformed. SEM analysis also indicated that a significant percentage of the impacted particles rebounded off of the steel surface and left craters in the substrate. These craters inspired further research focused on the empirical determination of the critical velocity required for particle adhesion. Although this type of study has been done for cold-sprayed crystalline metals, the cold-spray critical adhesion velocity had not been previously defined for amorphous metals. Material strength and stiffness needed for the critical velocity analysis were determined using nanoindentation. This empirical modeling work is ongoing, and results are expected to define a window of gas temperatures and pressures for the successful cold-spray deposition of Fe-base metallic glass particles.

398158 Combined Modelling and Experimental Investigation of Mechanical Behaviors of Metallic Glasses

Abstract:

Metallic glasses are innovative in the unique combination of steel-like strength and plastic-like formability. Their applications, however, are limited by a catastrophic failure due to the formation and propagation of shear bands. The 'bulk-scale' concerns are mitigated by processing metallic glasses as low-dimensional thin films using the near atomic scale control. In this work, we combine computational modelling and experimental testing to gain an *understanding of the mechanical behaviors of thin film metallic glasses (TFMGs), particularly the initiation of shear bands.* Experimentally, we conducted various nano-indentation tests, and the initiation of the shear bands were reflected by the discrete ripples on the measured load-displacement traces. Further, with increased strain rates of indentation, the shear band initiation was suppressed, leading to increased yield loads. Additionally, the measured yield loads vary with sample location as a result of the structural and chemical inhomogeneity of metallic glasses at nano-scale. The computational simulation results capture the experimental observations, and also provide insight into how the atomistic process leads to the formation of shear bands. This experimental-computational collaboration will give way for a broader grasp of the deformation mechanisms that ultimately yield catastrophic failure.

Synthesis of Molecular Electronic Components for Self-Assembly onto Metal Electrodes

Kindle S. Williams, Joseph E. Meany, and Stephen A. Woski
Department of Chemistry
The University of Alabama, Tuscaloosa, AL 35487

Introduction

The field of molecular electronics centers around the synthesis of single-molecule electronic components. These components have the potential to one day replace their silicon-based counterparts, resulting in both smaller and more efficient devices which can be assembled from the ground up rather than with the inefficient top-down approach taken today. Our present goal is to design and synthesize a carbon-based molecular diode. A diode – also called a rectifier – functions to convert alternating current (AC) to direct current (DC) by allowing unidirectional electron flow. Our current work is centered on molecules bearing an electron-rich dimethoxybenzene donor ring and an electron-poor quinone acceptor ring separated by a single bond. We have successfully synthesized a group of such molecules, which we call hemibiquinones (HBQs). Each side of these molecules features an attachment site (at the 4,4′ positions) onto which a functional group may be added. The purpose of such groups is to bond to metal electrode surfaces. Attached to an electrode at each side, electrons could travel from one electrode, through the molecule, to the next electrode unidirectionally, thereby making rectification possible.

(a) (b)

Figure 1. Aviram & Ratner molecule and Br,Br-HBQ
Above are the Aviram & Ratner template molecule (a) and dibromohemibiquinone (Br,Br-HBQ) (b), our precursor molecule. Each exhibits the acceptor and donor features (green and pink, respectively) as well as substitution sites (white).

It was posited by Aviram and Ratner in 1974 that such a molecule, if synthesized, could rectify current[3]. Their hypothesis was that an electron-rich ring and an electron-poor ring separated by a long enough insulator would allow a molecule to conduct electricity. While their theoretical molecule possessed an ethylene bridge as an example of such an insulator, our molecules of interest possess only a single bond (shown in yellow in (b) above). The torsion angle between the two rings serves to insulate the molecule.

Objectives

Our goals were to:

- Optimize the synthesis of Br,Br-HBQ from 1-bromo-2,5-dimethoxybenzene
- Model highest occupied molecular orbital (HOMO) and lowest unoccupied molecular orbital (LUMO) characteristics
- Acetylate aminohemibiquinone (NH_2,Br-HBQ)
- Prepare 4,4′-dithiodibenzoic acid for addition to NH_2,Br-HBQ
- Prepare symmetrical molecules for electrochemical and spectral comparison with HBQs
- Append NH_2,Br-HBQ with moieties capable of bonding directly to atomically flat metal surfaces

Synthesis and Methods

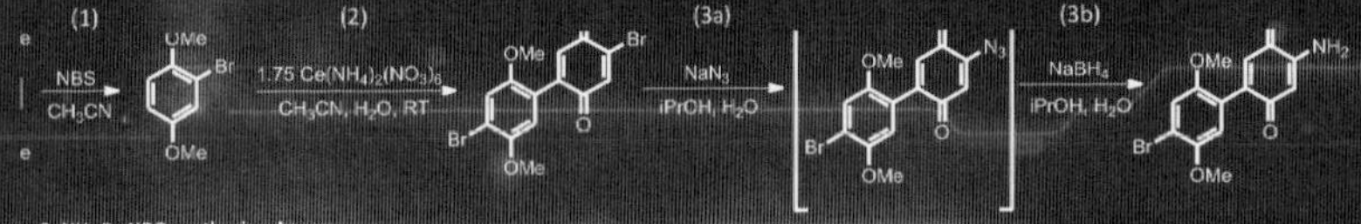

Figure 2. NH_2,Br-HBQ synthesis scheme
This synthesis of NH_2,Br-HBQ, a molecule to which many functional groups may be appended, has been optimized. Reaction (2) is now achieving yields around 75%. Reactions (3a) and (3b), which occur in the same pot, consistently yield over 90% pure NH_2,Br-HBQ.

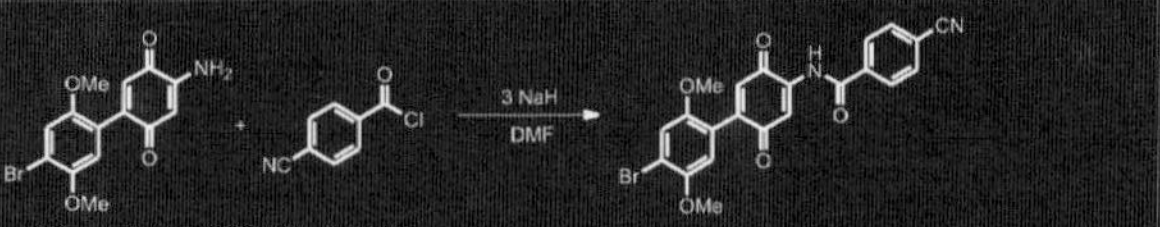

Figure 3. Addition of Benzonitrile group to NH_2,Br-HBQ
The above reaction was carried out under nitrogen atmosphere and yielded a product which will be subjected to surface studies.

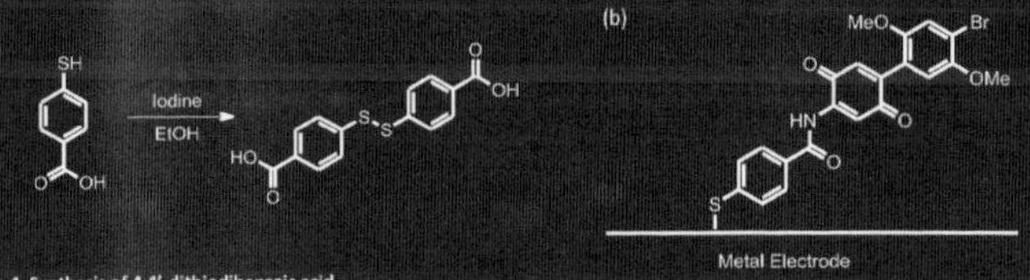

Figure 4. Synthesis of 4,4′-dithiodibenzoic acid
Some particularly useful groups for bonding to metal surfaces are the thiols. In order to append a thiol group to the hemibiquinone framework, we prepared 4,4′-dithiodibenzoic acid as shown in (a)[5], achieving a yield of 92%. Activation of the acid group will then allow for acylation of the aminohemibiquinone molecule, following the same procedure used in Figure 3. Once this is completed, surface studies will be conducted in order to determine the geometry of a monolayer of self-assembled molecules on a gold surface. (b) depicts how the resulting molecule might attach itself to an atomically flat metal surface.

HOMO/LUMO Modeling

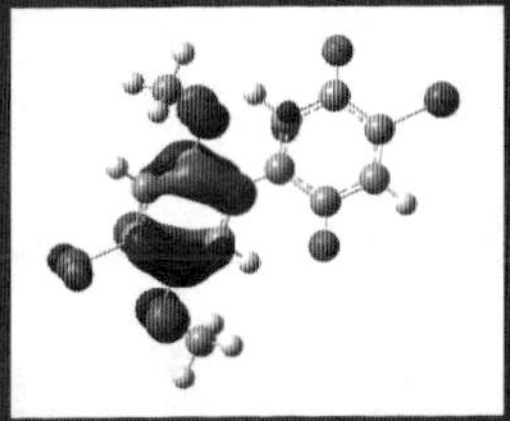

Figure 5. Br,Br-HBQ HOMO

Calculations of molecular properties such as HOMO/LUMO energies and orbital distributions were performed using Gaussian 09. As hypothesized by Aviram and Ratner, for each HBQ, the HOMO resides largely on the dimethoxyphenyl ring, and the LUMO is found on the quinone. This is exactly what is needed for rectification to occur. HOMO/LUMO orbitals for Br,Br-HBQ are shown here. The depicted shape is typical of all HBQs calculated so far.

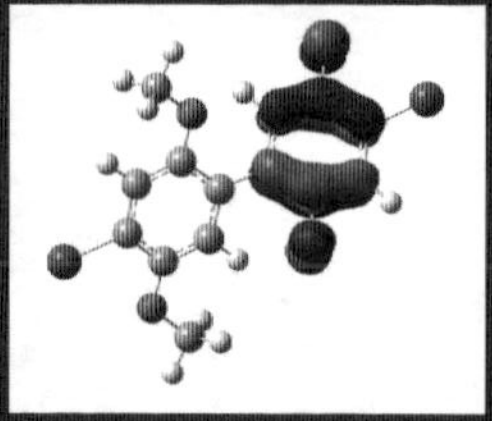

Figure 6. Br,Br-HBQ LUMO

HOMO/LUMO Modeling (cont'd)

Useful derivatives of the HBQs have also been modeled with Gaussian 09. Show below are the HOMO and LUMO of the benzonitrile derivative of acetylated NH2,Br-HBQ. The torsion angle of this reduced HBQ backbone is 50°, as opposed to the angle of 38° when oxidized.

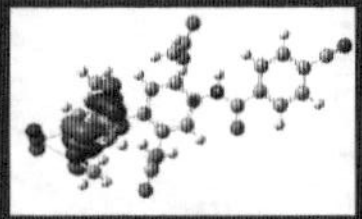

Figure 7. Acetylated benzonitrile derivative HOMO

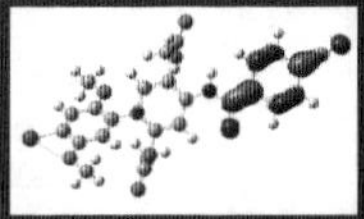

Figure 8. Acetylated benzonitrile derivative LUMO

Conclusions/Moving Forward

- Optimized Br,Br-HBQ and NH_2,Br-HBQ reactions are now more efficient and allow us to synthesize starting material with relative ease.
- A nitrile-containing HBQ has been synthesized via an acylation reaction with NH_2,Br-HBQ.
- A similar reaction will be attempted in order to synthesize a thiol-containing compound.
- The resulting molecules will be subjected to surface studies in order to determine their geometries when attaching to a metal surface.
- When the spectral and electrochemical properties of the HBQs are determined, they can be compared with symmetrical molecules that have been isolated, such as dibromotetramethoxybiphenyl.
- Since acetylation reactions have begun working, it may be possible to add different substituents to the dimethoxyphenyl side of the HBQs.

References

1. "Does molecular electronics compute?" Editorial. Nature Nanotechnology. 8, 377 (2013).
2. Aradhya, S. & Venkatamaran, L. Single-molecule junctions beyond electronic transport. Nature Nanotechnology. 8, 399- 410 (2013).
3. Aviram, A. & Ratner, M. A. Molecular Rectifiers. Chemical Physics Letters. 29, 277-283 (1974).
4. Ratner, M. A brief history of molecular electronics. Nature Nanotechnology. 8, 378-381 (2013).
5. Takeda, K., et. al. Molecularly Imprinted Tunable Binding Sites Based on Conjugated Prosthetic Groups and Ion-Paired Cofactors. J. Am. Chem. Soc. 131, 8833–8838 (2009).

Acknowledgements

Special thanks to the University of Alabama, the Computer-Based Honors Program, the College of Arts and Sciences, the Department of Chemistry, and the Howard Hughes Medical Institute.
Woski Group Fall 2014: Joseph Meany, Wenzhao Dong, Lynda Truong, Shanna Phillips, Kindle Williams, Julianna Betbeze, Sarah McClees, Brice Morpeth, Falkner Werkhaven

Abstract for AICHE Conference Undergraduates Student Poster Competition

Due to their very small size, nanomaterials can be used for versatile applications. Numerous nanomaterials have found uses in the catalytic, photonic, mechanical, biomedical and electrical industries. They are used every day in the production of commercialized products such as sunscreen, composites, catalysts, electronics and medical devices. However, their small size can also be detrimental since it grants them the ability to cross brain-blood and cellular barriers and thus affect cellular functions. Recent publications have shed light on the mechanisms of cellular responses to exposure to bioactive nanoparticles. For example, silver nanoparticles have widely documented cytotoxic properties that are mediated, in part, by oxidative stress.

The mechanisms by which nanomaterials induce cellular toxicity are complex, as nanomaterials exhibit a wide range of physical and chemical properties. These variable properties are based on their core material and their interactions with extra/intra–cellular proteins which adsorb and desorb from their surface. The functionality and consequently the toxicity of these nanomaterials are affected by factors such size, surface chemistry, cell type and environmental factors such as temperature. Current literature includes research assessing the toxicity of nanomaterials based on size, surface chemistry and cell type.

Our work focuses on characterizing nanoparticle-cell interactions using particles with variable surface chemistry. Silver nanoparticles were coated with a thermo-responsive copolymer of the formula MeO_2MA_x-co-OEGMAy. This thermo-responsive copolymer exhibits a variable Lower Critical Solution temperature (LCST) which is dependent on the composition of the more hydrophobic monomer MeO_2MA or the more hydrophilic monomer OEGMA. The resulting coated Silver nanoparticles exhibited a temperature dependent aggregation with the critical aggregation temperatures increasing with the co-polymer composition of OEGMA. Most notably, polymer coated silver nanoparticles particles with OEGMA composition around 5% aggregated at 37 degrees Celsius.

Using absorbance based cytotoxicity assays, we evaluated the toxicity of free copolymers solutions and silver nanoparticles coated with thermo-responsive polymers containing 80, 90 and 95% of MeO_2MA. The nanoparticle concentration was varied ranging from 0 to 6 nM. Free polymers in solution did not initiate potent cytotoxic responses at concentrations used for nanoparticle coating (< 1 mg/mL). Uncoated silver nanoparticles exhibited a dose and time-dependent cytotoxicity with 100% cytotoxicity attained at a particle concentration of 6 nM under 12 hours. Similarly, the polymer coated particles exhibited a dose-dependent cytotoxicity with a 100% cytotoxicity reached under 6 nM for an OEGMA composition of 5%. The cytotoxicity also increased with decreasing OEGMA composition. We also correlated cellular uptake of polymer coated silver nanoparticles with their cytotoxicity by quantifying partitioning of nanoparticles between the cellular monolayer and the supernatant media as a function of OEGMA composition.

Our future experiments will investigate molecular mechanisms of cytotoxicity, gene regulation and ROS generation.

Sputter Deposition of Metal Oxide Thin Films and their Plasma-assisted Transformation into Metal Dichalcogenides

Trevor Haak, Rachel Morrish, Colin Wolden

Department of Chemical and Biological Engineering

Colorado School of Mines

Abstract

Transition metal dichalcogenides such as WS_2 and MoS_2 are of great interest for applications ranging from electrocatalysts to sensors and semiconductors. Here we describe a two step approach for their synthesis. First, metal oxides were deposited in a reactive RF sputtering process. These films are then removed and transformed to dichalcogenides in a second reactor that employed an inductively coupled plasma source to dissociate H_2S. The application of plasma generates radicals such as atomic H and S which enable the process to proceed at significantly lower temperature. In this poster we describe the synthesis and characterization of both the oxide precursor and chalcogenide product as a function of process parameters.

Due to its two-dimensional geometry, surface chemistry, and high surface area-to-mass ratio, Graphene Oxide (GO) has been found to be highly effective at stabilizing pickering emulsions. This project has looked to investigate the barrier properties of these emulsions. One aspect of this is oxidation of the liquid in the emulsions. In an oil-in-water (o/w) emulsion, oil droplets are vulnerable to oxidation by oxidants in the bulk aqueous phase. GO in emulsion acts as both a physical and chemical barrier to oxidants, and therefore reduces the rate of oxidation of oil in these emulsions. This rate was measured, and compared to various other emulsifying agents, namely chemical surfactants (SDS) and silica nanoparticles, which forms a pickering emulsion like GO. Another property of the emulsion that can be affected by the barrier properties of the emulsifying agent is rate of evaporation. The rate of evaporation of hexane in an o/w GO-stabilized emulsion was measured and compared to rates for SDS and silica nanoparticles. In addition to this, the rate of evaporation was measured to investigate correlation with concentration of salts with different valence, in order to investigate cross-linking properties of GO and their effects on its barrier properties. Ongoing research includes the reduction of GO and the ability of this reduced GO (rGO) to form stable emulsions, particularly emulsions of two miscible liquids.

398934 Direct Fabrication of Mesoporous Graphene from Graphene Oxide By Magnesiothermic Reaction

Abstract:

$CaCO_3$ acts as the foaming agent to prevent the newly formed graphene sheet from agglomerating. The surface area increases from 66 m^2/g to 603 m^2/g with a high yield at the same time.

Stephen Gibbs
University of Florida - Dr. Rinaldi Lab 2/21/14

Synthesis of Manganese Zinc Ferrite Magnetic Nanoparticles for Biomedical Applications

Stephen Gibbs[1], Lorena Maldonado-Camargo[1], Dr. Carlos Rinaldi[1,2]
[1]Department of Chemical Engineering
[2]J. Crayton Pruitt Family Department of Biomedical Engineering

Synthesis and suspension of magnetic nanoparticles to create a fluid that responds to a magnetic field, a so-called "ferrofluid," has opened the door to various biomedical applications. Among these are: targeted drug delivery, magnetic resonance imaging (MRI) contrast agents, and nanoscale energy delivery through heat dissipation, just to name a few. The goal of this project is to synthesize Manganese Zinc Ferrite (MZF) nanoparticles with narrow size distributions and surface coatings that make them suitable for biomedical applications. Of note, we are interested in exploring the dependence of the mechanism of magnetic relaxation on size of the particles, and how this affects performance in biomedical applications. When responding to an alternating magnetic field, a particle can relax by the Brownian Mechanism (physical rotation) or the Néel Mechanism (dipole rotation, no physical rotation). It has been observed that when the size of a magnetic nanoparticle increases above a critical diameter, it will switch from relaxation by the Néel mechanism to relaxation by the Brownian mechanism. Due to its particular crystalline structure, the MZF particle is expected to have a low anisotropy constant, which should increase the critical diameter for this transition in the relaxation mechanism.

Elemental composition analysis by ICP-MS of our samples confirms that we have successfully incorporated Manganese and Zinc into the nanoparticles. Results from TEM show highly monodisperse particles with physical diameters of ~20 nm. Dynamic magnetic susceptibility experiments indicate the particles respond through the Néel relaxation mechanism. In comparison, iron oxide nanoparticles of similar size are found to relax through a combination of Brownian and Néel mechanisms. Forthcoming results of performance characterization for applications in magnetically mediated energy delivery will be also presented.

398367 Modification of the Co-Precipitation Method to Synthesize Iron Oxide Nanoparticles with Higher Specific Absorption Rates

Abstract:

Thanks to advances in nanotechnology, it has become possible to synthesize, characterize, and especially modify the surface of nanoparticles for biomedical applications. Due to their promising properties, several investigations have been carried out in the biomedical field using nano-scaled iron oxide magnetic nanoparticles, usually called ferrofluids, such as magnetite (Fe_3O_4). The properties of iron oxide nanoparticles are remarkable, particularly for their promising role in the biomedical field especially in diseases like cancer. However, the preparation method of these nanoparticles has become a challenge primarily because their properties are strongly influenced by particle size, shape, agglomeration and size distribution. These nanoparticles have demonstrated superparamagnetism characteristics at room temperature, in which is obtained an accurate ratio and pH range when compared to the conventional protocol. The hypothesis of this project is that by optimizing the conventional co-precipitation synthesis protocol, nanoparticles with a smaller hydrodynamic diameter can be produced along with less agglomeration, higher magnetization properties and a higher Specific Absorption Rate (SAR). To this end, an experimental design was developed to investigate the effect of temperature and peptization on the resulting SAR values. Focus was given to temperature at a range of 20°C-30°C and 80°C along with the condition of peptization, incorporating the use of an ultra-sonicator probe. To increase the magnetization properties of the particles sodium hydroxide (NaOH) was added in a scaled down co-precipitation synthesis with two different experimental set ups. Nanoparticle characterization techniques like Dynamic Light Scattering (DLS) and UV-Vis spectroscopy were carried out to measure the hydrodynamic diameter and the iron concentration, respectively. Preliminary results indicated a decrease in the diameter of the nanoparticles at room temperature and a more effective peptization. Lately, by understanding and optimizing the conventional co-precipitation protocol, our findings will improve the heating efficiency of the nanoparticles which will increase their potential for biomedical applications.

398557 Nanoparticle Photoresists: Synthesis and Characterization of Next-Generation Patterning Materials

Abstract:

Hybrid metal oxide nanoparticle photoresists are prominent candidates for next-generation photolithography due to their exceptional sensitivity to extreme ultraviolet (EUV) radiation. To improve the resist's resolution, roughness, and sensitivity (RLS) performance, we explore new compositions for this nanoparticle system. In this study, a series of new nanoparticle resist compositions are synthesized and characterized by solubility and radiation dose tests, as well as deep ultraviolet (DUV) and electron-beam patterning.

Comparative Study of Gold Nanoparticle Structure Modifications Induced by Aggregation Reagents

Elizabeth Stafford, Siyam M. Ansar, Ganganath S. Perera, Dongmao Zhang

Gold nanoparticles (AuNPs) have attracted tremendous interest in recent years for their applications in areas including plasmonics, surface enhanced Raman spectroscopy (SERS), biosensing, drug delivery, and catalysis. AuNP aggregation is implicated in many of these applications. AuNPs have to be aggregated in SERS in order to produce high electromagnetic enhancement for the molecules adsorbed onto the AuNP surfaces. On the other hand, AuNP aggregation is often undesirable in biomedical and catalytic applications. In this presentation, we will compare and contrast the AuNP structure modifications induced by various aggregation reagents that include a series of salts (KF, KCl, KBr, KI, KNO_3, $NaBH_4$, $NaBO_2$). Scanning electron microscopic (SEM) measurements were used to study the structural and morphological variations of the AuNP aggregates over time in the presence of the aggregation reagents. 2-mercaptobenzemidazole, an organothiol, was used as a probe molecule to determine the surface area of the aggregated AuNPs. In addition, the AuNP aggregation processes were studied with SERS, Inductively Coupled Plasma Mass Spectrometry (ICPMS), and UV-Vis spectroscopic techniques. This study offered a series of new insights about mechanisms of the structural evolution of AuNPs induced by the aggregation reagents.

Creating Monodisperse Particles Using Membrane Emulsification Technique for use in Environmental Sensing Applications

Shelby Thies
C. Duval[A], S. Husson[A] and T. DeVol[B]
[A]Department of Chemical and Biomolecular Engineering
[B]Department of Environmental Engineering & Earth Sciences
CLEMSON UNIVERSITY

Introduction

❖This research is part of a larger effort to streamline EPA testing processes by creating an environmental sensor that will detect ultra-low-levels of α- and β-emitting radionuclides in water

❖Sensor comprises of a transparent column filled with resin particles that contain fluorophores and selective ligands. Radionuclides present in the water bind to the ligands, and the release of α- and β-particles cause nearby fluorophores to emit light. Photomultiplier tubes placed around the column detect light and quantify the amount of radionuclides present.

Objectives

❖Produce resin with a uniform size of 200-300 μm to aid in the modeling of this detector as well as to improve the detection efficiency.

❖Use membrane emulsification as a pre-treatment technique to create monodisperse resin.

❖Determine the roles of critical parameters such as membrane pore size and pitch; stir speed; surfactant type; and salt type on particle size distributions.

Experimental Methods

Resin component	Porogen	Bulk material	Crosslinker
Compound	Toluene	Styrene	Divinylbenzene

Table 1. Composition of the monomer phase of the emulsion made using the emulsification unit.

Figure 1. Membrane emulsification setup, with membrane shown beside it.

❖After the droplets are formed, they are transferred to a 500 mL reactor and polymerized in a 10 hour suspension polymerization reaction.

❖It was necessary to use a reliable method to measure and analyze droplets formed from membrane emulsification as well as polymerized resin beads. Pictures were taken using the Digital Celestron Microscope Model 44341, and a program was written in MATLAB to automatically identify particles from these images to return statistics including number of particles, average radius, and standard deviation.

Determination of Optimal Stir Speed

To choose the best membrane and stir speed to obtain droplets in our desired size range, five membranes were tested at various stirring speeds. These membranes were as follows:

❖ Membrane A: Pore size 30 μm, pitch 200 μm, ringed membrane

❖ Membrane B: Pore size 30 μm, pitch 200 μm, standard membrane

❖ Membrane C: Pore size 20 μm, pitch 200 μm, ringed membrane

❖ Membrane D: Pore size 20 μm, pitch 200 μm, standard membrane

❖ Membrane E: Pore size 40 μm, pitch 650 μm, ringed membrane

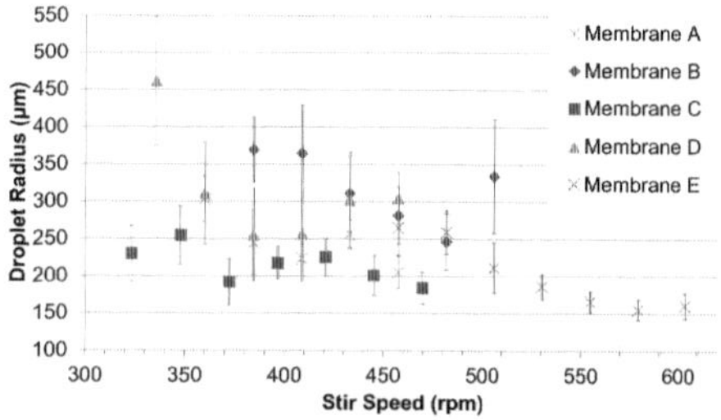

Figure 2. Membrane pore size, pitch, and type have great effects on droplet size and monodispersity.

Determination of Optimal Surfactant

Surfactants play an important role on the monodispersity of the suspension because they stabilize the monomer droplets in order to minimize coalescence or droplet breakup. Polyvinyl alcohol, sodium dodecyl sulfate, and Tween 20 are commonly used surfactants and were tested to determine performance in our particular system.

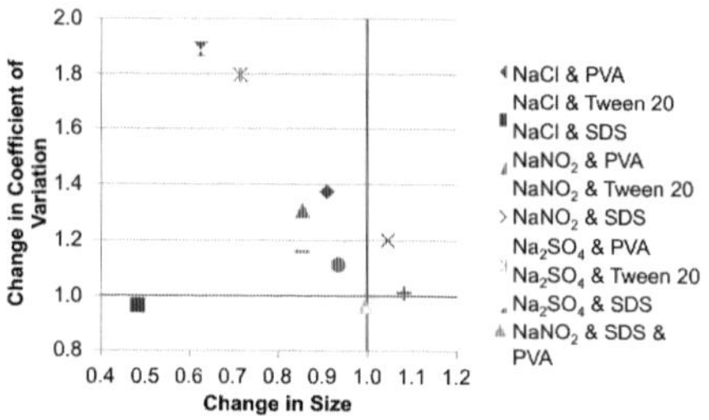

Figure 3. Since size and uniformity are important considerations in the making of our resin, data was compared from droplet samples taken just after formation and droplet samples taken after 30 minutes of stirring. The combination of sodium dodecyl sulfate and polyvinyl alcohol resulted in a nearly equivalent size and coefficient of variation after the stir period was complete.

Determination of Optimal Salt

The use of salt increases the ionic stability of the suspension, which is pivotal in during the suspension polymerization reaction. Salts were selected based on the varying anion position in the Hofmeister Series[1] as follows:

Increasing chaotropic effect ⟶

PO_4^{3-} SO_4^{2-} $CH_3{\bullet}COO^-$ Cl^- Br^- NO_3^- ClO_4^- I^- SCN^-

⟵ Increasing kosmotropic effect

	1 mg/mL Salt Concentration				8.333 mg/mL Salt Concentration			
Anion	SO_4^{2-}	Cl^-	NO_2^-	SCN^-	SO_4^{2-}	Cl^-	NO_2^-	SCN^-
Average Radius (μm)	C	88	30	C	D	45	>500	C
Standard Deviation (μm)	C	30	8	C	D	11	--	C
Coefficient of Variation	C	34	28	C	D	25	--	C

Table 2. In this table, C means that the resin droplets coalesced before the suspension polymerization was complete, and D means the particles disintegrated upon removal for the reactor. Droplets formed with the high concentration of NO_2^- were too large to accurately measure using the Celestron microscope.

Discussion.

❖After testing five membranes at different stir speeds, Membrane A at a stir speed of 425 rpm was selected for future experiments because of its suitable average radius and low standard deviation

❖Polyvinyl alcohol and sodium dodecyl sulfate are good surfactants to stabilize droplets formed in the membrane emulsification process. Using both surfactants together is highly effective and this combination was used in future experiments. Tween 20 is not an appropriate surfactant for our system.

❖Salt type has a complex effect on droplet stability, though it appears that highly chaotropic and highly kosmotropic salts do a poor job at stabilizing resin particles during suspension polymerization.

Conclusions

❖Monodispersity can be increased through use of membrane emulsification, although the relationship between monodispersity and the critical parameters discussed is complex

Future Work

❖ Further testing to determine optimal aqueous phase composition to create stable droplets and avoid droplet coalescence during polymerization

❖Explore the use of photoinitiators as a way to begin polymerization before transferring droplets to the reactor.

References

[1]GE Healthcare. Hydrophobic Interaction and Reversed Phase Chromatography. General Electric Company, 2006

Acknowledgements

This work was supported by the Defense Threat Reduction Agency, Basic Research Award #HDTRA1-12-1-0012, to Clemson University.

398108 Nanoporous Carbons As Gas Sensors: Exploring the Surface Sensitivity

Abstract:

Ammonia sensing properties of a poly(sodium 4-styrene sulfonate) derived carbon and its air oxidized counterpart were studied. The carbons were used to fabricate chips-like devices that were further used for electrical resistance measurements when exposed to various concentrations of dry ammonia. The sulfur-containing carbon after oxidation had a nine times greater response to ammonia than before oxidation. This response was evaluated from the point of view of reactive adsorption and physisorption. The initial reactions of ammonia with the sulfonic groups on the non-oxidized carbon led to an irreversible increase in resistance, while on the oxidized carbon, deprived of sulfonic groups, physisorption caused an opposite effect. After equilibration and reactive adsorption, a reversible decrease in resistance upon exposure to ammonia was found on both samples. The extensive characterization of the carbons' surfaces, and their performance as ammonia adsorbents, suggest that the sensing mechanism is governed by the physical adsorption of ammonia inside the micropores. Higher concentration of ammonia in air results in more pores being filled and thus, in the stronger response of the sensor.

398405 High Temperature Oxygen Sensors Based on Ruddlesden-Popper Type Oxides

Abstract:

There is a desire for oxygen gas sensors that are capable of high temperature operation and possibly in harsh environments of combustion processes. Sensor implementation would allow for combustion processes to be more efficient and more environmentally friendly. As a result, numerous studies have been done on different types of electrodes in an electrochemical sensor. In this study, praseodymium nickelate (Pr_2NiO_4) response to various oxygen environments at 750 °C was investigated. Pr_2NiO_4 was chosen due to its rapid diffusion and surface exchange coefficients of oxygen at elevated temperatures. Pr_2NiO_4 adopts a Ruddlesden-Popper structure and consists of oxygen octahedra between which there are rock-salt type layers allowing oxygen hopping via oxygen interstitial sites. Electrochemical impedance spectroscopy (EIS) was used to determine the transport resistance in an electrochemical cell. The evolution of electrode structure was determined by using X-ray diffraction (XRD). EIS results show that the electrode responds instantaneously to changes in oxygen partial pressure. Oxygen reduction reaction seems favored over oxygen evolution reaction at low potentials. In addition, EIS response was stable over the measurement period. XRD analysis shows the presence of phase transformation to higher ordered structures. Further research needs to be done in order to fully understand electrochemistry at low potentials.

398509 Explosive Thermal Reduction of Graphene Oxide Based Materials

Abstract:

Graphene Oxide (GO) – a single layer carbon atom sheet covered with various oxygen functional groups, is the main precursor for pristine graphene (G) synthesis in application of many commercial products such as transistors, touch screens, solar panels, and conductive inks. It has been observed that under certain conditions dry form multilayer GO could be energetic and may undergo violent decomposition, releasing large amounts of heat comparable to known hazardous chemicals or explosives such as TNT. The chemical decomposition reaction occurring is self-sustaining which allows even a small initial reaction to cause an entire bulk GO to react, making it hazardous. If large quantities of bulk GO were to suddenly react, it would pose a serious safety hazard, due to vast heat release. Therefore, it is imperative to understand the behavior before such an event can ever occur. The effects of common GO synthesis process contaminant ions such as Potassium and Sodium as well as treatment with both acid and base on relevant thermodynamic properties were evaluated. Treatment with base was found to destabilize the GO, shifting the violent decomposition reaction onset temperature by approximately 50C, but it was also found that the overall heat of decomposition decreased. GO treatments with acid and various salts were found to have no effect on thermal characteristics of the final bulk material. Our findings were backed by measurement of surface functionalities which showed changes in the relative abundances of surface functional groups for different treatments.

Band-gap Tuning of Colloidal Silicon-based Quantum Dots by Surface Functionalization using Conjugated Organic Ligands

Rebecca Pinals[1], Ryan T. Anderson[2], Dr. Alan Sellinger[3]
[1]Chemical Engineering Department, Brown University, Providence, RI 02912
[2]Materials Science Department, Colorado School of Mines, Golden, CO 80401
[3]Chemistry and Geochemistry Department, Colorado School of Mines, Golden, CO 80401

Silicon-based quantum dots (SiQDs) have been of recent interest due to their distinctive electronic properties that lead to applications in photovoltaics, fluorescent bio-labeling, organic light emitting diodes, and more. However, most quantum dots are presently based on heavy metal semiconductors that contain toxic and expensive elements such as cadmium, selenium, and lead. The purpose of this project was to connect conjugated organic ligands to SiQDs with vinyl linkages in attempt to improve energy absorption and energy transfer within and between the dots. The surface passivation of these environmentally benign SiQDs leads to significant orbital redistributions that can increase resistance to oxidation, improve solubility, and alter the band-gap.

The method presented here involves a palladium catalyzed cross-coupling reaction to synthesize the ligands followed by a hydrosilylation reaction to form a vinylic connection between the ligands and the SiQDs. This poster focuses on the synthesis and attachment of one ligand, 4-ethynyl-*N*,*N*-bis(4-methoxyphenyl)aniline (MeO-TPA), in comparison with dodecyl-terminated dots (used as a control). The photoluminescence emission spectrum of the new dot-ligand system was red-shifted by 50 nm, indicating energy transfer from the surface species to the SiQD core as incident photons excite electrons from the HOMO (localized on the ligands) to the LUMO (localized on the dot). Therefore, the MeO-TPA capping group effectively tuned the electronic band-gap of the SiQDs. Further, the dot-ligand systems were soluble and resistant to oxidation. This work demonstrated the feasibility of creating Si-C=C linkages to peripheral aromatic groups using mild chemistry.

Seebeck Coefficients and Electrical Conductivities of PEDOT:PSS/Te Nanowire Organic/Inorganic Composites

Cooper Vajner,[1,2] Shilin Ma,[1,2] Kimberly Anderson,[2,3] Greg Szulczewski,[2,3] and Hung-Ta Wang[1,2]

[1]Department of Chemical and Biological Engineering, [2]Center for Materials for Information Technology, [3]Department of Chemistry, The University of Alabama, Tuscaloosa 35487, United States

Thermoelectric generators have drawn much attention due to their capability of harvesting waste heat to generate electricity. A new class of thermoelectrics is the organic-inorganic composite because it is cost effective, and can be manufactured by a facile solution-based process. In this work, a microcircuit was fabricated to characterize the thermoelectric properties of the PEDOT:PSS/tellurium nanowire composite from 10 to 400 K. The Seebeck coefficient and the electrical conductivity were found to increase with increasing temperature through a complex, thermally activated mechanism. It is also shown that the Seebeck coefficient doubled and the electrical conductivity slightly decreased due to an oxidation process of the Te nanowires. This suggests that encapsulations of the composite will be necessary towards practical applications.

Hungta Wang 10/13/2014 12:24 AM
Deleted: for a purpose such as powering space exploration vehicles. An example would be a radioisotope thermoelectric generators in the Voyager 1, Voyager 2, or the Mars rover.

Hungta Wang 10/13/2014 12:25 AM
Deleted: microribbons of

Exploring the Synthesis and Characterization of Nanoscale Multifunctional Surfaces

Erick S. Vasquez, Andres F. Chaparro, and Keisha B. Walters

Dave C. Swalm School of Chemical Engineering

Mississippi State University

Surface modification of material—whether physisorbed or covalently-bound species—can dramatically alter surface chemistries and ultimate performances without requiring alterations to the underlying bulk substrate (Rakhmatullina, 2007). We are interested in grafting polymers, specifically stimuli responsive polymers, to change the surface properties, and also build-in a way to manipulate these properties in response to an external stimulus. In this study, a pH responsive material, poly(methacrylic acid) (PMAA), is grafted onto gold-coated silicon substrates and then combined with amine-functionalized magnetic nanoparticles to create a multifunctional nanocomposite surface. The substrates have potential for applications involving heavy metals removal, corrosion resistance, heat conduction and dissipation, and bacteria inhibition/removal.

Other research studies have examined stimuli responsive polymer brushes and magnetic nanoparticles for removal of heavy metals. Specifically, stimuli-responsive polymer brushes have been used to selectively remove metal ions (Azzaroni, 2012) and improve metal ion sensitivity for sensor applications (Schuwer, 2011). Magnetic nanoparticles have been used previously for the removal of heavy metals due to their high surface areas and, if porous, high pore volumes, uniform pore structures, and structured porous channels (Xin, 2012). By combining these polymer brushes and magnetic nanoparticles, a novel nanocomposite could result allowing for enhanced loading and efficiency in sensing and/or removing heavy metals. Moreover, polymer-magnetic composites can be used for bacterial growth inhibition/removal and cancer treatment via hyperthermia. These multifunctional nanocomposites containing both polymer brushes and magnetic nanoparticles have unique potential for novel properties in a wide range of application areas.

In this study, we explore the synthesis and characterization of PMAA polymer brushes on a planar substrate combined with magnetic nanoparticles to form a unique multifunctional nanocomposite. First, gold (Au)-coated silicon wafers were functionalized with a self-assembled monolayer (SAM) using a thiolated bromine-based initiator, (bis[2-(2-bromoisobutyryloxy)undecyl] disulfide). The bromine-based initiator then provided the site for growth of the PMAA polymer chains surface-confined atom transfer radical polymerization (ATRP). ATRP is a controlled polymerization technique that allows for the formation of well-controlled polymer molecular weight (Matyjaszewski, 2009). In this study, the SAM deposition and ATRP polymerization reactions were performed under different mixing conditions to determine the effect of mixing on the thickness and uniformity of the polymer brushes and incorporated nanoparticles. PMAA was chosen due to the pendant carboxylic acid functionality, since when exposed to different pH values these carboxylic acids protonate/deprotonate producing a change in thickness and a swelling/contracting response in the polymer brush.

Once the PMAA polymer brushes were produced, amine-functionalized magnetic nanoparticles (A-MNPs) were introduced. These A-MNPs have shown promising results for high heavy metal adsorptions due to the strong affinities between nitrogen and the metal ions (Huang, 2009). Modifications take place following a carbodiimide crosslinking reaction between the amine-functionalized group of the magnetic nanoparticle (–NH_2) and the corresponding –COOH groups from poly(methacrylic acid). The –COOH groups were activated by combining N-hydroxysuccinimide (NHS), and 1-ethyl-3-(3-dimethylaminopropyl) carbodiimide hydrochloride (EDAC) in water. Then, A-MNPs were added to the solution containing the Au-PMAA substrate and allowed to react (Zhang, 2008). The AMNPs were expected to bind to the PMAA via the COOH groups present in PMAA through the carbodiimide reaction.

To confirm the synthesis and reaction steps of PMAA and the binding A-MNPs and PMAA polymer brushes, multiple techniques were used at each step of the process. These techniques included dry and in situ ellipsometry to find thickness, Fourier-transform infrared spectroscopy (FTIR) to determine different chemical functional groups, contact angle to measure the wettability, and atomic force microscope (AFM) to determine changes in morphology.

Results show that, for the first time, A-MNPs nanoparticles were combined with a polymer brush tethered to a surface. IR absorbance characterization showed that when the amine-functionalized nanoparticles were added to PMAA by carbodiimide activation, the carbonyl absorption peaks shifted from 1575 and 1700 cm^{-1} to 1760 and 1805 cm^{-1} indicating COOC=NR, and RCOOR' esters bonds present in the nanocomposite structure (Kozlovskaya, 2006). Additionally, AFM images showed morphology changes in the PMAA polymer brush after the nanoparticles were reacted to form the nanocomposite structure. Height sensor and in-phase images confirmed the attachment of nanoparticles to the polymer brushes. Ellipsometry characterization showed an increase in thickness confirming attachment of nanoparticles. Finally, contact angle measurements showed an increase in hydrophobicity for the newly created multifunctional surface resulting from a decrease in the capability of the surface to create hydrogen bonds with water after the elimination of carboxylic acids during the carbodiimide reaction. Details of the synthesis and characterization of novel these pH-responsive, multifunctional nanocomposites will be presented.

References

Azzaroni, O. Polymer brushes here, there, and everywhere: recent advances in their practical applications and emerging opportunities in multiple research fields. *J Polym Sci A*: *Polym Chem*, (2012), 50: 3225-3258

Huang, S.H., Chen, D.H., Rapid removal of heavy metal cations and anions from aqueous solutions by an amino-functionalized magnetic nano-adsorbent. J Hazard Mater, 163 (2009), pp. 174–179

Matyjaszewski, K.; Tsarevsky, N. V., Nanostructured functional materials prepared by atom transfer radical polymerization. Nature Chemistry, (2009), 1 (4), 276-288.

Kozlovskaya, V., Kharlampieva, E., Mansfield, M.L., Sukhishvili S.A., Poly(methacrylic acid) hydrogel films and capsules: Response to pH and ionic strength, and encapsulation of macromolecules, *Chem. Mater*. (2006), *18*, 328-336.

Rakhmatullina, E., Braun, T., Kaufmann, T., Spillmann, H., Malinova, V. and Meier, W. Functionalization of Gold and Silicon Surfaces by Copolymer Brushes Using Surface-Initiated ATRP. Macromol. Chem. Phys., 208: 1283–1293, (2007). doi: 10.1002/macp.200700119

Schuwer, N.B., *Synthesis of Responsive Polymer Brushes for Sensing Applications.* (Doctoral dissertation). École Polytechnique Federale de Lausanne, (2011), Theses No. 5077.

Xin, X., Wei, Q., Yang, J., et al., “Highly efficient removal of heavy metal ions by amine-functionalized mesoporous Fe3O4 nanoparticles,” Chemical Engineering Journal, vol. 184, (2012), pp. 132–140.

Zhang, F., Zhang, Z., Zhu, X., Kang, E.T., and Neoh, K.G., “Silk-functionalized titanium surfaces for enhancing osteoblast functions and reducing bacterial adhesion,” Biomaterials, vol. 29, no. 36, (2008), pp. 4751–4759.

Modulated Solvothermal Synthesis of Large (3 mm) Single Crystals of $Cu_3(BTC)_2$ Metal-Organic Framework

William Nunn, Junjie Zhao, and Gregory N. Parsons
Dept. of Chemical and Biomolecular Engineering, NC State University

Metal-organic frameworks (MOFs) are a new class of crystalline, porous materials that consist of metal units linked by organic ligands. These materials have gained huge attraction due to their ultra-high surfaces areas, large porosity, tunable framework construction, and availability for post-synthetic modification. With these advantages, MOF materials are promising candidates for applications including hydrogen storage, capture of CO_2, separations, and catalysis. However, MOF powders (μm-scale crystals) obtained via conventional methods may not be the desired configuration for some applications, and large MOF single crystals are in demand for the bulk properties yet to explore. Here, we report a new synthesis method to dramatically increase the size of $Cu_3(BTC)_2$ MOF (or HKUST-1) crystals. Taking advantage of the competition effect between acetic acid (monocarboxylic acid) and 1,3,5-benezenetricaboxylic acid (ternary linker), we were able to control the rate of homogeneous nucleation at low temperature (55°C), making heterogeneous nucleation and growth favored. Large $Cu_3(BTC)_2$ crystals (up to 1mm) were successfully obtained within 3 days of synthesis using this approach. In addition, liquid-phase epitaxial growth was achieved on MOF crystal seeds allowing for $Cu_3(BTC)_2$ crystals greater than 3mm in size to be obtained after multiple reaction sequences. Optical, SEM, and TEM images and analysis of single-crystal XRD data will be presented and discussed for these large $Cu_3(BTC)_2$ MOF crystals.

Fabrication of Non-Wetting Copper Surface via Solution Immersion Process

John B. Votaw, Won Tae Choi, Dennis W. Hess

School of Chemical & Biomolecular Engineering, Georgia Institute of Technology, Atlanta

Strongly liquid-repelling materials are of significant interest to many industries and scientific communities for the unique anti-fouling, contamination-resisting, and self-cleaning properties. Generally, metals exhibit a number of favorable properties, such as their malleability, ductility, and electrical conductivity. By creating a liquid-repelling metal surface, a technologically useful material with a combination of these unique properties can be formed. In this work, we discuss a method to control the surface topography of copper by chemical treatment. Control of the copper surface topography was achieved by immersion of copper foil into a solution of hydrochloric acid and hydrogen peroxide. Variation of the immersion time and temperature in the solution allowed further control over the copper foil topography: by increasing either variable, there is an increase in pit density in the copper surface, which was confirmed by scanning electron microscopy. A fluorocarbon film was then plasma-deposited onto the copper foil to modify the surface chemistry. When the appropriate combination of surface chemistry and topography were achieved, a strongly liquid-repelling copper surface was formed. The contact angles of water and hexadecane on these surfaces were as high as 166.2° and 104.3° respectively.

Enhanced Infrared Molecular Sensing via Localized Surface Plasmon Resonances (LSPRs) in Doped-Silicon Nanowires

Emily Tucker

Faculty Advisor: Dr. Michael Filler

Engineering localized surface plasmon resonances (LSPRs) – the collective oscillations of charge carriers induced by electromagnetic radiation – provides an exciting avenue for low-cost enhanced molecular sensing. LSPRs supported in heavily impurity-doped Si nanowires absorb strongly in the infrared region (5-20 μm) and hold particular promise for probing infrared-active modes. We hypothesize that the coupling between a molecular vibrational modes and the strong, local electric fields setup by the LSPR will enable dramatically increased sensitivity (i.e., larger absorption strengths). A uniform and consistent oscillator density is essential to systematically study this effect. By limiting our limiting ourselves to a single monolayer (i.e., SAM), we only need to focus on the density and reproducibility of the surface layer. To this end, we first synthesize self-assembled monolayers of 3-(triethoxysilyl)propionitrile (TESPN) on planar Si substrates to determine monolayer quality with an easily characterizable platform before performing studies on more complex nanowire substrates. This species also contains a nitrile functional group, whose vibrational mode near 2250 cm^{-1} lies within the range of our LSPRs. Samples are characterized via transmission infrared spectroscopy, attenuated total reflectance spectroscopy, and contact angle measurements as a function of adsorption time. Our studies show that TESPN self-assembles on the Si surface, but atmospheric water inhibits pristine monolayer formation. Future work will focus on applying this monolayer to Si nanowire arrays and observing any LSPR-adsorbate coupling.

398699 Highly Conductive Silver Thin Films Synthesized at Room Temperature Using a Continuous Flow Deposition Process

Abstract:

Abstract

Silver has the highest thermal and electrical conductivity of all metals making it useful in various applications. The conductive properties of silver allow it to be useful in the production of electrical devices such as thin film transistors. Silver also has optical properties that make it optimal for use in various sensors. In this project conductive silver thin films were synthesized at room temperature using a continuous flow microreactor deposition process. The process utilized the silver mirror reaction to fabricate the thin films. There was no sintering process after the deposition or stabilizing agent added to the reactants. The thin films were found to have thickness that varied depending on the deposition period from under 100 nm up to 150 nm. Deposition times were collected between 1 and 15 minutes. The highest conductivity was measured at the same order of magnitude as bulk silver.

Nate Gomes (nategomes@my.uri.edu)
AIChE - URI Student Chapter
Abstract for AIChE National Student Conference
Fall 2014

Microstructural Analysis of Colloidal Silver Precipitates

This investigation focuses on the characterization of colloidal silver particles precipitated from an ammonium nitrate solution by neutralization of silver nitrate using ammonium hydroxide. Pellets of silver were reacted with nitric acid to form silver nitrate solution and allowed to cool and dry. Ammonium hydroxide was added to the silver nitrate crystals, forming an ammonium nitrate solution with nanoscopic silver colloids dispersed throughout the liquid phase. These preliminary reaction steps were necessary to produce the desired final product, i.e. a fine silver precipitate having the desired particle size and particle size distribution. These precipitated powders are routinely made in industry for use in several high performance electrical applications. We are investigating methods for fine-tuning the particle size distribution of the silver powders produced by this precipitation process. We believe if we can form uniform size silver colloids, we may be able to use these as seed crystals to grow the desired silver precipitates with the desired particle size and size distribution.

Silver colloids were isolated and imaged for microstructural and chemical analysis using Scanning Electron Microscopy and EDS (Energy-Dispersive analysis of X-rays). Elemental analysis of the colloids indicated they were largely silver and the resulting silver colloids appeared to form a monodisperse with a mean particle size of one micrometer (in diameter). This confirms our hypothesis that the uniformity of the silver colloids was sufficient to act as a template for subsequent growth of the silver particles with the desired particle size and particle size distribution.

398805 Fabrication and Optimization of Deep UV Photonic Crystals

Abstract:

We are investigating different synthetic conditions of highly charged silica nanospheres which we use to create photonic crystal wavelength selecting devices. These photonic crystals consist of a crystalline colloidal array of small (30-45 nm diameter) silica nanospheres which are highly monodisperse in size and charge. The current goal is to optimize the size and charge of the particles in order to improve the diffraction properties of our photonic crystals i.e. strong diffraction with a narrow bandwidth. These particles were functionalized with strongly acidic groups, resulting in a high negative surface charge. This high charge leads to electrostatic repulsion between nanospheres and self-assembly into face centered cubic photonic crystals which Bragg diffract a narrow wavelength region of UV light.

Dispersion of Carbon Nanotubes in a Complex Biological Medium

By Arthur Sloan, Alinne Pereira, Joyanta Goswami, Mark R. Liles, and Virginia A. Davis

The purpose of this research is to examine the ability of tryptic soy broth (TSB), a complex bacterial growth medium, to act as a dispersing agent for single-walled carbon nanotubes (SWNT). There have been notable advances in understanding how specific proteins interact with nanotubes. However, relatively little is known about SWNT dispersion in complex media commonly used in microbiological research. The dispersions examined in this study were prepared by probe ultrasonication of a 0.1 wt% mixture of SWNT in several varieties of TSB at 60W for 30 minutes. These dispersions were subjected to centrifugation at 17,000 x *g* for one hour. Ultraviolet-visible spectroscopy of the resulting supernatants was used to characterize ability of each type of TSB to disperse SWNT. These results suggest that SWNT-tryptophan interactions are a key component of TSB's ability to disperse SWNT, and provide a starting point for additional research to assess SWNT interactions with pathogenic bacteria without interference from the medium.

Thermal Degradation of Lead Sulfide Quantum Dots

Lead sulfide (PbS) quantum dots (QD's) are an inexpensive technology that can be implemented in thermoelectric applications and solar cells. Thermoelectric devices are important as they can recycle some of the waste heat back into electrical energy. Recently, PbS QD's have been shown to be air stable[1], making them even more valuable for such application. However, thermoelectric devices along with solar cells can reach incredibly high temperatures. Up to this point, PbS QD's have not been characterized at such high temperatures. In this experiment we use thermal gravimetric analysis (TGA) and photoluminescence (PL) testing to characterize PbS QD's. We have found from TGA testing that there appears to be two significant mass losses from 100 to 600 degrees Celsius, whereas the PL degrades at 120 degrees Celsius with no recovery of the PL at lower temperatures.

398945 Characterization of Material Properties Using Multiple Wavelength Interferometry

Abstract:

In microgravity, the performance of a constrained vapor bubble heat pipe is limited by the effect of Marangoni forces. This limit can potentially be overcome by using a mixture, rather than a pure fluid, in order to create a concentration gradient which offsets the temperature gradient. However, it is difficult to fully understand this change in concentration along the length of a pipe. In this project, multiple wavelength interferometry was used to examine the thickness and composition of the liquid film in a constrained vapor bubble heat pipe. A new application of multiple wavelength interferometry to determine concentration of a fluid at any point was derived. An algorithm to apply this method and to calculate the liquid film thickness was created.

398966 Hierarchical Assembly of DNA Bio-Templated Nano-Wire and Nano-Tube Films for Lightweight Multi-Functional Materials

Abstract:

DNA hydrogels can serve as 3-dimensional bio-templates for tunable nano-porous materials that serve both as electrochemical power sources and structural material. We present a general approach to 1) form a DNA hydrogel; 2) sensitize the DNA bio-template using catalytic palladium ions to mediate the electroless deposition; 3) synthesize 3-dimensional networks of copper and nickel nano-wire films via electroless deposition; 4) synthesize inorganic salt templated square cross-section nano-tubes that can be integrated into a hierarchal nano-material assembly. Such multi-functional electro-mechanical materials are envisioned to decrease the systems mass across a broad range of Army platforms and serve as a nano-architecture for other applications such as photovoltaics, catalytic systems, sensors, and energy absorption.

398967 Preliminary Characterization of Al-Mg/Al-12B Alloy from Al-Mg Found in Soda Cans

Abstract:

Aluminum beverage cans are typically dispensed at the end of their life cycle, contributing to an accumulation of waste. Al-Mg alloys are part of the cap of this product; they have unique charactristics: lightweight and hard. However, they lack characteristic properties that would make them a viable material for other applications. By forming a composite consisting of an Al-Mg alloy as matrix and AlB12 particles as reinforcement the materials properties can be optimized and their usable life, extended. This study seeks to characterize the Al-Mg/AlB12 composite mechanical properties by bulk indentatio and microindentation tests in order to determine the effect of chemical composition in the materials hardness. Understanding these properties would help optimize costs and fabrication processes in industries where aluminum alloys are vastly used. Some of industries with interest in alternative use of recycled aluminum can alloys are automotive, aerospace, gas and other transportation industries. The understanding of this material can yield cheaper raw material from where the manufacturers can derive their products while maintaining the same level of reliability on their products. Apart from reducing operation costs, recycling those alloys would help reduce their environmental impact

398924 3D Printing Using Extrusion of Micro-Particles Dispensed in a Liquid

Abstract:

3D printing is a multipurpose additive manufacturing (AM) technique that allows for the creation of free standing structures that can be fabricated solely from an appropriate building material. Conventional manufacturing methods either require a mold or subtractive manufacturing (SM) which severely limits the efficiency by providing excess material to make the desired structure. In turn, this makes for a bulky product as not all of the support structure can be removed. For example, copper is currently etched onto the surface of circuit boards, but if it were 3D printed, only the exact amount of copper or conductive material would be needed to make the board. Also, if there is no desire for the board, a three dimensional design can be used creating a 'floating' circuit as the medium required to start the structure, but not necessarily required to continue the structure. In all, 3D printing produces out-of-plane structures with compact and manageable options for support. We have been developing new materials for 3D printing based on the extrusion of pastes composed of micro-scale particles dispensed in a liquid.

The Studies of Short-chain Phosphatidylcholine Effect on the Spontaneous Lipid Transfer in Phospholipid-based Vesicles using Differential Scanning Calorimetry

Kamil Charubin [a], Yan Xia [a], Mu-Ping Nieh [a, b, c]

[a] *Department of Chemical and Biomolecular Engineering, University of Connecticut, Storrs, CT 06269, USA*

[b] *Polymer Program, Institute of Material Science, University of Connecticut, CT 06269, USA*

[c] *Department of Biomedical Engineering, University of Connecticut, CT 06269, USA*

Phospholipids are the building blocks of biological membranes, and thus have drawn much attention in their structural and dynamic studies. Due to their amphiphilic characteristics, phospholipids can self-assemble into a variety of bilayered structures, including multi-lamellar vesicles (MLVs), nanodiscs, lamellae, and unilamellar vesicles (ULVs, also known as liposomes), which were of particular interest in this study. Since liposomes resemble the structure of cells and biological vesicles, they are commonly used to understand the interplay between biochemical and biophysical properties of membrane-associated molecules (e.g., proteins) and their biological functions. Moreover, liposomes have been used as carriers for therapeutic agents due to their ability to encapsulate both hydrophilic and hydrophobic molecules. Recently, it was found that liposomes can form spontaneously in mixtures of long-chain and short-chain lipids, paving a new avenue for large-scale manufacturing of uniform liposomes. However, it is unclear how the presence of short-chain lipids affects the dynamics of liposomal membrane, which is of great importance in gaining a better understanding of the relative stability of lipid bilayers. One of the dynamic properties of lipid assemblies is the rate at which individual lipid molecules transfer between the bilayers through the aqueous phase, a process known as the spontaneous lipid transfer.

In this study, the effect of short-chain lipid 1,2-dihexanoyl-*sn*-glycero-3-phosphocholine (DHPC) on the rate of lipid transfer in liposomes was examined using differential scanning calorimetry (DSC). Liposomes were prepared using either long-chain 1,2-dimyristoyl-*sn*-glycero-3-phosphocholine (*h*-DMPC) or its deuterated counterpart 1,2-dimyristoyl-d54-*sn*-glycero-3-phosphocholine (*d*-DMPC), at various DHPC compositions. Since DSC can detect the phase transition temperature, T_M and *d*-DMPC exhibits lower T_M than *h*-DMPC, the DSC thermogram of the mixture of *h*-liposomes (*h*-DMPC/DHPC) and *d*-liposomes (*d*-DMPC/DHPC) yields two transition peaks. Over time, as lipid molecules transfer between liposomes, both peaks move toward each other and the temperature difference between them decreases. Hence, the rate of lipid transfer can be determined by measuring the rate at which the temperature difference, ΔT_M decreases as a function of the incubation time. The lipid transfer rates of individual lipids were determined in gel phase and liquid-crystalline phase, respectively. The results show that in the gel phase, the rate of lipid transfer increased as the amount of short-chain lipid increased, at a constant temperature. On the other hand, in the liquid-crystalline phase, the rate of lipid transfer was found to be faster, and independent of the short-chain lipid amount. Presumably, the DHPC molecules induce the 'melting' of DMPC molecules adjacent to them, and facilitate the dissociation of DMPC molecules from the bilayer. However, at temperatures higher than the T_M of DMPC, all lipids are in a liquid-crystalline state, thus the introduction of DHPC will not affect the lipid transfer rate. Furthermore, the enthalpies of phase transition, ΔH_M (from gel to liquid-crystalline) at different lipid compositions were determined and the results indicate that increased DHPC content reduces the required ΔH_M, consistent with the notion that DHPC induces the 'melting' of gel-phase DMPC.

398240 Underground Water Pipeline Corrosion

Abstract:

An underground 44" pipe carry fresh water and surrounded by sand and 3m underground faces corrosion issues; leakages occur at 2011 and 2014. In 2011, CP system install (Cathodic Protection) after that in 2014, the problem return but much worst. A team assemble to investigate these issues. Utilizing some old cases and data provided by our plant, we found the root of the problem is the installation of the CP after the leakage happened which affect the electrochemistry behavior of the pipeline metal.

High-energy and high-power electrical energy storage (EES) devices are in need of constant improvement due to the rapid expanse of technology in today's society. Electrical Double Layer Capacitors (EDLCs) are one type of high-power EES devices that store energy as charge at the electrode-electrolyte interface. This research focuses on developing an EDLC based on solid-state gel electrolytes by optimizing an ionic liquid electrolyte in Succinonitrile matrix, which expands the applicability of conventional liquid EDLCs.

In order to do this symmetric cells with Activated Carbon (AC) electrodes were fabricated with the following three ionic liquid electrolytes (soaked in glassfiber separator); Ionic Liquid (IL) 1-butyl-3-methylimidazolium tetrafluoroborate ($BMI\text{-}BF_4$), IL to Propylene Carbonate (PC) mixture in a 9:1 (v/v) ratio, and IL : Propylene Carbonate (PC) mixture in an 8:2 (v/v) ratio. These previous cells comprised the liquid electrolyte that was optimized. The gel polymer electrolytes were comprised of symmetric cells fabricated via drop-casting gels over AC electrodes. Gel electrolytes were prepared using Poly(vinylidene fluoride co-hexafluoropropylene) (PVdF-HFP)/ Succinonitrile (SN) matrix with $BMI\text{-}BF_4$ IL as ionic component in a 1:4 ratio of PVdF-HFP to SN/IL mixtures where the SN and IL ratios were; SN:IL, 3:1, SN:IL, 2:2, SN:IL, 1:3 and SN:IL, 0:4.

The above cells were all characterized using Electrochemical Impedance Spectroscopy (EIS) to 1 mHz, Cyclic Voltammetry(CV) from 10mV/s to 200 mV/s and Charge Discharge (CD) from .5mA/cm^2 to 10 mA/cm^2. Using the data from these tests specific capacitance values were calculated for each cell in order to determine the best ratio of IL:PC and SN:IL respectively. The gel cells were additionally characterized with long cycle charge discharge testing at 4 mA/cm^2 for over 3000 cycles. Lastly energy and power calculations were done for the optimized liquid electrolyte, gel electrolyte and control electrolyte and graphed on Ragone Plots.

Based on the experiments it was found that the addition of PC reduced the viscosity and hence the ESR of Ionic Liquid $BMI\text{-}BF_4$ and also increased the specific capacitance and rate performance of the electrochemical liquid electrolyte cells. The addition of SN to IL also improved the performance of the solid-state supercapacitor, with the gel electrolyte based on the combination of SN: IL (2:2; w/w) showing the best performance in terms of specific capacitance and long cycle life. The last conclusion of important is that the gel electrolyte outperforms the comparison with the liquid electrolyte due to lower ESR, higher specific capacitance (and hence higher specific power and energy) and longer cycle life.

Effect of Methanol on Intra- and Intermolecular Bonds of Ionomer Solutions

Abstract:

The effect of methanol on intra- and intermolecular bonds in sulfonated polystyrene (PS) ionomer, carboxylated PS ionomer, and poly(methyl methacrylate)(PMMA) ionomer solutions were studied through viscosity experiments. Ionomers were dissolved in a mixture of benzene and methanol and viscosity measurements were conducted at different concentrations. At low concentration, the intramolecular bonds between ionomer chains resulted in a decrease in solution viscosity, while at high concentration the intermolecular bonds between different chains increased solution viscosity. Methanol concentration in solvent was also varied to study the change in viscosity as a function of alcohol concentration. At low ionomer concentration, solution viscosity increased as methanol content increased, indicating a higher degree of breakage of ionic bonds by methanol. The effects of ionic content and type of ionomer on the viscosity behavior in the mixed solvent were also studied.

CENTER FOR CHEMICAL SENSORS DEVELOPMENT

CHEMICAL IMAGING CENTER / ALERT-DHS-NEU-UPRM-COE

Characterization and Identification of α and β-RDX Polymorphs on Stainless Steel Substrates

Amanda Figueroa-Navedo, José L. Ruiz-Caballero and Samuel P. Hernández-Rivera*

amanda.figueroa@upr.edu, jose.ruiz9@upr.edu, samuel.hernandez3@upr.edu*

Abstract

The highly energetic material (HEM) studied was the cyclic nitramine hexahydro-1,3,5-trinitro-s-triazine, which is also known as RDX or cyclonite. RDX has two stable polymorphs stable at room temperature (known as the α- and β- phases of RDX) which were generated by a spin-coating methodology on stainless steel substrates. These spin-coated substrates differ in their spectra recorded per increasing angular velocity during each deposition. β-RDX has been reported to be difficult to obtain at room temperature since α-RDX is the most stable form. Reflection-absorption infrared spectroscopy measurements for these coated surfaces were able to distinguish these conformational polymorphs in a large surface area of the substrates. Raman spectroscopy was employed to image a small area within three different crystal structures reported in literature of RDX: islands, dendrites and prisms. By employing principal component analysis, spectral findings were validated both for Infrared and Raman spectra of these RDX polymorphic structures. Our results from these spectra provide good correlation with calculated values reported in literature. Finally, characterization of these structures by atomic force microscopy is provided due to a marked intensity variation in Raman spectra for all samples. Consequently, it is important to optimize the conditions required to generate only β-RDX films at all angular velocities on these substrates.

Keywords: Energetic materials polymorphism, spin coating methodologies, reflection-absorption infrared spectroscopy, Raman spectroscopy, atomic force microscopy

398331 Dynamic Rheological Properties of Field Responsive Nanomaterials for Application in Transtibial Prosthesis

Abstract:

STFs (shear thickening fluids) are non-Newtonian fluids that display increases in viscosity at high rates of shear. This unique behavior of STFs can be utilized in various types of protective gear applications. Ultimately, the goal of this project is to design and implement an STF device into a transtibial prosthetic. Those who wear transtibial prosthesis are prone to degenerative joint disease due to elevated loading in the intact extremity of the leg. An STF damper located in the ankle of the prosthetic will absorb gait stresses and protect the intact extremity.

With this goal in mind, the current work seeks to understand the mechanism behind shear thickening and shear thinning at the microstructure level, characterize the STF through steady state and dynamic rheology, and to investigate the dynamic response of the STF to time-scale dependent deformations. These objectives were accomplished through conducting steady state sweeps, amplitude sweeps, frequency sweeps, and large amplitude oscillatory shear experiments to characterize the sample. The experiments probed the STF response to both constant and dynamic deformations, and allowed for a greater understanding of the material as well as an evaluation of STF effectiveness in a transtibial prosthetic.

The experimental results provided further evidence for the hydrocluster mechanism of shear thickening as well as confirming that the mechanism was universal across all compositions. Furthermore, the STF was shown to increase in both energy dissipation and elasticity with an increase in oscillatory strain. This unique property motivates for further research into the damper capabilities of STFs.

Fluorescence nitro-explosive detection through electrospun Pyrene-PES nanofiber

George Shaw, Xiangcheng Sun, Yu Lei

The electrospinning process was used to synthesize fluorescent nanofibers. The solution composition, concentration and solvents were experimentally determined through the optimization of the morphology of the nanofiber film. The film was then used to detect various amounts of common explosives such as picric acid (PA), 2,4,6-trinitrotoluene(TNT), 2,4-dinitrotoluene (DNT) and 3,5-trinitroperhydro-1,3,5-triazine (RDX) in aqueous phase. The non-linearity of the Stern-Volmer (SV) plot for the pyrene eximer exciting can be attributed to photo-electron transfer and energy transfer for piric acid. The SV plot for TNT, DNT, and RDX display linearity in the concentration range. Its application for the detection of nitro-explosives by naked eye was also demonstrated through fluorescence color changes. In addition, the explosive detection in vapor phase was also studied. The pyrene-PES electrospun film shows high sensitivity and selectivity, low cost and great reusability. These characteristics make explosive detection through fluorescent pyrene-PES nanofiber a very promising method.

There is a huge demand for small-diameter vascular grafts, as the majority of vascular disease cases involve small-caliber blood vessels. Recently, electrospinning has gained attention as a valuable technique for the fabrication of scaffolds for blood vessel engineering, as electrospinning produces nanofibers that closely approximate the structure of native extracellular matrix (ECM).

Accordingly, electrospun scaffolds were fabricated in a 3D tubular structure from a blend of the synthetic polymers, viscoelastic and durable polycaprolactone (PCL) and relatively fast-degrading shape memory poliglecaprone (PGC). The scaffolds were coated with a physiological biomatrix, HuBiogel^{TM}. The biohybrid graft was found to exhibit mechanical properties comparable to those of native blood vessels, and the HuBiogel^{TM} coating imparted bioactivity. The coating was crosslinked using EDC and the natural crosslinker genipin (Gp) to improve its stability in physiological conditions. This study evaluated the effects of EDC- and Gp-crosslinking on the scaffold mechanical, structural, and morphological properties. Additionally, coating stability was studied to assure the presence of collagenous biomatrix on the scaffold for effective cell-matrix interactions.

Mechanical testing showed little difference between EDC- and Gp-crosslinked scaffolds; both retained mechanical properties in the range of native human arteries (tensile strength 1-2 MPa, tensile modulus 9-12 MPa). SEM imaging revealed that while crosslinking with EDC resulted in an increase in fiber diameter compared to uncrosslinked scaffolds, Gp-crosslinking did not affect fiber diameter; the majority of fibers in EDC and Gp-crosslinked scaffolds had diameters ranging from approximately 0.9 – 1.4 microns and 0.7 – 1.2 microns, respectively. This is in the upper range of fiber diameters in native extracellular matrix. Coating stability studies using picrosirius red (PSR) stain showed that EDC-crosslinked scaffolds were more effective than Gp-crosslinked scaffolds in enhancing the stability of the biomatrix coating.

RDX is a major component of plastic bonded explosives. It is widely used in military applications. The most common polymorphs of RDX are α- and β-RDX, where α-RDX is the most stable at room temperature. In β RDX, all NO_2 groups are believed to be oriented in axial positions. In α RDX, one NO_2 group is located in equatorial position and two NO_2 groups are in axial positions with respect to the triazine ring. This difference in the position of the nitro groups grants the RDX crystals different orientations in space. At first the RDX deposited on stainless steel plaques was considered to adopt filmlike properties, but since the RDX solutions are not homogeneous, they will be regarded as deposits. This lack of homogeneity can be explained through the thermodynamic processes that cause RDX to nucleate, or precipitate in ever larger conglomerations as time passes. As opposed to what is generally thought, angular velocities are inversely proportional to thickness, which is related to the concentration of RDX in the solution being deposited. For this purpose, acetone proved to be the best solvent, due to its high evaporation rate. Among the objectives of the study is to find regions where we can identify which structures are favored and correlate them with the concentration of these deposits on stainless steel surfaces. Concentration will be determined by High Performance Liquid Chromatography (HPLC). The angular velocities examined were varied from 3000 to 7000 revolutions per minute (rpm), with increments of 1000 rpms. Calibration curves from concentration data will be built to predict the thickness of these deposits.

Danielle Lizarazo

October 27,2014

Abstract: AICHE conference 2014

Effect of Solvent Induced Dipole in ALD of Al_2O_3 on HOPG

Authors: Danielle Lizarazo[1]-Dlizarazo@ufl.edu, Greg Mordi[2]-gregstar155@gmail.com, Lanxia Cheng[2]-lanxia.cheng@utdallas.edu, Jiyoung Kim[3]-jiyoung.kim@utdallas.edu

The motivation of this project was that graphite was found to be a very interesting 2D material, but its inert surfaces do not let it chemically react well due to its hydrophobic nature which mostly causes reactivity only on the step edges. Devices made from graphene have many issues. One of these issues is that dielectric deposition techniques tend to cause damage to graphene's surface. An ideal technique would react with graphene without damaging its properties and deposit uniformly across the graphene surface. Previous studies have shown that TMA-O_3 shows better Atomic Layer Deposition (ALD) coverage than TMA-H_2O because of the high chemical reactivity of ozone in comparison to water. This reactivity causes the HOPG to be more hydrophilic which is more favored when performing ALD (B.Lee et al.,Appl.Phys. Lett.92,203102 (2008).) In this work we attempt to study the effect of a large variety of solvent treatments with different dipole moments on HOPG's surface and the solvents corresponding role in ALD coverage.

We began by soaking freshly cleaved HOPG (Highly Ordered Pyrolytic Graphite) in the solvents for 2 hours. The non-polar solvents included Hexane, Toluene, and

Chloroform. For the polar-aprotic solvents we used Acetone, DMSO, and Propylene Carbonate. Lastly, for the polar-protic solvents we used H_2O, and Acetic Acid. After they were soaked we performed Atomic Layer Deposition of Al_2O_3 on the treated HOPG. Lastly, an Atomic Force Microscope was used to see the amount of coverage on the HOPG. The results showed that polar-aprotic solvents (high dipole moments) and non-polar solvents (low dipole moments) did not have much uniform coverage. However, the polar protic solvents (around the dipole moment of water) showed much better, uniform coverage. After we tested the HOPG for uniformity, we exfoliated the HOPG again and tested the contact angle on the surface of the graphite. The results showed that the contact angles did not show any form of pattern depending on the dipole.

We concluded that since higher dipole moments do not lead to more uniform coverage, differences in dipole moments are not the reason why ALD on HOPG works better with some solvents as opposed to others. Our next thought was that maybe the reason why the polar protic solvents had better coverage may have dealt with the larger regions of O-H densities. These O-H densities could possibly react with each other or the dangling bonds on graphene.

Fabrication of Amphiphobic Paper with Improved Mechanical Strength

Georgia Tech | Chemical & Biomolecular Engineering

Hanyang Li, Zhenguan Tang, Lu Jiang, Victor Breedveld and Dennis W. Hess

School of Chemical & Biomolecular Engineering, Georgia Institute of Technology

Motivation

Cellulose-based paper products

- Inexpensive
- Globally available
- Environmentally friendly

Amphiphobic paper attracts extensive interest

- Self-cleaning surface material
- Paper-based microfluidic devices
- Impermeable textiles

Introduction

Requirements for amphiphobic surfaces

- Low surface energy
- Nano-scale roughness
- Reentrant angle structures[4]

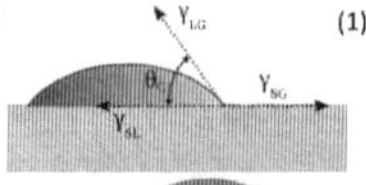

(1)

(2)

Low surface tension liquids wet surfaces more easily

- Water (γ_{sl} = 72.5 mN/m)
- Hexadecane (γ_{sl} = 27.5 mN/m)

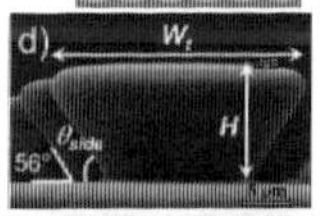

(3)

Prior work

Superamphiphobic paper (oil contact angle >150°)

(6)

- **Low surface energy coating**
  - Fluoropolymer deposition
- **Nano-scale roughness**
  - Oxygen plasma etching
- **Re-entrant angle structures**
  - Organic solvent rather than water for paper formation: increase spacing between fibers[5]
  - Refining: reduce the fiber diameters

The optimum ratio L/D ~17[2]

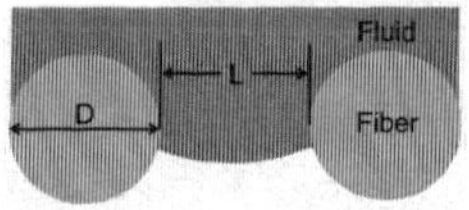

Paper displays very low mechanical strength due to large spacing between fibers

Paper Formation Procedure

Pulp → (7500 Revolutions) → Refined pulp → (Organic Solvent Exchange) → Paper → (Plasma Etching) → Etched Paper → (Fluoropolymer deposition) → Amphiphobic Paper

Objectives

Improve mechanical strength by using more polar solvents than sec-butanol

→ More amphiphobic

	Water	Methanol	Ethanol	Sec-butanol
Polarity	1.0	0.8	0.6	0.5

Higher strength ←

Results: Morphology

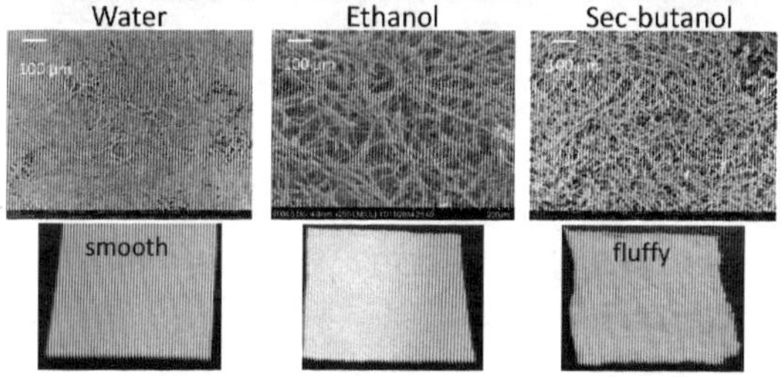

Ethanol paper has similar morphology as sec-butanol paper, but with a more smooth surface

Results: Wetting behavior

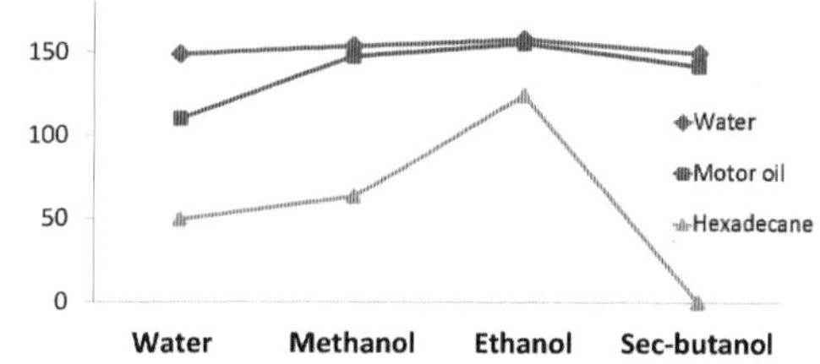

Ethanol -paper displays water and motor oil CA> 150° and hexadecane CA > 120° with improved mechanical strength

Results: Etching Time

Etching time is varied to determine the optimum conditions for fabricating amphiphobic ethanol paper

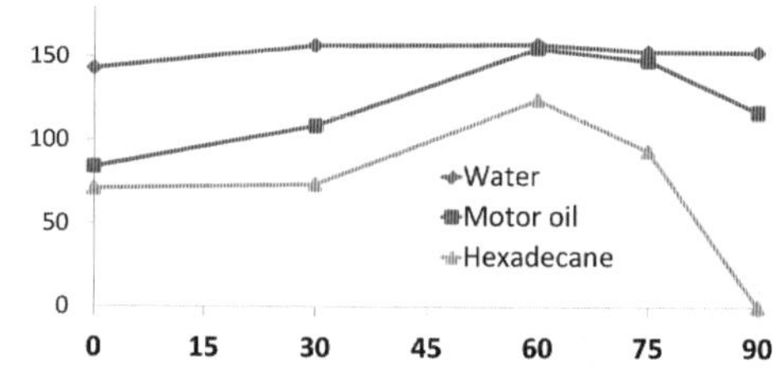

Ethanol paper etched for 60 min displays highest oil CA

Summary

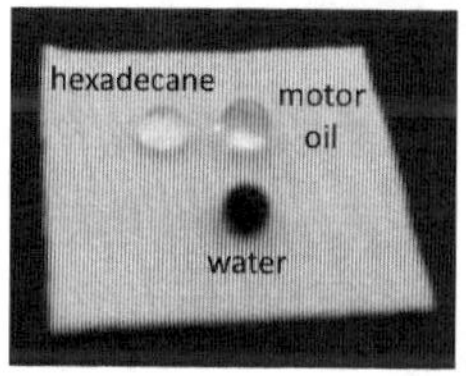

Amphiphobic property

Water CA	157.5°
Motor oil CA	155.0°
Hexadecane CA	124.0°

Improved mechanical strength based on results from simple tearing test

Future Work

- Quantify the mechanical strength of paper made with different solvents
- Determine the optimum plasma etching time for other organic solvents

References

1. rame-hart intrument co.
2. http://www.cchem.berkeley.edu/amsgrp/wettability.html
3. M. Im, H. Im, J. H. Lee, J. B. Yoon, Y. K. Choi, Soft Matter 2010, 6, 1401..
4. A. Tuteja, W. Choi, M. L. Ma, J. M. Mabry, S. A. Mazzella, G. C. Rutledge, G. H. McKinley, R. E. Cohen, Science 2007, 318, 1618
5. Mantanis, G.I.; Young, R.A.;Rowell, R.M. Swelling of Compressed Cellulose FIBER WEBS IN ORGANIC LIQUIDS. Cellulose *1995, 2, (1), 1-22*
6. Li, L.; Breedveld, V.;Hess, D.W. Design and Fabrication of Superamphiphobic Paper Surfaces. ACS Appl. Mater. Interfaces *2013, 5, (11), 5381-5386.*

398515 Synthesis of Tunable Spongy PVDF Membranes for Water Treatment Applications

Abstract:

Functionalized membranes are gaining increased use in many applications and often it is desired for such membranes to have a maximum degree of functionalization. In this study, flat sheet poly(vinylidene fluoride) (PVDF) membranes are synthesized with a thick cross section and spongy interior morphology by an immersion precipitation phase inversion technique. This results in a greatly increased membrane interior surface area allowing for a high degree of membrane functionalization. Using dimethylformamide (DMF) as solvent and an elevated membrane casting temperature (35-40°C) the equilibrium relationship between polymer and solvent is able to be controlled to create a thick and spongy membrane. Membrane morphology was examined using scanning electron microscopy (SEM) where a thick membrane top layer and open substructure were found. Membranes were functionalized in this study using poly(acrylic acid) (PAA) and it was found a higher degree of PAA functionalization was possible in spongy membranes over non-spongy membranes. The membrane flux can be modulated by pH variation. Spongy membranes have also shown highly enhanced divalent calcium ion pick up over non-spongy membranes. The iron nanoparticle functionalization capacity of these membranes is also being studied and it is expected that a high degree of membrane functionalization will be possible. Through this study, it has been shown that the synthesis of spongy PVDF membranes allows for increased functionalization with PAA allowing for tunable performance in any number of applications. This project has been supported by NIH-NIEHS-SRP, and by NSF EPSCOR program. Technical support of ULTURA Corporation is highly appreciated.

Validation of High Throughput Electrochemial Gas Sensing Screening System

Zixuan Wang
Arizona State University

Advisor: Dr. Jason Hattrick-Simpers, Dr. Jochen Lauterbach, Ben Ruiz-Yi

The development of solid-state, high temperature gas sensors for detecting combustion byproducts is essential to improve energy efficiency, reduce toxic emissions, and meet new environmental regulations. However, gas sensors for such applications must be capable of exhibiting high sensitivity and selectivity to analyte gases at high temperatures. At those temperatures, materials tend to inter-diffuse between layers and react with each other or the environment, negatively impacting sensor performance. Most current state-of-the-art gas sensors involve the doping of a base oxide with aliovalent or isovalent metals, here subatomic differences in composition substantially affect the ultimate sensor figure of merit. Optimizing sensor performance by testing the effect of changes to the dopant concentrations via one-sample-at-a-time methodology is a time consuming effort. The application of the high-throughput methodologies to probe the composition-microstructure-device performance phase space can expedite the search for materials with enhanced figure of merit. This project focuses on developing and validating new high-throughput sensor synthesis and characterization techniques. The approach uses a high-temperature electrochemical screening cell capable of simultaneously quantifying the sensitivity and selectivity of up to 30 individual gas sensing materials arranged in a planar geometry. To validate the electrochemical system, a uniform standard sample was synthesized and tested under NO gas flow at 600 °C. Preliminary electrochemical tests were performed on the standard sample while alternating the dosing of N_2 and different concentrations of diluted NO. The potential, measured using a multimeter, changed reproducibly during NO dosing, providing initial validation of the use of a planar sensor design for screening.

Colleen Lawlor
Chemical Engineering Student at Rochester Institute of Technology

Effects of Chemical Dopants on the Electrical Conductivity of Carbon Nanotubes for Novel use in Wire Applications

Carbon nanotubes (CNTs) are being investigated as an alternative to conventional metallic conductors. Individual CNTs have measured conductivities around $1x10^8$ S/m, but conductivities this high in bulk materials have not yet been achieved. Of the numerous strategies employed to increase the bulk conductivity of CNTs, chemical doping is a prominent pathway. The present work focuses on doping efficacy and dopant stability as part of making multifunctional wires that are both electrically conductive and mechanically strong. In addition, an investigation of conductivity as a function of purification was performed to understand how material processing impacts the conductivity. CNT sheet material was doped with a series of compounds of varying structure and functionality, including metal halides, nitrates, hydroxides, acids, halogen gases, and ionic liquids. Of the chemical dopants investigated, potassium tetrabromoaurate salt, gold (III) chloride, and iodine gas showed the greatest increases in electrical conductivity, with iodine gas reaching $2.98x10^5$ S/m, 8 times higher than the average purified sheet value. Metal halides, bases, and ionic liquids all showed limited doping results on the CNT sheet material. A qualitative analysis of the extent of dopant plating was performed, and it was found that the CNT samples immersed in the gold halide and alkali gold halide solutions showed the highest extent of dopant plating, while most other dopants did not have a visible change after doping. Densified CNT wires were fabricated using each of the top three dopants and tested for both electrical conductivity and mechanical strength. The wire doped with potassium tetrabromoaurate was able to withstand the most stress, while a control sample densified with deionized water was able to withstand the least amount of stress. Potassium tetrabromoaurate and iodine gas were selected for additional environmental experiments because they exhibited the highest potential for multifunctionality. Doped sheets were soaked in deionized water and aqueous sodium chloride and changes in electrical conductivity of each sample were measured before and after soaking. The iodine doped samples had an initial decrease in conductivity after a 15 minute soak, after which conductivity remained stable over a 5 week soaking period. The general trend for the potassium tetrabromoaurate doped samples in both deionized water and saltwater was a continuous decrease in electrical conductivity. These experiments served as a preliminary assessment of CNT wire stability for conventional cable applications, and demonstrate the viability of this developing technology. The present work indicates that potassium tetrabromoaurate and iodine gas dopants have emerged as the best for the fabrication of multifunctional wires.

Landon Tyler

Abstract for Research Project

The Effects of GXP Annealing on a PCBM/P3HT Bulk Heterojunction Layer's Surface Roughness

Organic photovoltaics (OPV) are promising devices which convert solar energy into electrical energy, fit for applications which require flexibility or light weight, and feature ease of processing. Process parameters which can affect in device efficiency include (1) layer thickness (2) layer uniformity and (3) proper dispersion of the nanoparticle in the polymer. One way to measure layer uniformity is surface roughness. The photoactive layer, called a bulk heterojunction (BHJ), consists of a photoactive polymer (P3HT) electron donor and a nanoparticle electron acceptor (PCBM). In order to achieve proper dispersion/morphology the BHJ layer must be annealed. At Tennessee Technological a new type of annealing termed GXP (gas expanded polymer anneal) using high pressure CO_2 has been developed. Herein we discuss how this GXP process affects the surface roughness of the flexible substrate which the thin (~100 nm) BHJ layer is applied to. Flexible substrates studied include a polyimide. Spin coating was used to apply the BHJ, followed by annealing. Surface roughness was quantified using atomic force microscopy (AFM). Early results indicate some contribution to surface roughness from the annealing process.

Small molecule diffusion through membranes for use in flow battery and fuel cell applications

Mary McBride[1], Samuel St. John[1], Gabriel Goenaga[1], Thomas A. Zawodzinski, Jr.[1,2]

1. Chemical and Biomolecular Engineering, University of Tennessee, Knoxville, TN, USA.
2. Materials Science and Technology Division, Oak Ridge National Laboratory, Oak Ridge, TN, USA.

Membranes are used in both flow batteries and fuel cells to create a medium for ion conduction. We are developing a class of membranes to use as a polyelectrolyte membrane separators in fuel cells and flow batteries with increased hydroxyl anion conductivity, relative to proton conductivity in Nafion membranes, and improved stability in alkaline electrolytes at elevated temperatures. In addition to hydroxyl anion conductivity, we are interested in the diffusion of small molecules as a proxy for the membranes' barrier properties for potential application to direct alcohol fuel cells. We use conventional fuel-cell-stack hardware to measure the diffusion coefficients of methanol, ethanol, isopropanol, and formate ion through commercial and experimental membranes. Various 2M solutions of small molecules were dissolved in 0.1M potassium hydroxide and pumped through one side of a fuel cell. On the other side, 0.1M KOH was circulated as a blank. Pumping time varied between 1 and 6 hours depending on the anticipated barrier property of the membrane being investigated. The concentration of the small molecule that diffused through the membrane was found using cyclic voltammetry. Diffusion coefficients were then calculated by applying Fick's law of diffusion for one-dimensional flux. Diffusion coefficients found for commercial Nafion membrane are comparable to literature values (0.105×10^{-6}, 0.0534×10^{-6}, 0.0475×10^{-6}, and 0 cm^2/s for methanol, ethanol, isopropanol, and formate ion respectively), validating our experimental approach. Diffusion coefficients for the various commercial and experimental membranes varied significantly, but were all higher than those for Nafion. Values for diffusion coefficients are higher for experimental membranes. Investigating these values gave us information on how to improve the barrier properties of experimental membranes for use in direct alcohol fuel cells.

A Synthetic Route for High-Yield Phospho-olivine Cathode Material for Use in Lithium-Ion Batteries through Liquid-state Salt Medium

Darren W. Kwee[1,2], Taehoon Lim[1,2], Eric Gonzalez[1,2], and Alfredo Martinez-Morales[1,2]

[1]Winston Chung Global Energy Center, 1084 Columbia Ave., Riverside, CA 92507
[2]College of Engineering - Center for Environmental Research and Technology, 1084 Columbia Ave., Riverside, CA 92507
University of California Riverside, 900 University Ave., Riverside, CA 92521

Abstract

$LiFePO_4$ (LFP) particles were synthesized via an ionothermal route (through use of 1-ethyl-3-methylimidazolium trifluoromethanesulfonate ionic liquid—'[EMI][TfS]'—a liquid salt solvent). Characterization of time-dependent phases has led us to propose a possible scheme for particle formation.

Due to the capability for synthesis at atmospheric pressure, open-vessel synthesis (alternative to considered bomb-type synthesis reactions[1,2]), low temperature, little monitoring requirement, environmental benignity, high yield, and overall simplicity in synthesis, this method could be suitable for large-scale commercial implementation.

The synthesized material was characterized by scanning electron microscopy (SEM), energy dispersive spectroscopy (EDS), and x-ray diffraction (XRD). Reported is an analysis on particle morphology and crystallinity of LFP particles synthesized through our ionothermal process.

Introduction

There has been high interest in lithium-ion batteries for energy storage due its wide variety of application; they hold high promise in both small devices—such as in cell phones and laptops–and larger ones—such as those in electric vehicles (EV), hybrid electric vehicles (HEV), and battery energy storage systems (BESS) for grid use. Their suitable energy density makes capable efficient light-weight energy storage systems[3]. Within lithium-ion batteries, cathode material remains intensively studied due to its strong influence in li-ion battery energy density, rate, and its direct correlation to overall cost.

Since its introduction in 1997 by Goodenough and group[4], $LiFePO_4$ (LFP) has attracted much attention due to its enhanced electrochemical performance when used as a Li-insertion/desertion compound for lithium-ion batteries. LFP had been extensively studied due to its relative environmental benignity (when compared to Co, Ni, and Mn—transition metals used in competitive alternative cathode material) and its structural stability.

This cathode material carries charges through the following reactions[4], involving the extraction of lithium ions on charging:

$LiFePO_4 - xLi^+ - xe^- \rightarrow xFePO4 + (1 - x)LiFePO_4$

Where the cathodic LFP material acts as a Li^+ donor, transferring Li^+ ions (during charge) to the anode to stabilize the oxidation of iron(II) to iron(III). Li^+ ions, alternatively, are transferred to the cathode to stabilize reduction of iron(III) to iron(II) in the cathode material during discharge:

$FePO_4 + xLi^+ + xe^- \rightarrow xLiFePO_4 + (1 - x)FePO_4$

In this approach, an ionothermal process -- involving the use of an ionic liquid solvent-- had been explored. Ionic liquids hold high promise as functional solvents due to their attractive properties as liquid-state salts[5]. Use of ionic liquids allow for a wide solvent operating temperature, further addressing environmental issues associated with solvent vaporization, a significant issue in scalable solvent-based reactions. Due to the relatively high boiling points of salts, ionic liquid solvents have potential for re-use in future reactions, further decreasing costs. Originally explored for this material by Recham and group[1], it has been discovered that ionic liquids maintain the ability to act as structural directing agents in the formation of crystalline LFP, further controlling directional growth.

Experimental

$LiFePO_4$ cathodic active material was prepared through an ionothermal approach using 1-ethyl-3-methylimidazolium trifluoromethanesulfonate ([EMI][TfS]) liquid-state salt, as described in Figure 1. Iron oxalate dihydrate ($FeC_2O_4{\cdot}2H_2O$) and lithium phosphate (Li_3PO_4) had been used as iron and lithium/phosphate precursors, respectively. Precursors had been suspended stoichiometrically in [EMI][TfS] and thoroughly mixed. Precursor / [EMI][TfS] system was then heated at constant temperature of 260°C for reaction times at 12, 18, and 24 hours. $LiFePO_4$ material had been synthesized under continuous nitrogen flow in efforts to prevent deterioration of ionic liquid[6] and to prevent the formation of potential impurities formed with oxidation of Fe^{2+} to Fe^{3+} through heating. $LiFePO_4$ product had been purified through extensive centrifugal/sonication treatment.

PANalytica Empyrean had been used to identify phases and product crystallinity through X-ray Diffraction. Particle morphology and composition had been observed through Scanning Electron Microscope (SEM) and Energy-dispersive X-ray Spectroscopy (EDS) using Nova NanoSEM450.

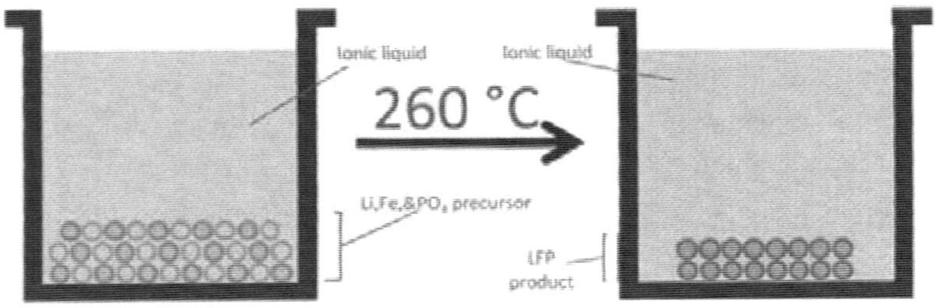

Figure 1: Schematic for used ionothermal synthesis reaction

Results and Discussion

Synthesis of LFP

Figure 2 shows the XRD patterns of synthesized LFP material at (a) 12 hours, (b) 18 hours, and (c) 24 hours from the start of reaction. All XRD graphs below had been indexed by orthorhomic olivine structure type, characteristic of LFP. Observing the transition between product at 16 hours of reaction and 18 hours—and more so in the transition toward 24 hours— we can see distinct differences in crystallinity of product.

Although these results still indicate the presence of other phases—precursor and intermediate—we have observed increased reaction time contributing to further improvement of LFP product crystallinity for this particular ionothermal approach. Further LFP crystallization with increased reaction time, decreases intermediate phases with further reaction progression. This leaves potential for a nearly complete yield.

Evident in Figure 3 is the presence of embedded 1-dimensional (1D) structures. Analysis of these structures has led us to observe these to be attributed to a multitude of intermediates in the formation of LFP material. It can be seen in Figure 4 that regions along these 1D structures differ in composition, though portions of these 1D structures (Figure 4b) are well-consistent with particle-like structures (as in Figure 4c).

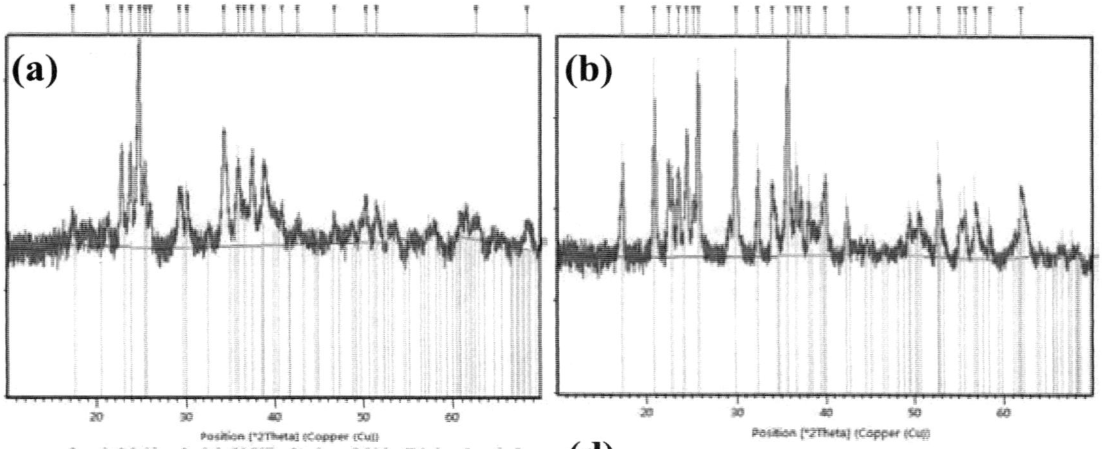

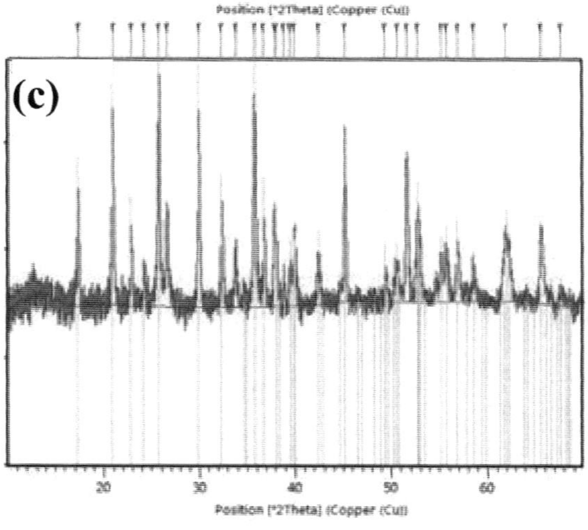

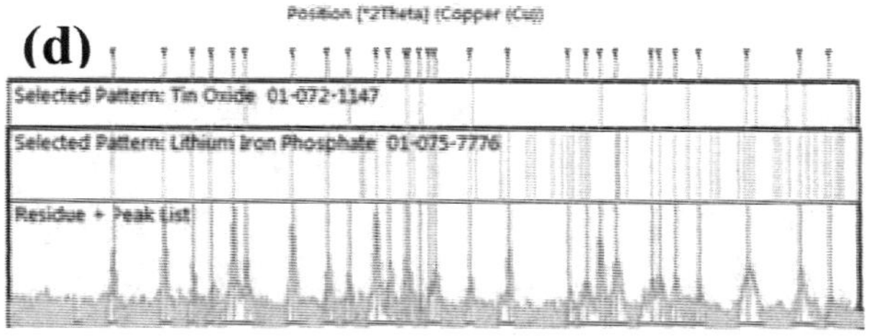

Figure 2: XRD data for LFP synthesis at (a) 12 hours, (b) 18 hours, and (c,d) 24 hours. (d) more detailed crystalline peak index; indexing with tin oxide (substrate) and LFP structures.
It is observed changes in crystallinity based upon time for reaction.

Figure 3: synthesized LFP morphology at (a) 12 hours, (b) 18 hours, and (c) 24 hours. There can be seen a presence of 1D structures in clumps of particles at the 12hr and 18hr synthesis. With increased reaction time, there can be seen more uniformity in particles.

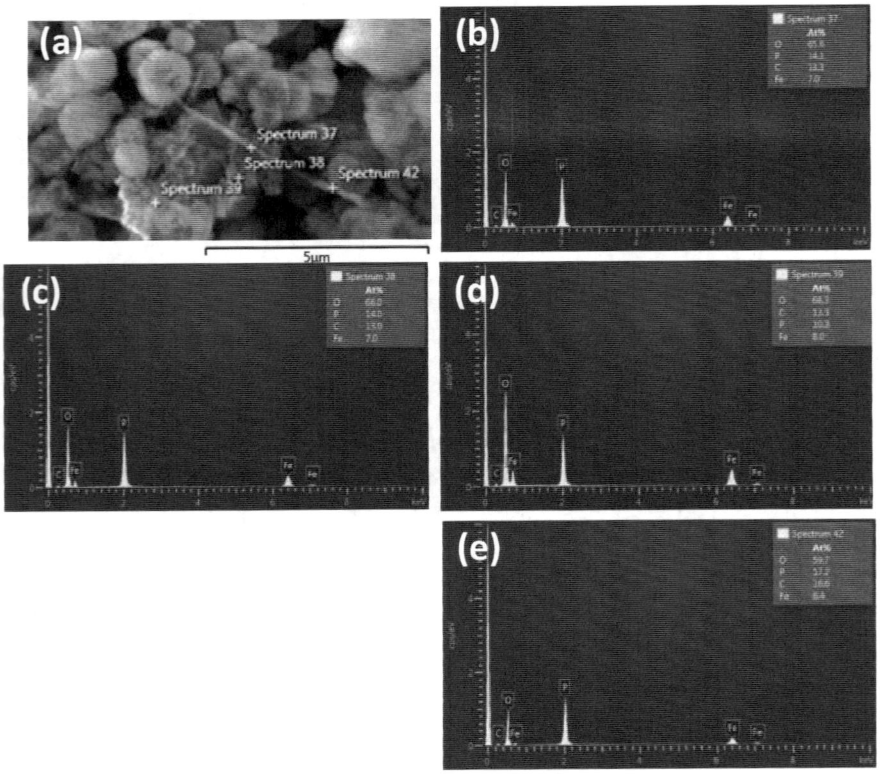

Figure 4 (left): (a) Analysis of 1D structure and surrounding composition. (b) Spectrum 37—located on 1D structure—indicates composition consistency with (c) spectrum 38—more developed particle-like structure—with a 1:2 Fe:P. (d) Spectrum 39 appears to have even closer Fe:P ratios and may be a intermediate closer to desired LFP product. (e) Spectrum 42 has considerably differing Fe:P ratios than (b), indicating differences in composition between regions of 1D structure.

SEM and EDS analysis further led us to propose growth of LFP material in our system as seen in Figure 5. Li undetected by analysis due to its light weight. The 'cluster' observed in Figure 5a can be seen to be composed of similar 1D-like structures. The consistency in At% of this 'cluster' with our Fe precursor leads this to be a very promising candidate for precursor. We've considered the observed 1D structures to have originated from the Fe precursor 'chunk', as visualized in the schematic in Figure 6.

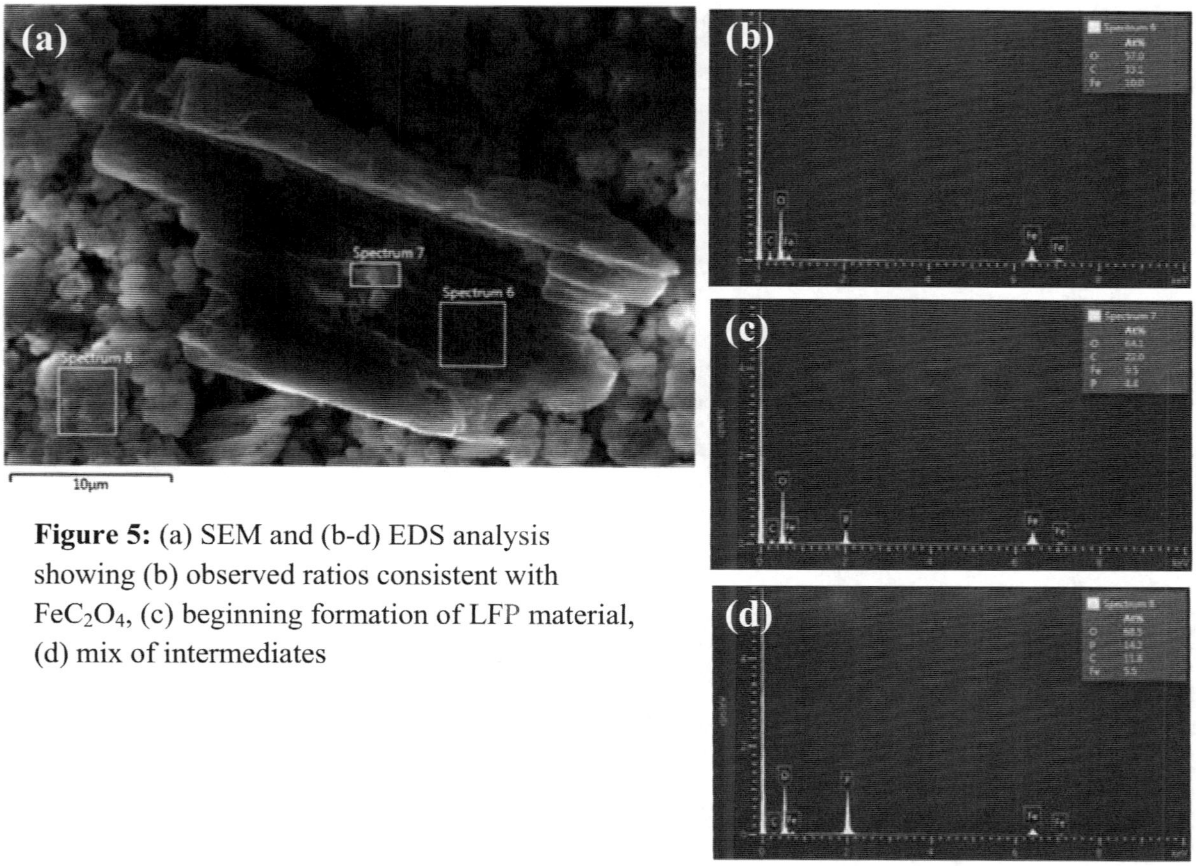

Figure 5: (a) SEM and (b-d) EDS analysis showing (b) observed ratios consistent with FeC_2O_4, (c) beginning formation of LFP material, (d) mix of intermediates

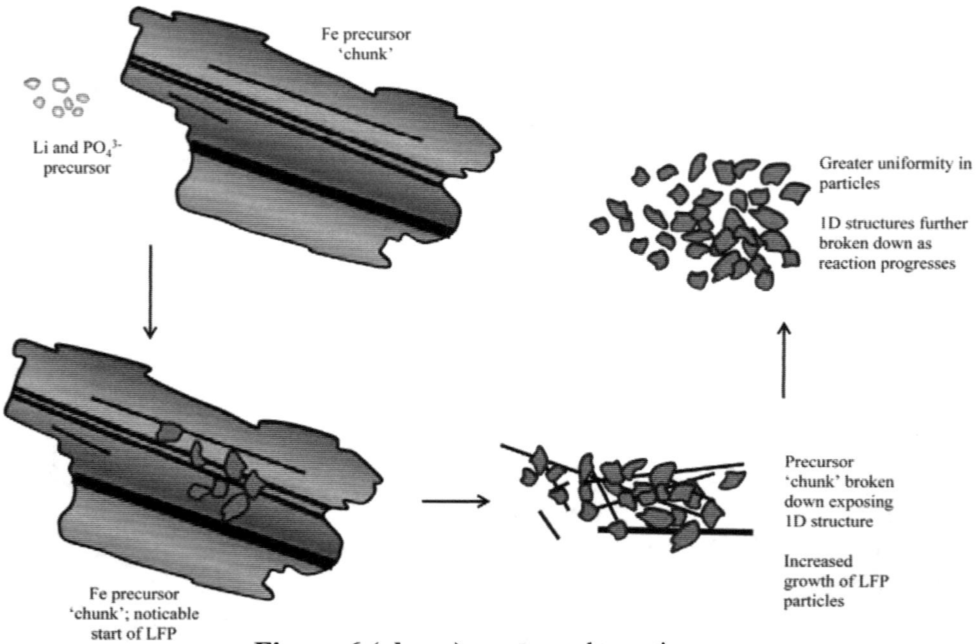

Figure 6 (above): proposed reaction scheme from precursor for this process

In this synthesis it is also important to note the development of an ionic liquid 'film' on synthesized product, as seen in figure 5. Without sufficient purification, such an ionic liquid 'film' develops a covering of much of the synthesized LFP particles. This film may negatively affect battery performance of active LFP material as it may cover much of the surface area of active LFP regions, blocking Li^+ effusion/diffusion sites.

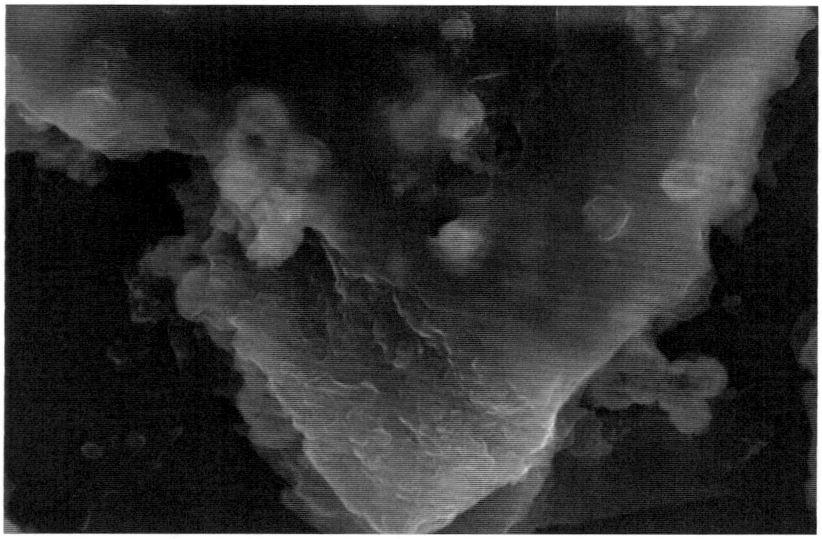

Figure 5 (left): Presence of ionic liquid 'film' covering synthesized LFP particles.

Conclusion

In this work, [EMI][TfS] ionic liquid was used as solvent to produce LFP particles. It was shown that increasing reaction time directly affects particle crystallinity; we believe that, with optimized reaction time, single-phased crystalline LFP material can be achieved through our synthesis approach. With shorter reaction time, one-dimensional structures were observed; this led us to propose a schematic for a possible formation of these 1D structures and, eventually, uniform particles.

Acknowledgements

This work was partially funded by Winston Battery Ltd. and the American Public Power Association (APPA) Demonstration of Energy & Efficiency Developments (DEED) Program. SEM characterization was conducted at the Central Facility for Advanced Microscopy and Microanalysis (CFAMM) at UC Riverside. We would like to thank Mr. Fred Lyn and the Rancho Cucamonga Municipal Utility for their support.

References

1. Recham N. *et al.* (2009), "Ionothermal Synthesis of Tailor-Made $LiFePO_4$ Powders for Li-Ion Battery Applications," *Chem Mater.*, 21, pp. 1096-1107
2. Recham N. *et al.* (2009), "Ionothermal Synthesis of Sodium-Based Fluorophostphate Cathode Materials," *J. Electrochem. Soc.,* 152, pp. A993-A999
3. Kodama T., Sakaebe H. (1999), *Journal of Power Sources,* 81-82, pp. 144-149.
4. Padhi A. K., Nanjundaswamy K. S., and Goodenough J. B. (1997), "Phospho-olivines as Positive-Electrode Materials for Rechargeable Lithium Batteries," *J. Electrochem. Soc.*, 144, pp. 1188-1194.
5. Mallakpour S., Dinari M. (2012), "Ionic Liquids as Green Solvents: Progress and Prospects," *Green Solvents II: Properties and Applications of Ionic Liquids*, pp. 1-32.
6. M. Kosmulski, J. Gutafsson, J. B. Rosenholm, (2004), "Thermal stability of low temperature ionic liquids revisited" *Thermochimica Acta,* 412, pp. 47-53.

The number of tendon transfer surgeries performed in the U.S. has been steadily increasing over the past few years. The Robotic and Human Control Systems Lab focuses on engineering a device that can be implanted into the human forearm to improve post-operative hand mobility in tendon transfer surgery patients. Typical tendon transfer surgery makes use of suturing techniques, which results in restricted hand motion. To implant this device, biocompatible materials must be selected or a body response would inhibit the new range of motion. Many other biomedical devices have been implanted in to the human body. By examining these devices, one can learn how other groups have dealt with the foreign body response and other body mechanisms that degrade implants.

A literature survey was conducted to determine the most biocompatible materials and coatings used in a forearm implant that reroutes the ECRL tendon through a pulley mechanism to improve postoperative grip following an upper arm tendon transfer surgery.

Articles published in English peer-reviewed journals between 1960 and 2014 looking at implants were found using the Oregon State University 1Search search engine without exclusion criteria. Scholarly journals and books were also loaned from the Valley Library on campus. Many implants were analyzed, including stents, total joint replacements, breast implants, and tendon implants. The review process involved finding articles that discussed the body's responses to implant materials and summarizing the biocompatibility of materials. These summaries were then compiled. Article references were used to search for more sources. The device consists of a pulley, a main body, screws to attach the device to bone, and a bag filled with hyaluronic acid surrounding the device upon implantation. Biocompatible materials and an anti-adhesion coating were selected.

UHMWPE with Vitamin E should be used for the device's pulley due to its mechanical strength and inertness in the human body. The main body of the device should also use UHMWPE with Vitamin E. This polymer has disadvantages. Fine wear particles released from this polymer due to oxidative damage[1] and wear cause adverse biological effects.[2] Vitamin E when added to UHMWPE reduces oxidative corrosion.[2] Fine wear particle release is a problem in total hip and knee replacements using UHMWPE.[3] Wear is less problematic in the forearm due to smaller force production than in the knee region.

Titanium should be used for the screws attaching the device to bone. The alloy Ti-6Al-4V should be used. Titanium is corrosion resistant and strong enough to withstand the forces occurring in the forearm. Titanium ions are nontoxic.[2] Titanium can passively dissolve but the presence of wear causes it to corrode more readily.[4] Titanium is not used when hardness or resistance to wear is the primary concern. Wear is not a concern for a forearm implant.[1] Titanium allergy has been noted but is uncommon.[1] For a Ti-6Al-4V alloy, Titanium is the only metal ion that has a significant serum concentration. Aluminum and Vanadium do not have significant serum

[1] Biomaterials science: an introduction to materials in medicine
[2] Principles of Human Joint Replacement by Buechel, Frederic F., and Pappas, Michael J
[3] Effects of UHMWPE Wear Debris Generated from Total Hip and Total Knee Replacements
Hulbert, S.F. and S. J. Megremis.
[4] Local accumulation of titanium released from a titanium implant in the absence of wear by Bianco, P.D. et. al. Vol 31. Issue 2. 1996. Page 227, 232, 233.

concentrations. Vanadium concentrations in serum and urine were uniformly low.[5] Vanadium is readily excreted while Titanium is not. Aluminum is cleared by the kidney ordinarily.[6] A study on the paraspinal muscles of rabbits suggested that Ti-6Al-4V is an appropriate metal for implants. [7] Ti-6Al-4V screws were used in an intervertebral disk replacement.[1] Cobalt-chromium alloys are possible alternatives. Co-Cr-Mo is the strongest, hardest, and most fatigue resistant of the orthopedic materials available. There is concern over use of nickel in the alloy due to its toxicity. Metal allergy is usually caused by nickel, cobalt, and chromium. Other studies support toxicity of cobalt.[8] Most cobalt ions are excreted but less than half of chromium ions are excreted.[9] Cobalt-chromium alloys should therefore not be used.[2]

Kevlar (aramid) cabling should be used. A study on the paravertebral muscle tissue in rabbits indicated that a low degree of irritation was noticed compared to PVC. Kevlar fibers have similar biocompatibility to silicone. [10] Stainless steel is also ductile enough to be cabling but has low corrosion resistance which is problematic in a physiological environment. Stainless steel is recommended for surgical instruments that do not enter the body.

The bag surrounding the device should be silicone due to the strong siloxane backbone. Silicones have been successfully used in breast implants. These implants have faced a gel bleeding problem.[11] Hyaluronic acid will likely be bled. This molecule induces tendon healing and is not damaging so bleeding is not problematic.[12]

Poly(ethylene oxide) should be used as a coating because it can inhibit protein adsorption and bacterial adhesion. [13] PEO was selected due to its common use. Other non-fouling surface coatings include hyaluronic acid, casein, albumin, mucin, and hydrogels in general.[1] Silver ions are primarily used due to their antibacterial properties but are disadvantaged by loading and release kinetics.[14]

---

[5] Metal Release and Excretion from Cementless Titanium Alloy Total Knee Replacements by Jacobs, Joshua J. et. al (1999). Pg 175-176.

[6] Toxicology of Metals: Biochemical aspects. Editors: Robert A. Goyer and M. George Cherian. Chapter 7: Biochemical mechanisms of aluminum toxicity by Jeffery, E.H. Page 139.

[7] Tissue reaction in rabbit muscle exposed to metallic implants (1967). Volume 1, Issue 1. Page 136-138,142, 148.

[8] Tissue Response to Implants of Pure Cobalt by McNamara, A. Engineering in medicine. Volume 10, Issue 1.

[9] Storage and elimination of titanium, aluminum, and vanadium salts by Merritt, K. et al (1992). Volume 26, Issue 11. Pg 1512-1514.

[10] Tissue biocompatibility of Kevlar aramid fibers and polymethylmethacrylate, composites in rabbits by Henderson, James D., Richard H. Mullarky, and Doreen E. Ryan (1987). Journal of Biomedical Materials Research. Volume 21. Issue 1. Page 59, 60, 62, 64.

[11] Biomedical Application of Commercial Polymers and Novel Polyisobutylene-based Thermoplastic

[12] Tendon Healing: Repair and Regeneration by Voleti, Pramod B. et. al

[13] Bioinspired Implant Materials Befuddle Bacteria
By Bryers, James D. and Ratner, Buddy D.

[14] A mechanistic study of the antibacterial effect of silver ions on Escherichia coli and Staphylococcus aureus by Feng, Q.L. et. al (2000).

Design, Synthesis, and Characterization of an Aqueous Alkaline Developable Molecular Resist for EUV Lithography

Ashten L. Fralick[1], Brandon L. Sharp[1], Richard A. Lawson[2], Hannah L. Narcross[1], Laren M. Tolbert[1], Clifford L. Henderson[1,2]

1) School of Chemistry and Biochemistry, Georgia Institute of Technology, Atlanta, GA 30332
2) School of Chemical & Biomolecular Engineering, Georgia Institute of Technology, Atlanta, GA 30332

Over the past 50 years, advancements in semiconductor fabrication technology, particularly in the scaling of device feature size, has enabled the production of microprocessors with a high density of circuit components, resulting in incredible processing power. Moore's Law suggests that the density of components will continue to double every two years, and a modern integrated circuit device contains features that are on the order of 30 nm in size (by comparison, a human hair is 50,000 nm and an *E.coli* bacterium is 2,000 nm). Features of this size are possible due to advancements in the lithographic technologies used to the print the circuit features during semiconductor device fabrication, however current methods have been unable to continue the trend observed in Moore's Law. Current lithography methods involve coating a substrate with a photosensitive material, and specified areas of the resist are exposed to radiation using a mask. The radiation exposure induces a solubility change in the material, and the differences in solubility between the exposed and unexposed regions of the resist allows for selective removal of portions of the resist in a developer to form a relief image. Following exposure, development in solvent, and drying, the substrate is etched using this relief image mask to form the desired device features. In order to maintain Moore's Law, lithography must overcome several challenges for producing sub-30nm features including: (1) development of resists that function at next-generation exposure wavelengths (i.e. EUV at ~13nm), (2) collapse of nanometer scale patterns due to capillary force during development and drying, and (3) decay in mechanical properties of molecular and polymeric resist materials as a function of dimension due to confinement and high surface-to-volume effects. In order to combat these challenges, this work focuses on the design, synthesis, and characterization of a negative tone resist based on a photo-induced cross-linking material. This particular resist was designed to include an environmentally friendly processing method as well as the ability to produce mechanically robust nanometer scale features. In order to maintain safe processing methods, the resist was designed so that aqueous alkaline developers could be used to develop the material in replacement of organic solvents. The resist is comprised of an aromatic core that is functionalized with epoxide and a phenolic groups. The phenolic groups render the resist soluble in an aqueous alkaline developer. The epoxide groups act as a site for cross-linking, and upon initiation with a photoacid, undergo robust polymerization. The epoxides in the exposed resist also react with the phenolic groups, greatly reducing the functional groups that contribute to the aqueous solubility of the exposed region. This polarity change, combined with the increased molecular weight resulting from polymerization of the resist, result in a large imaging contrast between exposed and unexposed regions of the resist. This presentation will review the design methods, the synthesis techniques, and the basic characterization and lithographic evaluation of this new resist.

398781 Composition Effects on the Optical and Mechanical Properties of Cellulose Nanocrystal Films

Abstract:

The objective of this research is to compare the optical and mechanical properties of cellulose nanocrystal (CNC) films with various compositions. CNC produced by the acid hydrolysis of natural cellulosic materials are increasingly being recognized for their potential in advanced materials applications. Sulfuric acid hydrolyzed CNC (CNC-SA) from wood pulp and acetic acid hydrolyzed CNC (CNC-AA) from cotton are the two types of CNC being used in this work. Aqueous dispersions of CNC can be sheared and dried to produce a solid film. However, CNC-SA films tend to be brittle. It was hypothesized that the addition of hydroxypropyl cellulose (HPC) or CNC-AA to CNC-SA dispersions could enable films with a broader range of mechanical properties. Aqueous dispersions containing a total of 12-13 wt% cellulosic material were prepared with ratios of 8:1, 9:1, and 10:1 CNC-SA to CNC-AA or HPC. Films were produced by doctor blade coating and characterized using cross-polarized optical microscopy and nanoindentation.

Full Spectrum Phosphor Based LEDs

Jannatul Ferdaous, City College of New York

Mentor: Partha Dutta, Rensselaer Polytechnic Institute

Abstract

Replicating sunlight for indoor lighting has been one of the many goals of researchers and companies since centuries. The goal of this research is to determine the ratio of a mixture of three to four different colors of phosphors, produced from rare Earth materials, which closely replicates the sunlight or produces a light which has a full spectrum of the visible range when excited by a blue UV light. After a ratio of phosphor powders was determined to give a high CRI as the sunlight, a multi-phase phosphor powder was to be produced from the rare Earth materials, consisting of the same ratio which was determined to best replicate the sunlight. The purpose is to produce a single phosphor powder which alone can replicate the sunlight instead of having to mix several phosphor powders to attain a high CRI. Several ratios of phosphors have been determined which emit light of CRI above 90. However, reaction of the Earth materials to produce one multi-phase powder of that ratio has not been successfully produced yet.

398886 Construction Board Made from Coconut Fiber

Abstract:

Coconut fiber (extracted from coconut husk) and coconut pith can be made into a biodegradable fiberboard when combined with an eco-friendly binder. Coconut fiber is inexpensive, abundant in many tropical countries, and has good mechanical properties. This makes it very promising for use in developing countries due to the large demand for fiberboard and makes its manufacture both feasible and profitable. By varying the amount of coconut fiber, the length of the fiber, the amount of coconut pith, and the percentage of binder, a fiberboard was created that passed the ANSI Standards for use as indoor furniture and construction board.

Enhancement of stability and superhydrophilicity of plasma-modified microfluidic materials

Yale SCHOOL OF ENGINEERING & APPLIED SCIENCE

UPMC SORBONNE UNIVERSITÉS

École nationale supérieure de chimie de Paris

Lea Winter[1,2], Bradley Da Silva[1], Guillaume Schelcher[1], Cedric Guyon[1], Daniel Bonn[3], and Michael Tatoulian[1]

(1)Institut de Recherche de Chimie Paris, **Chimie ParisTech/CNRD, Paris, France**, (2)Department of Chemical and Environmental Engineering, **Yale University**, New Haven, CT, (3)Institute de Physics, **University of Amsterdam**, Amsterdam, Netherlands

OBJECTIVE: Identify polymers and deposition method for enduring hydrophilic coating of microfluidic materials[1,2]

MOTIVATION

Conventional microfluidic materials exhibit **hydrophobic recovery** after treatment with hydrophilic coating.[3,4,5,6,7]

Hydrophilic, patterned[8] microfluidics necessary for:

- Biological research (on-chip PCR, cell culture)
- Chemical research (catalysis[9], kinetics)
- Physics research (fluid mechanics, droplets)
- On-chip Diagnostics

RESULTS

AGING OF POLYMER COATINGS

Water contact angle measured wettability over a two month period following deposition of the silica-like layer.

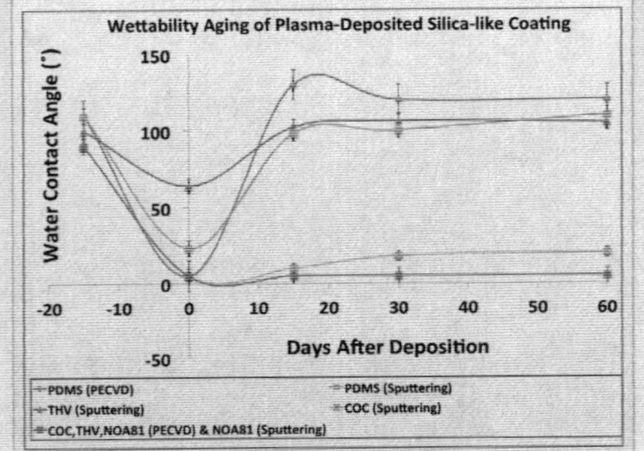

Fig. 2: Wettability aging of plasma-deposited silica-like coating on polymers.

- ➢ **PDMS recovered hydrophobicity** within two weeks for both deposition methods.
- ➢ **PECVD** was successful for depositing **enduring hydrophilic coating (<10°) on COC, NOA81, and THV.**
- ➢ **Sputtering** deposited an **enduring hydrophilic coating on COC and NOA81.**
- ➢ **Sputtering** was **ineffective** for rendering the **THV** surface hydrophilic.

MATERIALS

Microfluidics require:
Optically transparent thermoplastics that are easily molded, robust, with modifiable surface properties.

PDMS (polydimethylsiloxane)

- Conventional microfluidic material
- Extremely hydrophobic, fast hydrophobic recovery[3,4,5,6,7]

COC 6013S (Cyclic Olefin Copolymer)

- ideal for surface modification (T_g =130°C)[4] for functionalization[5]

NOA81 (Norland Optical Adhesive)

- Thiolene-based photocurable resin
- Dissolvable[6], biocompatible[7]

THV (fluorinated polymer)

- Low surface energy and high chemical resistance[10]

METHODS

Deposition of **silica-like layer** on surface of each polymer using two plasma methods:

1) Plasma-enhanced chemical vapor deposition (PECVD) in home-made reactor with hexamethyl-disiloxane precursor

2) Sputtering in commercial reactor with pure silica target

Fig. 1: Homemade reactor for PECVD.

SURFACE MODIFICATION

➢ **Successful incorporation of silanol groups** onto polymer surfaces (confirmed by FTIR and XPS)

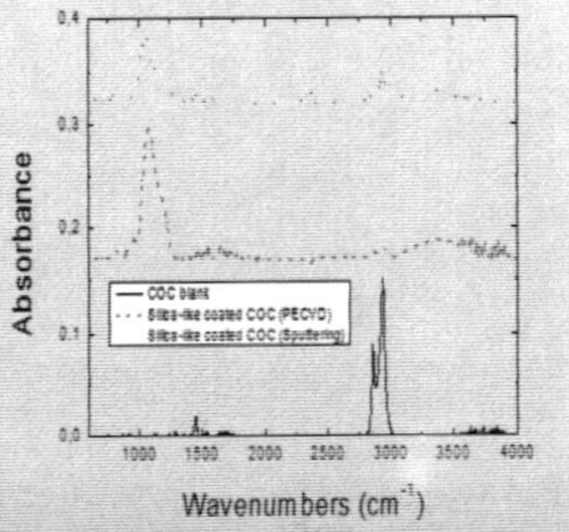

Fig. 3: FTIR-ATR spectra of blank COC surface (black), and silica coated COC surface functionalized using PECVD (red) or sputtering (blue).

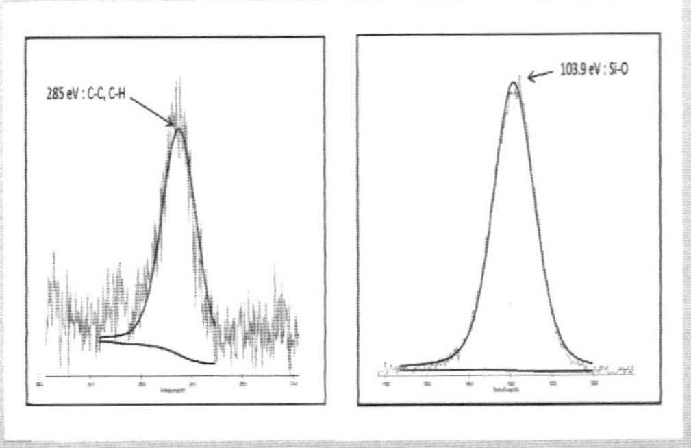

Fig. 4: XPS C_{1s} and Si_{2p} high resolution spectra of the surface of COC coated with sputtered silica-like layer.

➢ **Sputtered silica layer on COC is less hydrophobic than PECVD-deposited**

- ✧ Ellipsometry revealed **silica layer thickness** of 30 nm for sputtered layer and 1 µm for PECVD layer.

➢ **Sputtered silica layer on THV remains hydrophobic**

- ✧ Fluorine atoms in the THV polymer backbone (confirmed by XPS and FTIR) prevent sputtered silica layer from adhering.
- ✧ Combined with comparative thinness of layer, results in delamination of the hydrophilic coating.

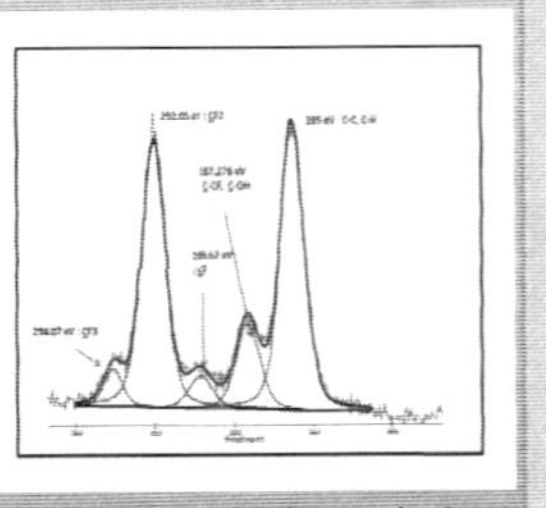

Fig. 5 XPS C_{1s} high-resolution spectra of blank THV surface.

CONCLUSIONS & FUTURE DIRECTIONS

Successful coating of new polymeric materials (COC, NOA81, and THV) with a durable silica-like hydrophilic coating provides an essential innovation for microfluidics studies and medical applications. A PECVD silica-coated COC microfluidic device is being tested for catalytic degradation of model pollutants.

References: **1**. Da Silva, B., Schelcher, G., Winter, L., Guyon, C., Tabeling, P., Bonn, D., Tatoulian, M. *European Cells and Materials* (2013) 26 (6): 105. **2**. Da Silva, B., Schelcher, G., Winter, L., Guyon, C., Bonn, D., Tatoulian, M. "Enhancement of Stability and Superhydrophilicity of Plasma-Modified Microfluidic Materials." *Submitted.* **3**. J. Zhou, D. A. Khodakov, A. V. Ellis, N. H. Voelcker, *Electrophoresis*, 2012, 33, 89-104. **4**. D. Bodas, C. Khan-Malek, *Sens. Act. B*, 2007, 123, 368-373. **5**. J. L. Fritza, M. J. Owen, *The Journal of Adhesion*, 1995, 54, 33-4. **6**. V. Barbier, M. Tatoulian, H. Li, F. Arefi-Khonsari, A. Ajdari, P. Tabeling, *Langmuir*, 2006, 22, 5230-5232. **7**. S. Bhattacharya, A. Datta, J. M. Berg, S. Gangopadhyay, *J. Microelectromechanical Sys.*, 2005, 14, 590-597. **8**. E. Sollier, C. Murray, P. Maoddi, D. Di Carlo, *Lab Chip*, 2011, 11, 3752. **9**. G. L. Chen, C. Guyon, Z. X. Zhang, B. Da Silva, P. Da Costa, S. Ognier, D. Bonn, M. Tatoulian, *Microfluidics and Nanofluidics*, 2014, 16, 141-148. **10**. R.K. Jena, C.Y. Yue, Y.C. Lam, *Microsystem Technologies*, 2012, 18, 159.

Thanks to the support of the Calhoun College Richter Summer Fellowship.

Design and Manufacturing of Microfluidic Devices

Kevin Nagle
Advisor: Brandon Vogel
Department of Chemical Engineering

Microfluidic devices and used to study the behavior of fluids on small scale. The microfluidic devices created were sent to the Geisinger Medical Center of Pennsylvania for use in cancer research. Using chemostaxis the researches there traced the movement of the drug that they were developing to the cancer cells. The devices are used by passing drugs or fluids through the channels. Using specifications from the hospital, a design for the devices was made. The devices were created using PDMS. SU-8 was used to create a master silicon wafer in order to leave the print in the PDMS.

398910 Effect of Stresses during Algae Growth on the Dimensions of Extracted Cellulose Nanocrystals

Abstract:

Cellulose nanocrystals are a valuable material resource that can be derived from algae with a wide range of applications, such as filtration and reverse osmosis (RO) membranes. If the dimensions of cellulose nanocrystals were controlled, one could tailor them for specific applications. It was hypothesized that, by stressing algae during its growth through high salinity media or air feeds with high carbon dioxide concentrations, there could be an effect on the biosynthesis of cellulose and, consequently, the dimensions of extracted cellulose nanocrystals. To test this, three algae, Nannochloropsis occulata, Desmodesmus abundans, and Scenedesmus obliquus were grown under ambient air, 10%, 20%, and 40% carbon dioxide concentrations as well as under different salinities. Cellulose nanocrystals were extracted from the algae using a bleaching pretreatment with 1.7 wt% sodium chlorite in an acetic acid buffer[1] followed by a 60 wt% sulfuric acid hydrolysis treatment[2]. Characterization was done through TEM and AFM. The work is ongoing.

1. Ma, L., Zhang, Y., Cao, J. and Yao, J. (2014). Preparation of Unmodified Cellulose Nanocrystals from Phyllostachys heterocycla and their Biocompatibility Evaluation. *BioResources*, *9*(1), 210-217.

2. Sèbe, G., Ham-Pichavant, F., Ibarboure, E., Koffi, A. L. C. and Tingaut, P. (2012). Supramolecular Structure Characterization of Cellulose II Nanowhiskers Produced by Acid Hydrolysis of Cellulose I Substrates. *Biomacromolecules*, *13*, 570-578.

Effects of Reaction Temperature and Aging Time on LTA Zeolites for Molecular Sieve Purposes

Julia King, Chemical Engineering
Mentors: Dr. Mary Laura Lind, Afsaneh Khosravi
School of Engineering for Matter, Transport, and Energy

act

For use as the selectively permeable component in a molecular sieve water purification purposes, Linde Type A (LTA) zeolites were synthesized while systematically nt parameters. The product nanoparticles were characterized using scanning electron microscopy and x-ray diffraction spectroscopy. It was found that a longer aging time of the precursor gel solution during synthesis and a lower reaction in smaller particle sizes of the LTA zeolites. It was also found that the aging time has no effect on the composition or · parameters will be evaluated.

Background

- A posite membrane for use in osmosis processes will /ater desalination and ɔf water-selective molecular sieve on-resistant, water-impermeable polymer
- a possible selectively permeable particle used. They ıeters in width with a pore size of contaminants.
- Since the calcination process is damaging to the particles and the energy expense is large, template- e particles will be made.

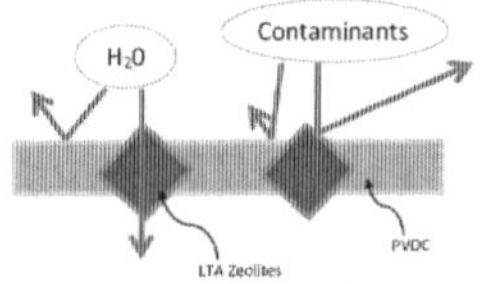

Figure 1: Schematic of membrane design – Only water is allowed to pass the membrane

Experimental Methods

LTA zeolites were synthesized while systematically evaluating specific · solution and crystallization The morphology of the synthesized zeolites was using scanning electron microscopy (SEM) and crystalline -ray diffraction spectroscopy.

Results and Discussion

Particle Size

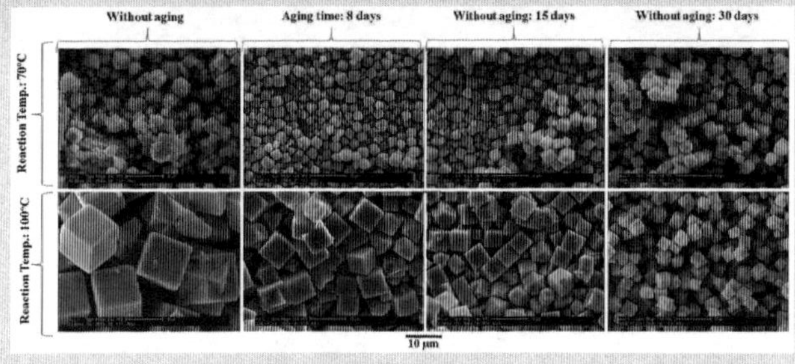

Figure 2: SEM images of LTA zeolites at different aging times and varying reaction temperatures

- The LTA zeolites formed at a reaction temperature of 100°C are much larger overall than the particles formed at 70°C. This is because higher temperatures increase the solubility of compounds and consequently the driving force of crystallization –supersaturation—decreases[1]. Under supersaturation, nucleation will terminate.
- The difference in particle size is greater for shorter aging times of the precursor gel.
- Many amorphous structures formed at a reaction temperature of 70°C without any aging of the precursor solution. Lower temperature produce smaller particles. However, lower temperatures suppress nucleation due to the drop in the atomic mobility. For this reason, at 70°C the reaction yield decreases significantly. The particles were more cubic after longer aging times.

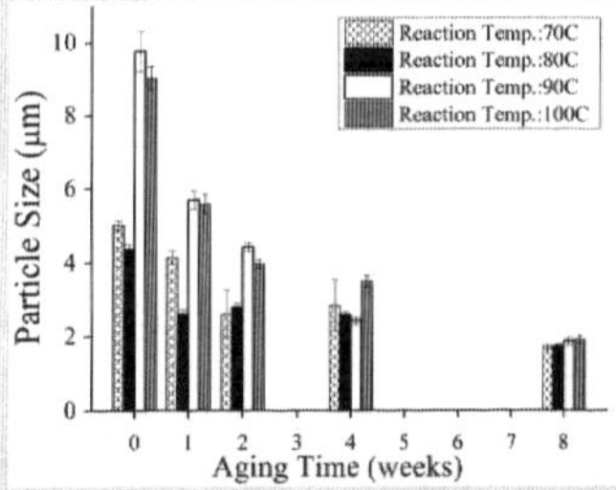

- The particle sizes of synthesized LTA zeolites decrease with longer aging times of the precursor solution.
- Higher reaction temperatures produce larger particles. This is because at higher temperatures, the rate of nucleation decreases significantly.

$$N \propto \exp\frac{-C}{(\Delta T)^2}$$

where N is number of nuclei [m^3/s], ΔT is the change in temperature and C is a constant[2]

Figure 3: Effect of aging time on the LTA particle size

Characterization

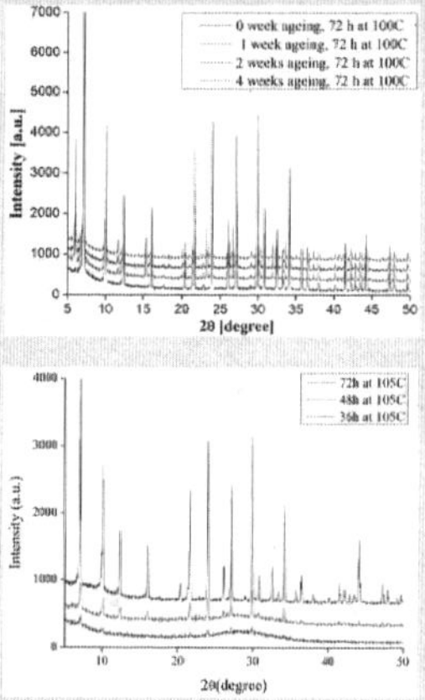

- the chemical structure of the synthesized LTA zeolites.

Figure 5: XRD of LTA zeolites

- The time of the precursor solution has an effect on the chemical zeolites. Shorter amorphous structures.

Conclusions

- Longer aging times produce s
- Higher reaction temperatures
- The aging time has no effect on the chemical structure but the reaction time does
- Further work will be done on the effect of other parameters
- Other zeolitic particles will be

Acknowledgements and References

- Fulton Undergraduate Research Initiative
- Lind Lab
- Arizona State University

[1] P. Jansens, J. Horst, S. Jiang, BIWIC 2006: 13th Crystallization, Sep. 13-15, 2006, Delft U Press, 2006
[2] A. Myerson, Handbook of Industrial Crystallization, Butterworth-

FURI research program

IRA A. FULTON SCHOOLS OF engineering
ARIZONA STATE UNIVERSITY

Fibrous encapsulation occurs as a result of the implantation of devices such as pacemakers, artificial breast implants, and micro-encapsulated islet cells used in the treatment of type 1 diabetes. The foreign body response (FBR) is responsible for the development of the fibrous capsule, which is often detrimental to the function of the implanted device. The FBR consists of a cascade of events involving many immune cells and cytokines, eventually leading to the differentiation of fibroblasts into myofibroblasts (MFs). MFs secrete and contract ECM components, including collagen, to form a fibrous capsule around the implant. In the case of encapsulated islet cells, this fibrous capsule prevents diffusion of insulin and other factors, ultimately rendering the treatment ineffective. The **objective** of our project is to significantly reduce the thickness of the fibrous capsule, allowing proper function of the device. It is important, however, that a small amount of fibrous encapsulation occur, as this will facilitate incorporation of the implant by providing wound closure. We will achieve our objective by using amidine-like modifications of poly-L-arginine (PLR) to coat alginate, which, in turn, will cover the implant. The amidine-like functional groups within these modifications inhibit matrix metalloproteinases (MMPs), which are part of the FBR that leads to the formation of the fibrous capsule. Inhibition of MMPs will impede cleavage of latent transforming growth factor beta (L-TGF-β) into transforming growth factor beta (TGF-β), an important cytokine responsible for stimulating fibroblast to MF differentiation. Complete prevention of TGF-β activity means that the presence of other cytokines, such as interleukin-4 (IL-4), that lead to MFs, will be important to provide the small amount of fibrous encapsulation necessary for incorporation of the implant. My presentation will focus on the effect of our materials on IL-4 induced fibroblast to MF differentiation. I will discuss the viability of NIH/3t3 fibroblasts on these biomaterials in the presence of IL-4, as well as evidence that IL-4 induced differentiation is not compromised on these surfaces. It is often important for the function of implants to be adjacent to vasculature; therefore, vascular endothelial growth factor (VEGF), which is produced by fibroblasts and MFs, will be analyzed on the surfaces of these materials to ensure that MMP inhibition does not prevent secretion of this molecule.

397334 Utilization of Optical Emission Spectroscopy in Characterization of Microwave Plasma

Abstract:

The patterning of electronic materials has been an essential technology in the manufacturing of integrated circuits. This investigation used a TEL Etch System paired with a 2.4 GHz microwave power source and a 30.56 MHz bias power source to create a novel plasma etching system. Critical operating parameters of this system were altered in order to determine their effect on fundamental plasma characteristics. These parameters included pressure within the processing chamber (5-100mTorr) and power from the microwave source (0.4-2kW). The dissociation of chlorine in Cl_2 plasmas was monitored via optical emission spectroscopy (OES) using emission ratios with trace rare gases, specifically the ratios Cl_2/Xe and Cl/Xe were monitored. Using nitrogen plasmas, spectra were analyzed in MATLAB with a simulation to provide the temperature of the plasma gas. It was found that the extent of chlorine dissociation depended on the power delivered to the plasma. The lowest Cl_2 concentration and the highest Cl-atom concentration were seen at the maximum power of 2 kW. Gas temperatures were found to increase with power, reaching up to 1150 K. Also, at higher pressures, higher temperatures were seen for the same power. A 'dim' mode of the system was discovered, in which instead of plasma forming near the upper power source, a standing wave formed between the wafer and the upper surface of the reactor. The resulting hysteresis of the system indicated by switching between 'dim' and normal operating mode was investigated.

Pre-Treatment By Common Anion Intercalation

This work was done in Dr. Alex Yokochi's Laboratory of Innovative Reaction Engineering for Materials & Sustainability (iREMS), in affiliation with NRGindependence, Dr. Alex Bistrika, and Dr. Pavel Mardilovich.

Carbon electrodes are widely used in industry today for electrochemical applications since they are cheap and relatively chemically inactive. Graphite Intercalation occurs where another chemical species becomes imbedded between two layers of graphene. This process is useful in many applications and can be done both chemically and electrochemically. This Project used the electrochemical method.

The idea behind this Project was to determine the extent of the improvement of electrical performance that sulphate intercalation has on sintered graphite electrode materials. Furthermore, it was hypothesized that there might exist an "ideal" current density to oxidize the graphite electrode so as to get the largest drop in running potential.

The experiments consisted first of baseline tests that measured an open circuit potential, and the potential needed to drive a current through an unaltered electrode at a certain current density in a potassium chloride solution. The next test was to intercalate sulphate into the electrode. This was done by running current at different current densities through the electrode while in a dilute sulfuric acid solution. The next test was a repeat of the baseline test.

After this series of tests, one of two tests were performed on each electrode. The first was testing for air stability, which was accomplished by taking baseline measurements after 12 hours and again after an additional 48 hours. The second test was conducted to rule out pitting as the cause of the potential drop since a visual alteration of the surface was observed. This second test was done by sanding the surface of the electrode and then taking the baseline measurements again.

These tests confirmed that sulfphate intercalation of electrodes drops the over-potential by about 150mV, and that lower current densities seem to be more effective at this task.

These tests also suggest that sulphate intercalation is air stable and that the drop in potential is not due to pitting.

Plans for future work include studying other intercalates and their effect on the behavior of graphite electrodes. Another possibility would be to test other stresses on the stability of the modified electrodes, as well as a study to relate the depth of intercalation for various carbon materials, e.g., cloths, felts, or composites.

Improving the Energy Density of Supercapacitors; Study of the Effects of Different Redox Couples and Electrodes

Marzieh Tousi[1], Brian Evanko[2], Galen Stucky[2], and Martin Moskovits[3]
[1]Department of Chemical and Environmental Engineering, UC Riverside
[2]Materials Department, UC Santa Barbara
[3]Department of Chemistry and Biochemistry, UC Santa Barbara

Background

❖ **Importance of Alternative Energy Sources**

✓ The U.S has been consuming almost 6 times as much fossil fuels per capita as the global average.

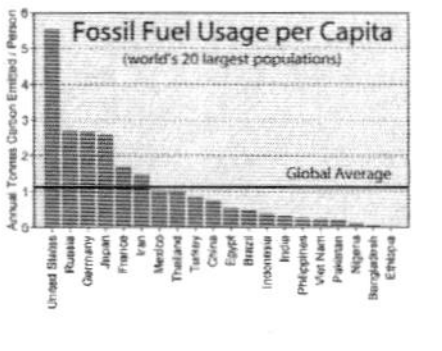

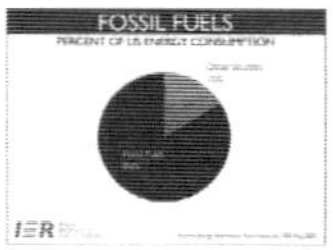

❖ **A Few Alternatives:**

Batteries Vs. Electrochemical Capacitors

✓ High power density compared to batteries
✓ High energy density compared to traditional capacitors

❖ **Among ECs are:**
- ➢ EDLCs store energy electrostatically
- ➢ Pseudocapacitors utilize redox reactions

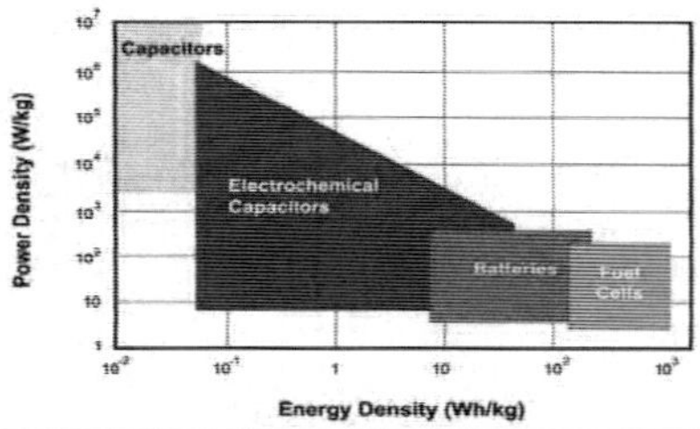

Other Applications

❖ **Regenerative braking:**

✓ Conversion of kinetic energy to electrical energy

✓ Capacitor charges when there is no acceleration

✓ Capacitor discharges when accelerating

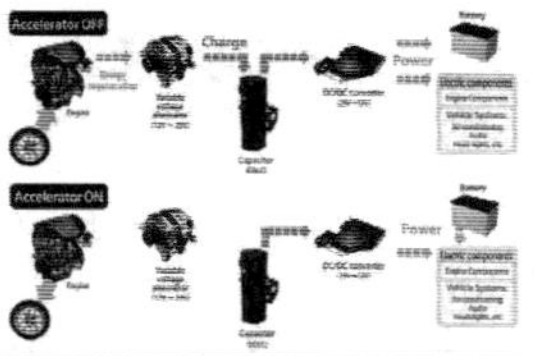

Traditional Capacitors Vs. Supercapacitors

$$C \propto \frac{A}{d}$$

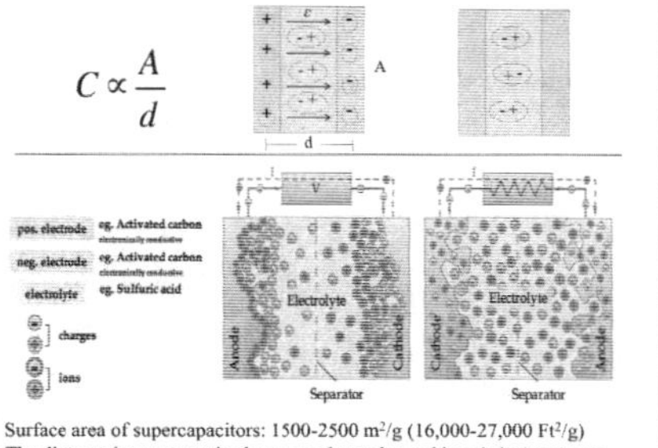

✓ Surface area of supercapacitors: 1500-2500 m^2/g (16,000-27,000 Ft^2/g)
✓ The distance between static charge on electrodes and ions is in Angstroms

Pseudo Capacitors

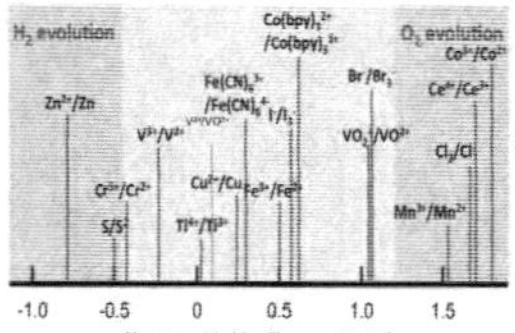

Chart provided by Shannon Boettcher

Electrolyte:
0.1 M $VOSO_4$ and 0.15 M KI

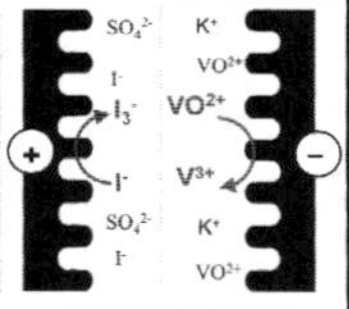

(+) $3I^- \rightarrow I_3^- + 2e^-$

(-) $2VO^{2+} + 2e^- \rightarrow 2V^{3+}$

Experimental Methods

❖ **Cyclic voltammetry:**
✓ Working electrode: glassy carbon
✓ Reference electrode: Ag/AgCl
✓ Counter electrode: platinum
✓ Supporting electrolyte: 0.1 M Sodium Sulfate
✓ Concentration of Analyte: 10mM

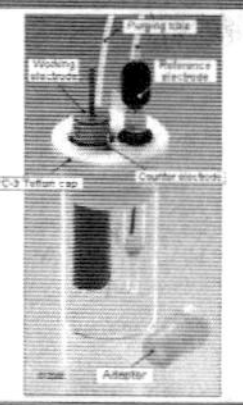

Results and Discussion

Cyclic Voltammograms With Different Electrolytes:

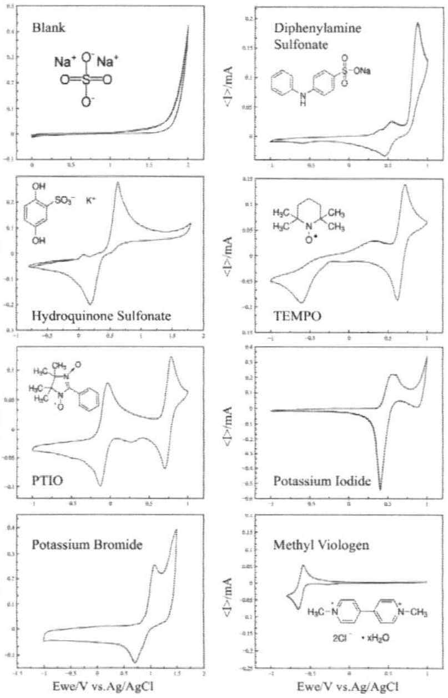

❖ **Other Criteria For Catholyte Selection:**

	HQS	DPAS	TEMPO	KI	KBr	PTIO
Cell Potential with MV	1.02	1.12	1.28	1.09	1.51	1.36
Cell Potential with Saf O	0.76	0.86	1.02	0.84	1.25	1.10
Molecular weight (g/mol)	228.3	271.3	156.3	166	119	233.3
Minimum solubility in water (M)	1.1	0.9	0.1	3.6	4.5	0.1

Binder for Activated Carbon Electrodes

❖ Cross linking poly-acrylic acid and carboxymethyl cellulose to obtain PAA/CMC, an insoluble binder

✓ PAA/CMC binder adheres well to the current collector and is cheap to manufacture

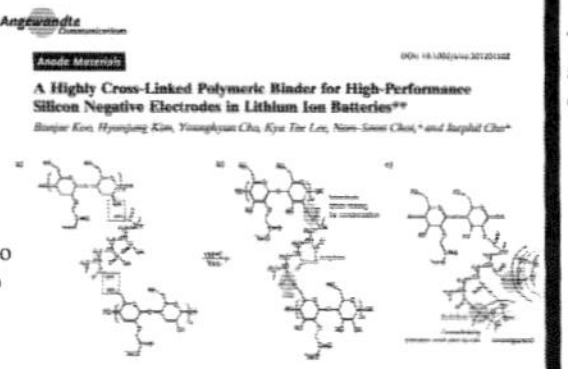

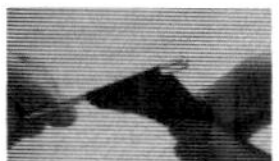

❖ Electrodes contain:
✓ 73 wt.% Norit A Supra Activated Carbon (1700 m2/g)
✓ 9 wt.% Super P conductive additive
✓ 18 wt.% CMC/PAA in 1:1 ratio
❖ Carbon appears hydrophilic
❖ The optimized cross-linking time is about 2-4 hours
❖ Binder can be decreased to 15 wt.%, possibly lower
❖ Ratio of CMC:PAA has not been optimized

Results and Discussion

❖ **Stability of PAA/CMC Binder**

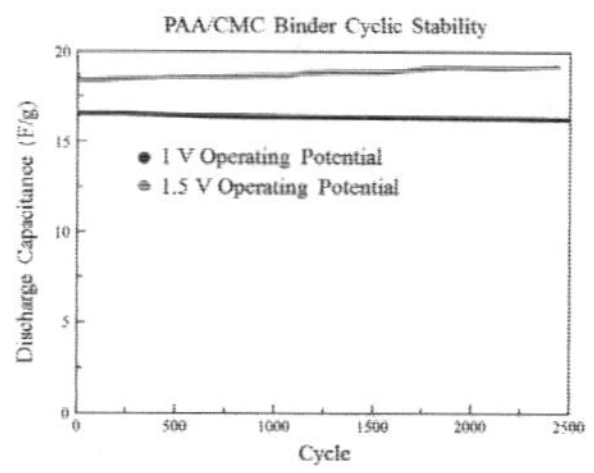

✓ Very stable cycling with coulombic efficiency > 98%
✓ Stable up to at least pH 8.6 (5 M LiI)
✓ Separates from current collector in 0.1 M KOH

❖ **Rate Performance Plots**

✓ Norit AC
1.9 mg/cm2

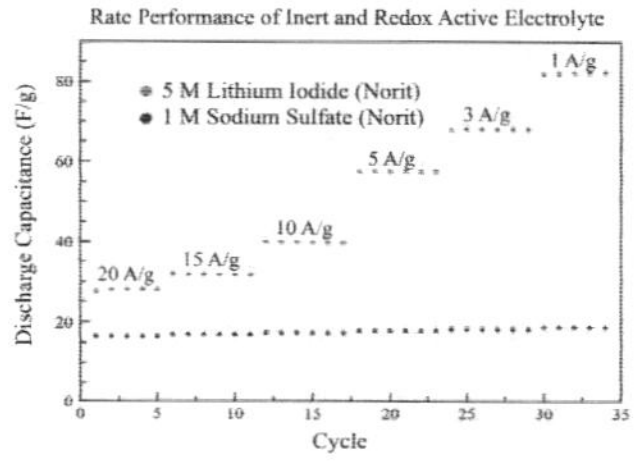

✓ Donacarbo AC
3.8 mg/cm2

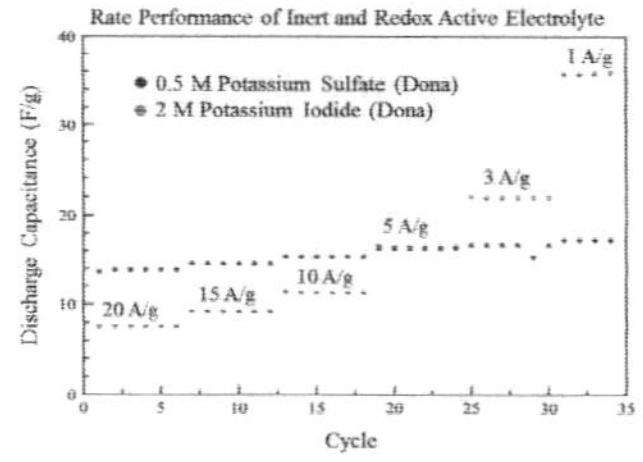

Future Work

✓ Investigating more catholytes
✓ Building cells with different analytes and catholytes to maximize the cell potential and performance
✓ Making PTFE (Polytetrafluoroethylene) electrodes and compare them to PAA/CMC with regards to efficiency and stability

Acknowledgments

The authors would like to thank the UC LEADS program and its coordinators for their support and funding for this project as well as the graduate division of the University of California, Santa Barbara.

References

✓ Energy Information Administration. MER. May 2009
✓ Gregg Marland, Tom Boden, and Bob Andres. Oak Ridge National aboratory. University of North Dekota
✓ Chijuan Hu , "Fluid Coke Derived Activated Carbon as Electrode Material for Electrochemical Double Layer Capacitor," 2008.
✓ P. Jampani, et al. "Advancing the supercapacitor materials and technology frontier" *Electrochem. Soc. Interface*, 2010.
✓ S. Roldán et al. "Towards a further generation of high-energy carbon-based capacitors by using redox-active electrolytes.," Angew. Chem. Int. Ed. Engl., vol. 50, no. 7, pp. 1699–701, Feb. 2011.
✓ B. Koo, et al. "A Highly Cross-Linked Polymeric Binder for High-Performance Silicon Negative Electrodes in Lithium Ion Batteries."*Angewandte Chemie.* 2012.

CMSE
MIT Center for Materials Science and Engineering

MPC
Making Matter, Matter

Investigation of Materials in Solid State Thin Film Lithium Ion Batteries

Rahul Kini[1], Chen Xi[2], Ahmed Al-Obeidi[3], Carl Thompson[3]

[1] Department of Chemical Engineering, Yale University

[2] Department of Chemical Engineering, Hong Kong University of Science and Technology

[3] Department of Materials Science and Engineering, Massachusetts Institute of Technology

Massachusetts Institute of Technology, Cambridge, MA 02139

Massachusetts Institute of Technology

Yale

LUX ET VERITAS

ɔn

ous systems, e ubiquitous with umer electronic and laptop m Ion batteries are s systems and novel materials for :trolytes[1].

work has gone into the investigation of cyclability and energy capacity. Additional nding electrolytes with high ionic conductivity e high mobility while in operation. Part of the h brings to the microbattery world is the use icrofabrication and integration of new timal conditions for each new material, roved significantly and programmable battery

research is to develop microbatteries that well as energy storage devices. By chargeable thin film lithium ion battery with capacity, this lays the foundation for batteries, batteries harvesting energy from a crucial step for efficient microbatteries.

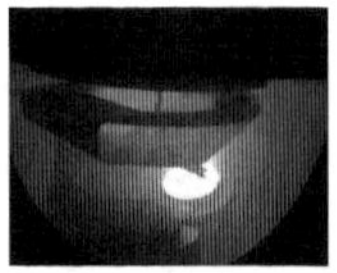

Deposition of thin film layers posed several problems: slow deposition rate and non-uniformity of deposited layers. Construction of a holder for the samples offset the gun and allowed deposition closer to the gun at higher rates and with greater uniformity. .

ɜen
ɔ
ɔn with
ɪduce
cy of

masks
ɪ and

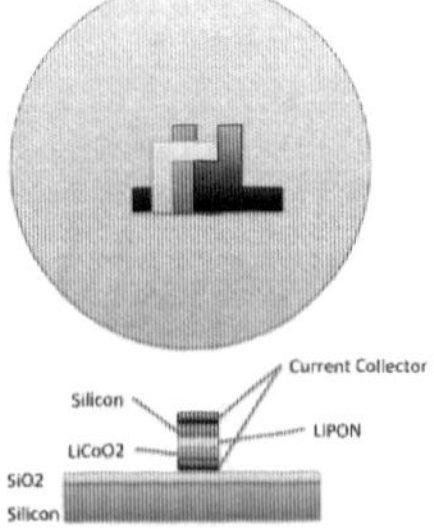

Results

Anode: Silicon

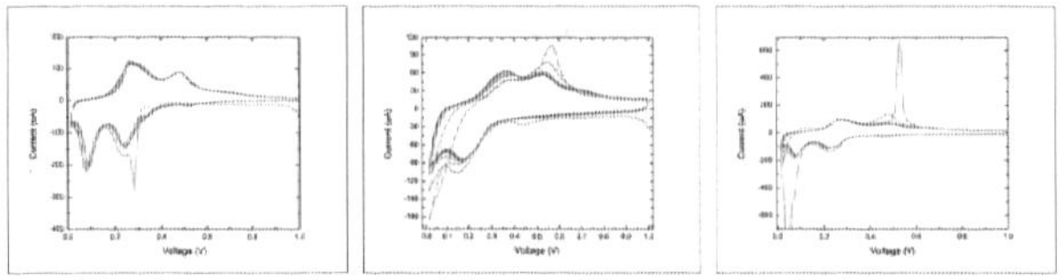

- Silicon has proven to be successful in terms of high energy capacity. As an anodic material it is both inexpensive and familiar in industry and efficient in charge mobility.
- The electrochemical tests carried out on the galvanostat proved to show a characteristic Current-Voltage curve with two characteristic peaks.
- Left to Right: Current-Voltage runs for Silicon with different liquid electrolytes
  - 100 nm PECVD aSi/TiN/glass LiTFSI/DOL electrolyte
  - 80 nm sputtered aSi/Pt/Ti/alumina LiPF6 EC:DMC electrolyte
  - 80 nm sputtered aSi/Pt/Ti/alumina LiTFSI DOL electrolyte

Electrolyte: LIPON (Lithium Phosphorus Oxynitride)

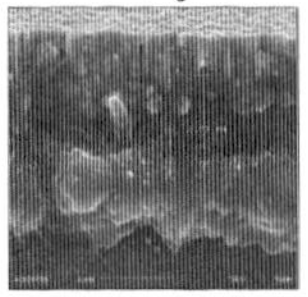

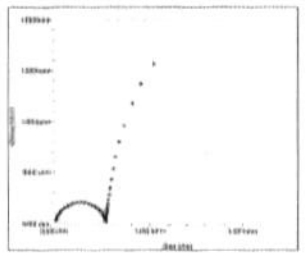

- LIPON is an amorphous glassy solid state electrolyte
- LIPON is incredibly reactive with ambient: A major problem encountered was instantaneous reaction with oxygen, thereby forming a thin layer of oxide on the top layer. Solution included quick shadow mask changes between depositions and deposition of a protective titanium layer when measuring ionic conductivity.
- Optimized deposition conditions proved to show strong ionic conductivity for LIPON from EIS spectroscopy. The testing proved that the ionic conductivity (κ) was close to theoretical values.
  - Theoretical: 2.00*10^(-6) S cm-1
  - Literature[2]: 3.10*10^(-6) S cm-1
  - MIT: 5.09*10^(-6) S cm-1

Cathode: Lithium Cobalt Oxide

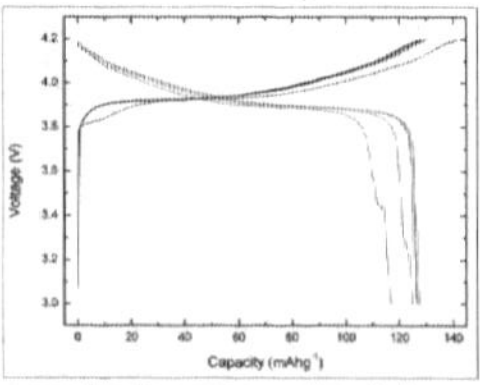

- Lithium Cobalt Oxide is theoretically and experimentally proven to have higher capacity than other lithium ion cathodes.
- Annealing the cathode films using Rapid Thermal Annealing (RTA) have been proven to induce desired crystalline structure[3].
- Electrochemical tests prove that there was stable cycling over 40 cycles for a 1.5 um film.
- By testing a half cell with Lithium Cobalt Oxide, we were able to prove experimentally constant voltage at higher capacities. In the picture above, note a flat voltage for range 20-90 $mAhg^{-1}$.

Conclusion

Optimal Conditions:

Material	Vacuum (Torr)	Gas (Flow Rate)	Power (W)	Time
Si	$2.51*10^{-7}$	Ar (25 sccm)	90	10 min
LIPON	$4.2*10^{-7}$	N_2 (16 sccm)	90	6 hour
$LiCoO_2$	$1.67*10^{-7}$	Ar (25 sccm)	150	3.5 hour
Ti/Cu	$3.14*10^{-7}$	Ar (20 sccm)	150	3 min/ 20 min

Challenges and Solutions:

- Lithium Cobalt Oxide delaminated during anr
  - We changed the one step ramp up to a tı 400° C and second step to 600°C .
  - Eliminated water vapor from the oxide su with ambient air. Long cool down to 200°
- LIPON target is sensitive to ambient, pre-cleı active) was conducted with Argon (25 sccm) sputtering with N_2 (16 sccm) for 15 minutes ı
- Improvements to the sputtering tool: new chiı plasma and a three way divider that prevents

Future Work

- Future investigation in the assembly and testing of the full battery is required and will provide the basis for improvement in design of novel thin film batteries.

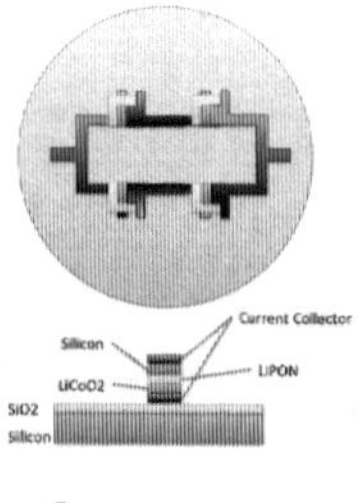

- Once construction and testing of the full battery is complete, optimization procedures can be employed to minimize resistance between thin film layers thereby increasing energy capacity and reusability.
- Other key electrolytes that warrant investigation include Lithium Sulfide and Molybdenum Chromium Oxide.
- This work sets the foundation for the design and testing of thermogalvanic batteries and programmable battery arrays.

References

(1)N.J. Dudney, *Materials Science and Engineering: B*, Volume 116, Issue 3, (2005)
(2)B. Fleutot, et al., *Solid State Ionics*, Volume 186, Issue 1 (2011)
(3)Yongsub Yoon et al., *Journal of Power Sources*, Volume 226 (2013)

This work was supported by the CMSE I Undergraduates Program, as part of the N National Science Foundation under grant and by the MIT Materials Processing Cente provided by the Singapore-MIT Allian Technology.

NSF

Elastic Properties of Bismuth Telluride Two-Dimensional Nanosheets

Haoming Yan, Lingling Guo, and Hung-Ta Wang

Department of Chemical and Biological Engineering, Center for Materials for Information Technology

The University of Alabama, Tuscaloosa 35487, United States

Two-dimensional (2D) nanosheets of bismuth telluride (Bi_2Te_3) have shown a potential for future applications in electronics, thermoelectrics, and heterogeneous catalysis. In this work, chemical vapor transport (CVT) method was used to synthesis Bi_2Te_3 2D nanosheets on a mica substrate. 2 to 10 nm-thick 2D nanosheets were then transferred onto a pre-fabricated substrate with holes (diameter~2 μm) using a poly(methyl methacrylate) (PMMA) film as a sacrificial layer and elastic poly(dimethylsiloxane) (PDMS) as a transfer stamp. Atomic force microscope in the contact mode or force-distance mode was used to probe the suspended 2D nanosheets. It was found that the 2D elastic constant of an 8 nm-thick 2D nanosheet is about 238N/m. Understanding 2D nanosheet mechanical properties will help to understand their thermoelectric properties under mechanical strains.

398697 Sol-Gel Synthesis of Clear Uncracked Silica Glass with Large Pore Sizes

Abstract:

Silica glass is a useful optical material because of its excellent transparency, good radiation hardness, high chemical stability and mechanical strength.In this study, clear uncracked silica glasses were manufactured via sol-gel synthesis. Sol-gel method is a wet chemical process that is suitable for manufacturing silica glasses at relatively low temperatures (1000-1300°C) without melting. The basic concept used in this experiment was assembling a mold containing the precursor solution of different elements mixed in a certain ratio in a low pH for the sol to convert to gel. The precursor solution was then heated in a small oven to form a dried gel followed by additional heating in a larger oven to 1300°C till a sample of clear uncracked glass was obtained. Large pore sizes are desired in the glass because it helps to prevent fractures or crack formations. A chemical named Tetraethyl orthosilicate/ Tetraethoxysilane (TEOS) was used as the silicon source provider that reacted with water under ethanol as a solvent. Besides these chemicals, the other two chemicals used were Hydrofluoric acid (HF) and Dimethylformamide (DMF). Hydrofluoric acid worked as an acid catalyst and also helped to form large pore sizes and DMF worked as a drying control chemical additive (DCCA) helping to prevent hydrogen bond network in the solvent in order to eliminate any crack formations.

Deposition of CuI for Thin Film Transistors via Inkjet Printing

Oregon State University Department of Chemical, Biological, and Environmental Engineering

Marshall Allen, Dr. Changho Choi, Dr. Chih-Hung Chang

Abstract

Since the late 1990's, a variety of Thin Film Transistors (TFT's) have been utilized in both commercial and academic applications. This work culminated in the emergence of transparent electronics around 2002 through use of ZnO as a negative carrier semiconductor (n-type). Since then, a large array of inorganic n-type semiconductors—including ZnO, In_2O_3, and SnO_2—have been applied, most notably in liquid-crystal displays (LCDs). While these have been successful, p-type semiconductors are required for expansion of transparent electronics. However, few examples of functioning inorganic p-type semiconductors have been developed. Of these, Cu_2O has shown promise with a relatively high band gap around 2.1-2.6 eV and a carrier mobility above 100 $cm^2/V*s$, however fabrication of quality thin films requires a high temperature synthesis above 500°C.

Of particular interest is Copper (i) Iodide, which has a high band gap in the range of 2.9-3.2 eV, giving it semiconductor properties useful for transistor applications. Importantly, the active crystal structure is formed at temperatures well below 300°C, allowing for the possibility of low-energy film synthesis. In this work we present a low-temperature fabrication method of CuI thin films with use of inkjet printing. Additionally, we study the film's properties as p-type semiconductor in thin film transistors. CuI was printed with an acetonitrile-based ink onto a SiO_2 substrates, forming thin films with thicknesses on the nanoscale. Inks were formulated at variable concentrations of CuI. The effect of adding the amine stabilizing agent TEMED to the ink was investigated. The printed films were annealed at variable temperatures ranging from 25 to 150°C. The electronic properties were then measured. Films were also analyzed with use of UV-Visible spectroscopy, X-ray diffraction, and scanning electron microscopy.

TFT performance was shown to be non-existent for prints contaminated with the stabilizing agent—indicating a possible breakdown in film formation in its presence. Additionally, XRD measurements showed an amorphous film, while UV-Vis spectroscopy revealed a material band gap much lower than the accepted value for CuI. Films made using no stabilizing agents, however, were shown to form a continuous film of CuI, which was verified using both XRD and UV-Vis spectroscopy. TFT performance was identified for films using voltage-current curves, which showed the low on/off ratio of around 2 with a V_{on} of 12V. The hole mobility for the film was found to be around .01 $cm^2/(V*s)$ The on/off ratio was lowered by a high film conductivity, caused by absorption of moisture.

Further work will focus on improving the TFT performance by fine-tuning film thickness, as well as lowering film conductivity.

An Interpretation and modification on Kapustinskii lattice energy equation

Mokhtar Shakerian*
Department of Chemical and Natural Gas Engineering,
Texas A&M University- Kingsville
Mokhtar.shakerian@gmail.com

Abstract:

Crystallization is one method of separation. The development of Catalyst and Solid state physics were depended upon the crystal studies.
The most developments of crystals are related to the physicists and chemists.
Great German Physicist, Max Born and great German Chemist, Fritz Haber had worked on crystals, and had formed Born-Haber Cycle for measuring the amount of Lattice Energy.
Born discoveries were not limited to Born-Haber Cycle. He discovered a theoretical equation to evaluate the amount of Lattice Energy along with Lande. Born also formed another equation to evaluate the amount of Lattice Energy with Mayer.
But interesting point is the Kapustinskii equation which can approximately give the Lattice Energies for crystals with out paying attention to their characteristics.

$$U = \frac{120200VZ(+)Z(-)}{R}(1 - \frac{34.5}{R}) \quad \text{KJ/mol}$$

However, as mentioned above, it is an approximate equation.
As chemical engineers we deal with crystals in many areas, such as crystal catalyst, separation, production and etc.
This abstract is allocated to an interpretation of the Kapustinskii's equation and a modification on that to include some important factors such as electronegativity and radial ratio of cation to anion.
The obtained formula gives more precise results for Lattice Energies compared to Kapustinskii equation.

Name	Born-Haber cycle	New equation	Kapustinskii equation
1.LiF	1049	1045	952.7
2.LiCl	864	861.8	803.7
3.LiBr	820	816	792.9
4.LiI	764	754.2	713.0
5.NaF	930	956.2	884.9
6.NaCl	790	790	752.9

“ Data in table are in KJ/mol

References:

1. Kapustinskii, A.F.Q.Rev., Chem. Soc. 1956, 10, 283
2. Kapustiskii, A. F.; Yatsimirskii, K. B. Zhur. Obshch. Khim. 1956, 26, 941. Yatsimirskii, K. B. Zhur. Neorg. Khim. 1961, 6,518.

3.Inorganic Chemistry, Huheey,E. Keiter, R. Keiter, Fourth edition, Harper Collins College publisher, 1993

Atomic Layer Deposition for Solar Energy Applications
Katharine Greco – Poster Presentation

Solar cells are a viable means of alternatrive electricity generation – if they can be made with environmentally methods and have long service lifetimes. Deployment can be suited for both individual and commercial application. However, hindering the widespread use of solar energy technologies is the poor efficiency and high materials cost of photovoltaic cells, which are the most common solar cells. Atomic layer deposition (ALD), a process of depositing ultra-thin films on metallic or ceramic substrates, can be used to improve the efficiency of photovoltaics. With ALD it is possible to grow uniform films that have precise thicknesses. By using ALD to deposit anti-reflective films and other optically enhancing coatings, we can enhance light absorption, thereby increasing efficiency. To determine how film thickness affects a substrate's optical properties, oxide films of varying thicknesses have been studied on both silicon and glass substrates. Precursor delivery and reactor conditions are studied to optimize the cycle time for ALD growth. Optical properties are measured with a spectroscopic ellipsometer. These experiments should increase the commercial lifetime of photovoltaics and make them more competitive with other means of electricity generation.

In addition to improving existing solar cells, ALD can also be used to produce an entirely new type of solar cell. These cells, called plasmonic solar cells, implement electron tunneling to generate electricity. Electrons travel through copper nanostrcutres, called "rectenna," across a gap to a collector electrode, generating a current. For tunneling to be possible, the rectenna must be within 1-2 nanometers of the collector electrode. Selective ALD can be used to create arrays of tunneling nanostructures, as the process grows copper exclusively on palladium seed layers, allowing growth of desired shapes. Unlike other growth techniques, selective ALD is capable of nanometer resolution, and no etching is required. To determine the best conditions for selective ALD growth of the rectenna, experiments varying partial pressure of precursors, pulse and purge times, and number of cycles have been conducted. Once ideal conditions for selective ALD have been determined, plasmonic solar cells will be closer to commercialization. Conducting these experiments will help solar energy become a more popular, sustainable choice for electricity generation.

Novel Silver/Carbon Material for 2, 4-Dinitrotoluene Detection

Bianca Davis[1], Kiara Moorer[1], Shamim Begum[1], Nader Vahdat[1], Samuel Hernandez-Rivera[2], Jonathan Mbah[1],

([1]) Department of Chemical Engineering, Tuskegee University, Tuskegee, AL 36088

([2]) Department of Chemistry, University of Puerto Rica, Mayaguez, PR 00681

ABSTRACT

The prevalent use of nitro-aromatic materials including, 2, 4, 6-trinitrotoluene (TNT) for explosive devices has made their detection crucial in the fight against act of terrorism and detection of military explosives. But rather than attempting to detect TNT, researchers focus on detecting 2, 4-dinitrotoluene (DNT) which is a common impurity in TNT –based explosive. DNT has a higher vapor pressure than TNT, allowing for more sensitivity and easy detection. In the production of munitions and explosives, DNT is used as a gelatinizing and waterproofing agent. High sensitivity and selectivity, combined with the ability to lower the deployment cost of sensors using mass production, is essential in winning the war on explosives-based terrorism.

In this study, the donor based electrode for the detection of DNT was synthesized by modifying the surface of carbon fiber material with zero valent silver (Ag) via chemical deposition. DNT detection was conducted electrochemically in aqueous media of pH 6.0. The cyclic voltammetric oxidation of DNT is diffusion controlled with detection sensitivity increasing with Ag loading of the electrode. The only oxidation product is benzoic acid, which suggests that the difference in performance of Ag-C electrode compared to other available materials is directly ascribable to the electrode fabricated in this study. Detection limit of DNT was 5μM with the electrode showing no signs of degradation after several cycles, and supported by surface behavior of characterized Ag-C electrode. Kinetic study was also investigated. In another study, the fabricated Ag nanoparticle on carbon fiber substrate was used as an efficient surface enhanced Raman spectra (SERS) substrate to detect the adsorbed DNT molecules with a detection limit of 50 ppm. In advent, our developed SERS substrates could have great potential in detecting other nitro-aromatic based-explosive materials, such as TNT molecules.

396383 Passivation of Silicon Using Aluminum Oxide: A Novel Spray Pyrolysis Method

Abstract:

The objective behind this research is to utilize spray pyrolysis as a cheaper, safer and easier method of manufacturing silicon solar cells. The first stage of research involves testing the method on soda lime glass for desired film quality and thickness. The second stage is aimed towards investigating the films effect on silicon. Thus far, aluminum oxide films have been deposited onto glass substrates. It has been found that, however, the film quality is poor and the thickness is too large. After refining experimental parameters, it is expected that the method will produce consistent quality results so that deposition can be conducted on silicon.

398540 Effects of Branched and Cycloalkyl Functionality on CO2 Separation Performance of Poly(IL) Membranes

Abstract:

The separation abilities of polymerizable ionic liquids (poly(ILs)) have been studied for the past several years, with a specific focus on their capabilities regarding CO_2 removal. The structure of ILs allows for a wide degree of flexibility regarding the addition of functional groups, important because the molecular composition of IL monomers directly correlates to the monomer's gas transport properties. Here, a variety of IL monomers were synthesized, with varying branched- and cycloakyl groups appended to the imidazolium ring. These monomers were used to form poly(IL) membranes. The permeabilities of each of these membranes to CO_2, N_2, and CH_4 were then analyzed, and their selectivites for CO_2 relative to other gases determined. Poly(ILs) with branched and cyclic functionalities exhibited ~20% larger CO_2/N_2 and CO_2/CH_4 selectivities than *n*-alkyl poly(ILs), while exhibiting a reduction in gas permeability of almost an order of magnitude. It is proposed that the greater selectivity of branched and cyclic functional groups is due to smaller fractional free volumes (FFV) than analogous *n*-alkyl chains. Computational analysis using COSMOTherm supports this.

397274 Synthesis and Gas Separation Performance of Poly(ionic liquid) Membranes Containing Branched and Cycloalkyl Group Functionalities

Abstract:

Abstract: The use of poly ionic liquids (poly(ILs)) for CO_2 separation has been studied in the past decade as the need for solutions to CO_2 emissions has risen. Alkyl-functionalized imidazolium-based poly(ILs) provide a tunable stage for gas separation. The synthesis and characterization of branched and cycloalkyl functional groups on the imidazole ring are reported in this investigation. Also reviewed, is the differences between straight-chain alkyl-functionalized imidazolium based poly(ILs) and their branched and cyclic equals. In the counterparts of the straight-chained poly(ILs), a 20% increase in the selectivity of the membrane was observed for CO_2/N_2 and CO_2/CH_4. However, a decrease in permeability of the cyclic and branched poly(IL) membranes is also seen. This is plausibly due to the lower free volume in cyclic and branched functionalities as compared to those of the straight-chained imidazolium based poly(ILs).

AIChE
Student Conference
Abstract
Gianna Credaroli

A New Thin Film Nanofiltration Membrane for Mixed Salt Rejection

Membrane technology is a promising platform for water treatment. The commercial grade, nanofiltration membrane (NF), which is manufactured by The Dow Chemical Company, is commonly used for water softening. However, 3M has a unique membrane whose support layer (Nylon-6,6) exhibits certain hydrophilic properties. The objective of this study is to test the performance of a nanofiltration membrane on a 3M support layer. This is the first nanofiltration membrane designed on a hydrophilic support.

Reverse osmosis tests were conducted using two different salt solutions, which were tested under commercial specifications. From this study, the pure water and salt permeance of the 2000-ppm magnesium sulfate solution of the Nylon-6,6 membrane was lower than the specifications provided by Dow. However, the salt rejection was similar to that provided by Dow. The salt rejection of the Nylon-6,6 membrane was much higher than the NF-270 membrane when tested experimentally.

Performance of Positively and Negatively Charged Nanofiltration Membranes under Various Operating Conditions

David A. Whitehead, Dr. Dibakar Bhattacharyya, Andrew Colburn

Abstract

Nanofiltration membranes are a continually expanding separation method of removing ions from a wastewater stream. Commonly used in the water softening industry, nanofiltration membranes have expanded to be used in the pharmaceutical and fine chemical industries. Created through interfacial polymerization of an amine and acyl chloride, they boast rejections higher than ultrafiltration membranes and operating costs lower than reverse osmosis membranes. Due to the surface charge present on these membranes, ion rejection of mixed salt feeds is not yet fully understood. The effectiveness of separation of a membrane changes with the operating conditions imposed on that membrane. The effects of different pressures showed an expected change in flux as well as a decrease in the ion rejection of these membranes at pressures less than 6 bar. Altering cross-membrane flow was seen to alter the ion rejection based on the valency of those ions—a result of the changing concentration polarization. Finally, the effects on auxiliary ions on the flux and specific ion rejection were studied, and it was found that even if the base ion concentration remained constant, the specific ion rejection decreased with increase ionic strength in the solution. The results found through these experiments will ultimately be used to develop a model for the performance of nanofiltration membranes. In order to fully understand the effects of various operating procedures, each individual effect needs to be isolated and studied. The effect of charge shielding from confounding ions should be studied in more depth in order to more accurately model the change in separation efficiency. In addition, different operating conditions have the possibility of altering the selectivity of certain ions over others. The research work done by this project was supported by the NFS EPSCOR program and by Southern Company.

DETAILED FINITE-ELEMENT CALCULATIONS OF PERMEATION IN MIXED-MATRIX MEMBRANES

Student: Tom Giordano, School of Chemical and Biomolecular Engineering

Advisor: Prof. Sankar Nair

ABSTRACT

Membranes offer an inexpensive and efficient route toward large-scale gas separations in the chemicals and energy sectors, with the potential to replace conventional energy-intensive processes such as distillation and crystallization. 'Mixed-matrix' membranes (MMMs) that incorporate nanoporous molecular-sieving filler particles in a polymer matrix are considered to be one of the leading membrane candidates for such applications. Existing equations that approximate permeability behavior in MMMs, including the Maxwell model, are limited in accuracy to low-pressure conditions and less than 25 vol% loading of the fillers. In reality, these membranes will be used at a wider range of pressure conditions and with very high loadings of filler particles. The purpose of this study is to perform accurate calculations of the permeability of MMMs via a detailed finite-element approach. We discuss the construction of detailed MMM models and permeation simulations using the COMSOL Multiphysics simulation package. These predictions of MMM permeation are used to examine the extent of applicability of approximate models with respect to the dependence of membrane permeability on filler loading, filler and matrix permeabilities, and membrane thickness.

398968 Transport Properties of Graphene Oxide Membranes

Abstract:

The purpose of my research project was to investigate the transport properties of graphene oxide-based material. The two-dimensional lattice structure of graphene oxide exhibit a layer-to-layer distance of approximately 0.34 nm. However, upon the exposure of polar compounds which are small enough to penetrate through the layers, such is water, the layer-to-layer distance increases beyond 0.34 nm. This expansion allows small polar compound to be co-transported with water through the layers. Moreover, it is hypothesized that an asymmetric transport of small compounds against water is not possible. Besides the asymmetric transport phenomena, graphene oxide pillaring with planar compounds, such as pyrene, is being currently investigated.

Membrane technology in co2 removal

Next to its environmental impact, CO_2 reduces the heating value of natural gas (mainly CH_4) in power plants and due to its acidic character, the presence of CO_2 can lead to corrosion in equipment and pipelines. This makes the removal of Co_2 from natural gas of crucial importance.

Traditional methods used to separate CO_2 from gas mixtures are pressure swing adsorption, cryogenic distillation and the most frequently used method is amine absorption; also, membrane processes are frequently used for gas separation.

Energy requirements on the other hand significantly benefit the use of membrane technology over other technologies: membrane technology uses 70-75 kWh per ton of recovered CO_2 compared to significantly higher values for pressure swing adsorption (160-180 kWh), cryogenic distillation (600-800 kWh) or amine absorption (330-340 kWh), making membrane technology an attractive alternative.

Fundamentals

Certain gases permeate or pass through the membrane more easily than others. This allows the more rapidly permeating "fast" components to be collected in one stream and the slower permeating "slow" components to be collected in a second stream. In natural gas treating, CO_2, hydrogen, helium, H_2S, and water vapor are highly permeable gases, sometimes characterized as "fast" gas molecules that permeate to a greater extent across the membrane to the lower pressure side than do the "slower" gases (i.e., hydrocarbons).

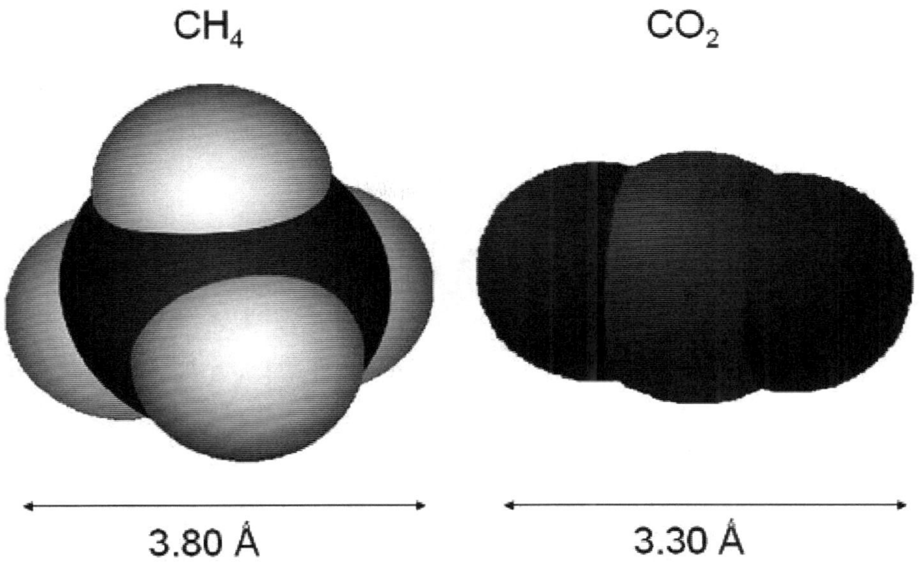

The driving force for permeation is the difference in partial pressure of a given component between the feed "high pressure" side and the permeate "low pressure" side of the membrane. The greater the partial-pressure difference, the greater the driving force.

The classical thermodynamics tells us nothing about the rate at which diffusion process takes place, but Fick's law for solution-diffusion membranes in rectangular co-ordinates do

$J_i=(S_i*D_i*P_i)/l$ where,

J » the flux of component; (the molar flow of component I through the membrane per unit area of membrane.)

S » solubility term.

D » diffusion co-efficient.

ΔP » partial pressure difference across the membrane.

The term Pi/L is membrane and operating conditions dependent.

ΔP is process dependent.

S, D depends on temperature and pressure.

Selectivity α 1-2, the ratio of two permeabilities, P1/P2.

A selectivity α of 20 for co2/CH4 means that co2 moves through the membrane 20 times faster than does methane.

The permeability coefficient, K, is a function of both the "solubility" and the "diffusivity" of the gas in the polymer matrix. Several mathematical models are available that derive the permeability coefficients.

Next to the permeability, the main factor to describe the performance of a membrane is its separation ability, called the selectivity α. The ideal selectivity using pure gases is given by the ratio of the permeability coefficients (Pi), which is, when assuming solution-diffusion to occur, composed of a diffusivity selectivity term (DCO2/DCH4) and a solubility selectivity term (SCO2/SCH4):

$$\alpha_{ideal} = \frac{P_{CO_2}}{P_{CH_4}} = \frac{D_{CO_2}}{D_{CH_4}} \cdot \frac{S_{CO_2}}{S_{CH4}}$$

For gas mixtures, the feed composition has to be taken into account as well, and the membrane selectivity is expressed as:

$$\alpha = \frac{Y_{CO_2} / Y_{CH_4}}{X_{CO_2} / X_{CH_4}}$$

Where Yi is the concentration of component (i) in the permeate stream and Xi is the concentration of component (i) in the feed.

Selectivity refers to the ratio of the permeation rates of a fast gas (CO2) and a slow gas (methane or other hydrocarbon). The selectivity of CO2 to methane determines the efficiency of the separation and the methane recovery.

EQUATION

$$\nu = KA\,(P_1 - P_2)$$

where:

ν	Volume of permeated gas, scf/hr
K	Permeability coefficient of permeating gas, ft/(hr-psi)
A	Membrane surface area, sq ft
P_1	Partial pressure of feed stream, psi
P_2	Partial pressure of permeate stream, psi

Membrane structure

Membranes can be used to separate other gaseous components such as H2S, N2, He, and water vapor from non-associated or associated gas stream under a partial pressure differential.

Typical membrane used	Category preferred polymer material	Component to be permeated
Cellulose acetate, polyimide	Glass	Co2
Cellulose acetate, ether-amide block Copolymer	Glass and rubber	H2s
Polyimide, perfluor polymers	Glass	N2
Many	Glass and rubber	Water vapor

The table shows typical membrane materials that can be used to separate the impurities from natural gas.

Membrane uses include dehydration, fuel gas conditioning and bulk co2 removal which is the most important application. The membrane market share is about 2%, essentially all for CO2 removal. This fraction will likely increase because other applications of membranes in the natural gas processing industry are under development.

Many different types of membranes have been developed for industrial separation, but for CO2 removal, the industry standard is presently cellulose acetate.

Membrane composition

A thin layer (0.1-0.5) ή m of cellulose acetate is on top of a thicker layer of a porous support material. Permeate compounds dissolve into the membrane, diffuse across it, then travel through the inactive support material. The membranes are thin to maximize mass transfer, minimize surface area and cost. The support layer is necessary to support the needed mechanical strength.

Commercial membrane configurations are either hollow fiber elements or flat sheets wrapped into spirally wound elements.

Spiral wound membrane structure

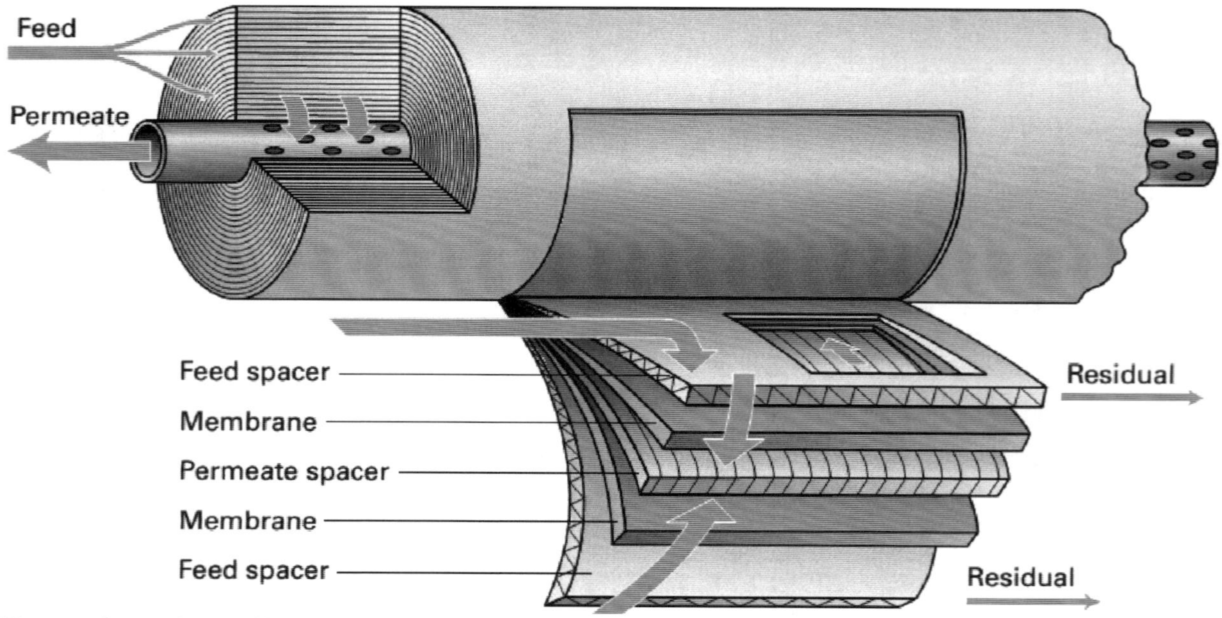

*Two membrane sheets with permeate spacer between; leaves are separated by feed spacers and wrapped around a permeate tube facing it with three open ends.

A membrane sheet has two layers: a relatively thick microporous layer that is in contact with the cloth support, and a thin active layer on top of the microporous layer.

The membrane sheets are separated by a permeate spacer and glued shut at three ends to form an envelope or leaf. Many of these leaves, separated by feed spacer are wrapped around the permeate tube, with the open end of the leaves facing the tube.

Feed gas travels along the feed spacer, the permeating species diffuse through the membrane and down the permeate spacer into the permeate tube, the residue gas exits at the end and the gas flow is cross flow. The spiral configuration is inherently more resistant than the hollow fiber membranes to trace components that would alter the polymer permeability. It also allows a wider range of membrane material to be used. The hollow fiber membrane is cheaper than the spiral one.

Each membrane unit contains a pretreatment unit and membrane skids. Each pretreatment unit consists of a filter coalescer, a guard bed, and a particle filter. Each membrane skid consists of membrane tubes containing membrane elements

398674 Formulating Novel Draw Solutions for Forward Osmosis

Abstract:

In recent years research has been done on the behavior and performance of membranes in different orientations (Forward Osmosis, Reverse Osmosis, Pressure Retarded Osmosis), although relatively few studies have been done analyzing the draw solutions for these systems. This study evaluates draw solutions for an FO system operating in PRO mode, with emphasis on the effects of the concentration and composition of the draw solution. Also, the system used was not a small bench scale, but rather a commercially available, full cartridge membrane system from Hydration Technology Innovations (HTI). Three electrolytes were chosen for the study: Sodium Chloride, Potassium Chloride, and Calcium Chloride, and the effects of changing the concentrations, as well as mixing the electrolytes, were studied. Since chloride is a common ion with all three electrolytes, the effects of the diffusivity coefficients and cation charge of the alkali metals could be studied. Increasing the concentration of the draw solutions was found to increase the flux, but the relationship was not linear. For example, doubling the concentration led to an approximately 35% increase in the output flux for all three electrolytes. Also, when graphing the flux of individual electrolyte solutions against the diffusivity coefficients it was found that there was an inverse relationship, counter to all other studies and theories. It should be noted that the velocity through the membranes was not held constant, as well as the temperature (since each test was conducted at room temperature).

Osmotic Pressure Generation Model for a Pressure Retarded Osmosis System

Torrie Sewell
Chemical Engineering, Undergraduate Student, New Mexico Tech
tsewell@nmt.edu

Frank Huang Ph.D., P.E
Department of Civil and Environmental Engineering, New Mexico Tech
huang@nmt.edu

Mike Riley Ph.D.
Department of Chemical Engineering, New Mexico Tech
mriley@nmt.edu

Oil and gas production is the largest energy production field in the world. When these products are extracted a large volume of fresh water is utilized to drill wells, lubricate drilling equipment, and extract fossil fuels. What is left over after these processes is called produced water, which is the largest waste stream generated by oil and gas industries [1]. This produced water can have extremely high levels of dissolved salts (TDS). In New Mexico, over 20 billion gallons of produced water are generated annually, most of it coming from the Permian Basin [2]. The produced water varies from 1% to 40% TDS, compared with seawater at 3.5% [2]. This water is composed of various salts at varying concentrations as well as biological impurities. Instead of storing this waste water in salt water disposal wells, it has been proposed to use pressure retarded osmosis (PRO) to generate power. PRO is a form of osmosis where a high salinity and a low salinity water source are separated by a semi-permeable membrane. Water naturally passes through the membrane from low salinity to high salinity via forward osmosis. In a PRO process, a pressure is also applied to the high salinity side retarding the flow across the membrane. This results in a pressurized exit stream from the high salinity side which will be used to generate power. This applied pressure also alters the system so it cannot reach equilibrium and runs continuously. In order to create a PRO system a theoretical thermodynamic model of expected osmotic pressures was developed. These osmotic pressures were dependent on the water activity and chemical potential difference between the various salts in the two water sources. To produce this model, the program ChemCAD was used to find the vapor pressure and saturation points of various salts, which was then incorporated into Van't Hoff's equation. The model displayed the maximum osmotic pressure to be approximately 46,000 psi for magnesium chloride, and the minimum osmotic pressure to be approximately 450 psi for potassium sulfate. The major salt components for a sample of produced water are sodium chloride with an expected osmotic pressure of approximately 5,400 psi and sodium sulfate with an expected osmotic pressure of approximately 2,300 psi. This theoretical model will be utilized by comparison to results acquired from a bench scale PRO system.

References

1. Coday, B.D., Pei Xu, Geaudry, E.G., Herron, J. , Lampi, K., Hancock, .N.T., Cath T.Y. (November 2013) The sweet spot for forward osmosis: Treatment of produced water, drilling wastewater, and other complex and difficult liquid streams, Desalination, 333, 23-35
2. Huang, F.Y.C., Rogelj, S., Leclerc, C., and Riley, M., Osmotic Power Development, In New Mexico EPSCoR RII (Energize New Mexico) Proposal

Computational Modeling of Filtration Fundamentals

Salamacha, Nicole, nms65@pitt.edu, and Joseph J. McCarthy, mccarthy@engr.pitt.edu
Department of Chemical and Petroleum Engineering
University of Pittsburgh

In the chemical industry today, filtration can play an important role, from clarifying a material to removing unwanted by-products. Knowing exactly how to filter a product and how to do it consistently can be an art that takes many trial and error attempts to perfect. This work aims to develop a computational model that combines the fluid flow dynamics of the Lattice Boltzmann Method (LBM) and the particle interaction schemes of the Discrete Element Method (DEM) in order to model filtration in lab-scale equipment. Results that include pressure versus flow as a function of filter cake height for a variety of filter cake geometries will be collected. In order to test the validity of the LBM-DEM method, follow-on work that will compare simulated results to an experimental system is being developed.

Understanding Behavior of Ionic Liquid and Water Interface via Molecular Dynamics Simulations

Shehab Alzobaidi, Shanmuga Venkatesan, and Neeraj Rai

Dave C. Swalm School of Chemical Engineering, Mississippi State University, Mississippi State, MS, 39762

Biomass, materials derived from plants and animals, is currently being explored as a carbon-neutral source of fuels and chemicals. Dissolution of biomass is a critical step in developing efficient and economical chemical processes for biomass conversion. Thus, there is a significant push to design solvents that not only dissolve the ligno-cellulosic material but also can be easily separated for recycling. One set of promising solvents are room temperature ionic liquids (RTILs). Due to millions of possible ionic liquids, it is experimentally daunting task to synthesize and evaluate all such possible solvents. In this work, we employ molecular dynamics simulations to explore the ionic liquid/water interface. Analysis of trajectories includes interface area, density fluctuations, and diffusion coefficient of water in IL, and structural details of the interface. Our simulations not only provide the interfacial properties for designing novel separation processes but also provide molecular level insights in the IL/water miscibility. Collectively, this information will allow us to come up with design rules for rational design novel ionic liquids for biomass dissolution.

Effects of the Ratio of Excipient and Solute on Amorphous Particle and Unique Polymorph Formation

Victoria M. Karakis and Ryan C. Snyder

Department of Chemical Engineering, Bucknell University, Lewisburg, PA

Introduction

In order to effectively deliver an Active Pharmaceutical Ingredient (API), pharmaceutical companies must employ a drug with a reasonable amount of bioavailability, the fraction of drug that dissolves appropriately in the patient and is delivered to the system to provide relief of symptoms. One way to do this is to utilize higher energy states than the most stable crystalline structure, specifically amorphous dispersions. In this work, we utilize monodisperse droplet evaporation technology, which has shown to be a process that leads to the formation of less stable or amorphous structures[1], to create amorphous suberic acid. Using polyvinylpyrrolidone (PVP) as an excipient to induce amorphous formation, we are able to determine the ratio of Suberic Acid:PVP in the solution that creates amorphous suberic acid. In Initial X-ray diffraction tests of pure suberic acid without the addition of PVP, it was found that the methods and conditions used in this study produced the metastable phase of suberic acid, denoted as β-suberic acid, one which had yet to be created at room temperature[2]. This study further investigates the uncommon and little known polymorph.

Goals

- Produce monodisperse droplets of pure suberic acid and with the addition of PVP
- Determine the size, morphology and internal structure of suberic acid with and without the addition of PVP
- Determine the polymorph of suberic acid produced
- Determine the ratio of suberic acid to PVP at which amorphous particles are produced
- Compare our work with previous work on succinic and adipic acid

Size and Morphology Results and Analysis

Scanning Electron Microscope (SEM)

- Suberic acid particles with the addition of increasing weight ratios of PVP were viewed under a scanning electron microscope (SEM).
- Their size is consistent with pervious work.
- The particles lose the dimple and become more smooth with increasing amounts of PVP added

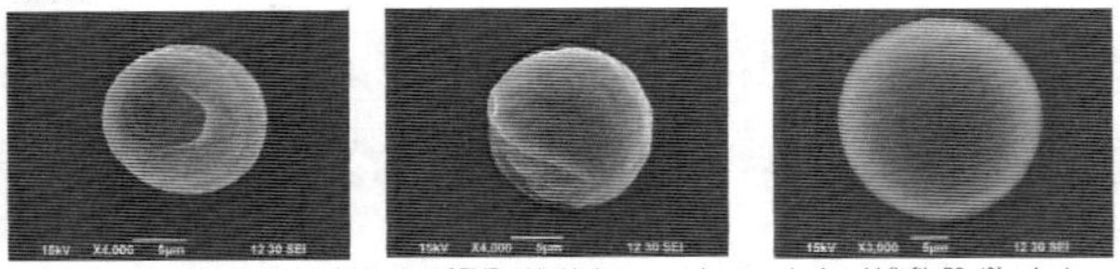

Suberic acid with increasing weight ratios of PVP added in isopropanol; pure suberic acid (left), 50wt% suberic acid 50wt% PVP (middle) and 31.25 wt% suberic acid and 68.75 wt% PVP

- Suberic acid particle without the addition of PVP were also analyzed.
- Confirms monodisperse droplet formation (left) and internal structure (right)

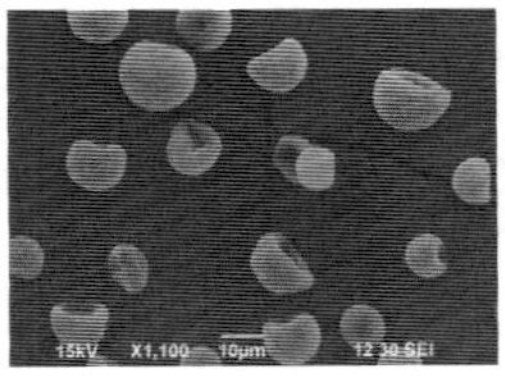

Overview of pure suberic acid particles

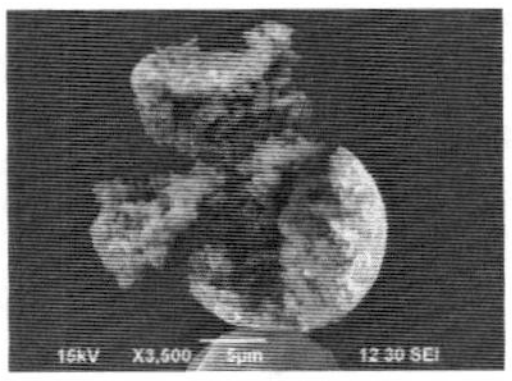

Cracked suberic acid particle

Internal Structure Results and Analysis

X-Ray Diffraction (XRD)

- Particles were analyzed using X-ray Diffraction (XRD) to determine internal structure (whether the particles are amorphous or crystalline) as well. The Powder XRD data shown below is for suberic acid with increasing weight percent ratios of PVP.

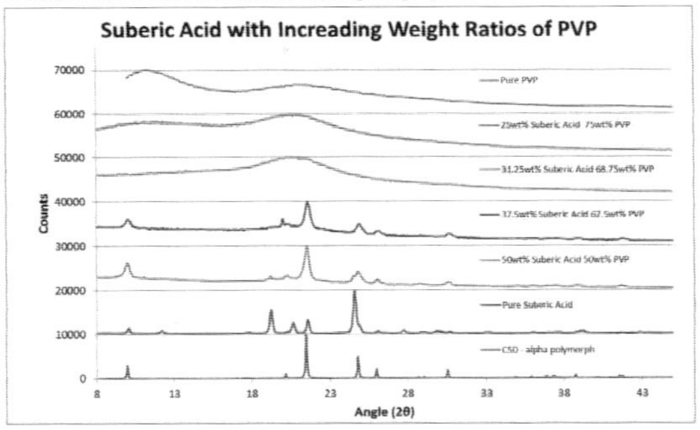

- Particles with less than or equal to 31.25wt% suberic acid show a distinctly amorphous feature, while particles with more than or equal to 37.5wt% suberic acid show a distinctly crystalline feature. Therefore, the ratio at which amorphous particles are produced is somewhere between 31.25wt% and 37.5wt% suberic acid and the correspond weight ratio of PVP
- This corroborates previous work with succinic and adipic acid.
- Analysis was confirmed by Differential Scanning Calorimetry (DSC).

Materials and Methods

- Suberic acid particles are produced via a modified Vibrating Orifice Aerosol Generator (VOAG)
  - Periodic mechanical disturbances (vibrations) of liquid jet allows for production of monodisperse droplets
  - Droplets are dispersed and diluted with air in a drying column
  - Particles evaporate as they move through the column and into the cyclone
- Solutions of pure suberic acid and suberic acid and PVP were prepared using isopropanol
- Operating parameters:
  - Syringe speed: 7.5×10^{-4} cm/sec
  - Vibration frequency: 45 kHz
  - Dispersion air: 1500 cc/min
  - Dilution air: 45 L/min
  - Liquid Pressure: 10-20 psi
- Dry particles were collected at the base of the column and cyclone for analysis

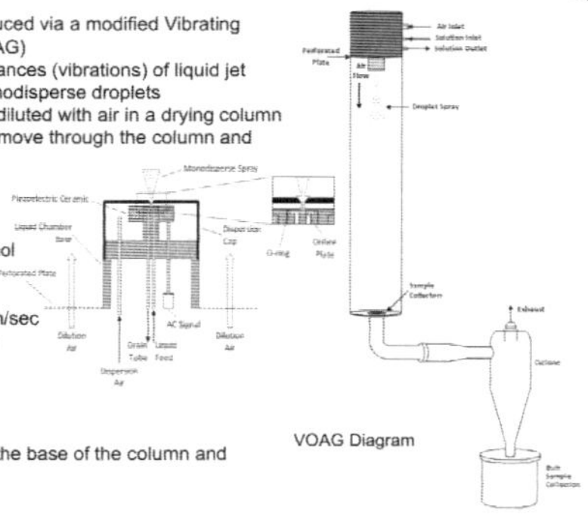

VOAG Diagram

Polymorphic Results and Analysis

X-ray Diffraction (XRD)

- Particles were analyzed using X-ray Diffraction (XRD) to determine the polymorph of suberic acid.
- XRD data shown below is the pure suberic acid in isopropanol data collected from the VOAG alongside the XRD data for suberic acid in the Cambridge Structural Database (CSD).

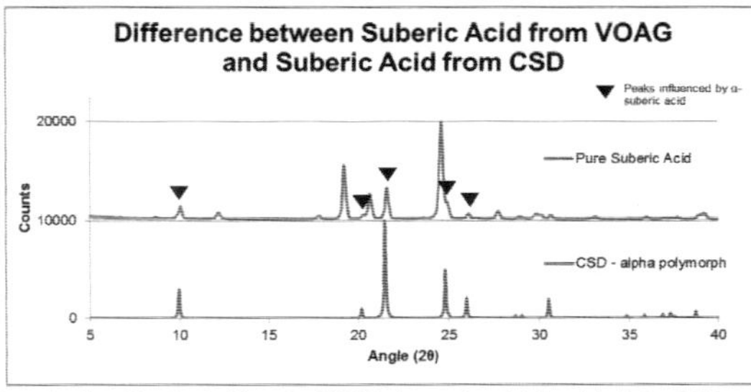

- The suberic acid collected from the VOAG does not match the data in the CSD, suggesting a different polymorph of suberic acid, β-suberic acid, was formed.
- The sample collected from the VOAG most likely has a combination of α- and β-suberic acid.

Conclusions and Future Work

Conclusions

- We were able to produce monodisperse droplets of suberic acid with and without the addition of PVP.
- The size and morphology of suberic acid with and without the addition of PVP is consistent with previous work[3].
- The polymorph of suberic acid was found to be β-suberic acid, one in which was only found in widely different conditions.
- The ratio of suberic acid:PVP at which suberic acid is amorphous was determined to be between 37.5 and 31.25wt% suberic acid and the corresponding weight ratios of PVP, which is the same as results found with succinic and adipic acid.

Future Opportunities

- Consider other dicarboxylic acids in order to find the ratio at which it becomes amorphous and compare to the results found from suberic acid
- Use other solvents like water, ethanol, or acetone to see if it changes the results
- Continue to experiment with β-suberic acid

References

1. Carver, K.C. "The Investigation of Uniform, Monodisperse Crystalline Particles via the Evaporation of Small Droplets." (2012).
2. Ha, J. Hamilton, B. Hillmyer, M. Ward, M. Phase Behavior and "Polymorphism of Organic Crystals Confined within Nanoscale Chambers." *Crystal Growth & Design Vol. 9 4766-4777.* (2009).
3. Webster, L. "Formation of Metastable Organic Structures Via Monodisperse Droplet Evaporation." (2014).

398666 Using Experimental Data to Model Small-Scale Batch Stills for the Purpose of Scaling-up the Column

Abstract:

Alcohol is produced by distilling a fermented corn mash made up of corn, sugar, and yeast. A batch still is used with varying column diameters, stages, and heights to produce high quality ethanol, in excess of 95% by volume. Craft distillation businesses are known to build small-scale, hobby stills capable of producing this high percentage alcohol, but at a low product flow rate. The purpose of this research is to scale-up these hobby stills to produce alcohol at a higher production rate. Experimental data obtained through the use of LabVIEW software such as temperature, pressure, and reflux flow rates along with still geometry measurements are used to accurately model the still in order to scale-up the column. In addition, an alternate method of heating the corn mash was examined. Through intensive heat-transfer calculations, steam coil geometric measurements were appropriately designed to heat up a known amount of corn mash as well as give the specifications of steam needed through the steam generator.

Title:

SEPARATIONS USING SUPERCRITICAL CO_2 DEPOSITED ADSORBENTS

Authors:

Seth Woodard, Jacob Deal, Dr. Christy Wheeler West, Dr. Kevin N. West

Abstract:

The separation of ethylene from ethane and other substances is very important in the chemical industry. This importance is evident in the production of polyethylene, which accounts for the consumption of more than half the world's ethylene production. A vast majority of the world's chemical industries have some type of production currently using a system to separate ethylene from ethane. This type of separation is complex and expensive. However, many metal salts complex ethylene rather easily. The down side is that this complex is unstable in the presence of H_2, which is formed during ethylene's production. Metal salts suspended in and stabilized by ionic liquids have shown great success in separating and protecting the ethylene, but the process degrades over time. Another approach is to support the salt on an inexpensive sub-structure, bringing about the question of how to effectively disperse the salt. Super-critical fluid deposition (SCFD) can be used to make highly disperse structures from metal organic salts. SCFD is useful for the fact that SCF's possess properties of both the gaseous and liquid phase. The liquid property that makes this process attractive is the allowance for a greater solvent solubility, while the attractive gaseous properties are high diffusivity and low viscosity. The goal of this research is to make supported, highly dispersed salts that are protected by a thin film of ionic liquid to reject H_2. This will consist of generating an efficient procedure for super-critical CO_2 deposition of $Cu(acac)_2$ and Ag(acac), which entails investigating this process over a variety of temperatures and pressures, while examining the amounts deposited. Key questions involve the whether the process is kinetically limited or equilibrium limited on practical deposition timescales and to what degree copper loading affects dispersion. In this work the investigators demonstrate the SCFD of copper salts on g-alumina and report both ethane and ethylene adsorption isotherms.

398711 Design of Bis(imidazole) Monomers for Step-Growth Polymerizations

Abstract:

Title: Design of Bis(imidazole) Monomers for Step-Growth Polymerizations

Frank Foley, J. David Roveda, Jason E. Bara

Imidazole is a planar five member ring. It is aromatic and amphoteric, and serves as a versatile building block for a variety of materials. Various substituents can be added at the 1,2,4 and/or 5s positions as a means of controlling the physical and chemical properties. We have produced a number of bis(imidazole) compounds via reaction of various imidazole cores with p-dicholorxylene. These compounds serve as "A-A" monomers which can then be polymerized via reaction with an additional stoichiometric equivalent of p-dichloroxylene to form imidazolium-based polymeric "ionenes". The properties of the resultant ionenes can be correlated with the structure of the imidazolium ring within the polymerbackbone. Initial experimental evidence and molecule modeling suggests that these imidazolium ionenes can adopt helical structures which may promote supramolecular assembly and provide unique properties for the separation of CO2 from industrial gas streams.

Using Vapor-liquid Equilibrium to Evaluate Solvent Selection in the Synthesis of a Model Ionic Liquid

Abstract

Ionic liquids (ILs) are a class of salts that are liquid at temperatures below 100° C. They offer a wide range of promising applications as designer solvents due to their highly tunable structures and are notable for their nearly nonexistent vapor phase. However, ILs are currently manufactured in small batch processes, and research which optimizes the selection of solvent for the synthesis methods is limited. Solvents are needed to safely disperse heat generated from the exothermic synthesis reaction, as well as to enhance the reaction rate. However, increases in reaction kinetics must be balanced against separation energy requirements; ionic liquids are frequently cited as "green" solvents in literature, but must be synthesized in an equally green manner to be truly environmentally friendly. In this study, the phase equilibrium of a model imidazolium-based ionic liquid, 1-hexyl-3-methyl-imidazolium bromide[HMIm][Br], is studied to collect thermodynamic data using two competing solvents: acetone and acetonitrile. From isobaric vapor-liquid equilibrium Txy data, we will elucidate binary interaction parameters which will allow subsequent thermodynamic modelling with NRTL and PR-EOS models to more accurately account for deviations from ideality. This project will quantify these interaction parameters for each binary system within a synthesis mixture and will later conduct separation energy analysis for each solvent using processing engineering software. This will establish the relative ease of separation for each system studied and allow experimental data to validate model correlations. Furthermore, the results will exhibit positive or negative deviations from Raoult's Law, which will show whether the separation has higher or lower energy needs than ideal estimates predict. The results of this study are intended to direct future research on ionic liquid synthesis so that an optimum solvent may be found which minimizes energy use during separation.

Amine Polymer Impregnated MCM-36 for Carbon Dioxide Capture

Justin Ramberger, Daniel Accetta, Christopher F Cogswell, Hui Jiang, Sunho Choi, Ronald Willey

Department of Chemical Engineering, Northeastern University, 313 Snell Engineering Center, 360 Huntington Avenue, Boston, Massachusetts 02115-5000

Solid adsorbents are currently being investigated as possible replacements for liquid amine systems for the capture of carbon dioxide from flue gas and dilute air streams, due to their low heats of adsorbtion and high surface areas. However, these systems cannot on their own reach the capture capacities and capture rates required to become economically feasible. Solid adsorbents generally are not capable of reaching the selectivity for carbon dioxide over other gas stream components possible in liquid amine systems, and often are found to have below the adsorbent capture capacity believed to be necessary for their feasibility as replacements for current practices[1–4]. In order to overcome these challenges, solid adsorbents which have been functionalized with amine groups either through physical impregnation, covalent binding, or in situ polymerization are being studied by multiple groups[5–10]. Despite this, a clear understanding of the kinetics and diffusion properties of these materials has not been established. In this work, we have attempted to elucidate these properties by the study of an MCM-36 porous support functionalized through physical impregnation with varying amounts of polyethyleneimine. This material was chosen due to the multiple porous channels present in the bare support. MCM-36 is composed of MWW type microporous layers, staked in the c-axis in a lamellar fashion[11–13]. Due to this, MCM-36 contains both microporous channels accessible to CO_2 in the a- and b-axis of the layers, as well as larger pore channels present in the pillared network created between each layer. This provides the ability to selectively functionalize the large pore space, because amine groups are too large to become stuck in the microporous channels in the layers. This allows for the creation of a material with open diffusion channels for carbon dioxide through the layers, as well as an amine functionalized zone in the pillared space between the layers. The material was studied systematically by varying the amine loading and investigating the carbon dioxide capture capacity, equilibrium adsorbtion half time and initial capture rate as well as physical properties such as the surface area and micropore volume, in order to develop a clear understanding of the ability to use this material as an industrial carbon dioxide capture system. It was found that with increasing amine content, the diffusion properties and capture capacity are greatly reduced. In fact, with higher amine loading the efficiency of each amine site toward the capture of carbon dioxide becomes greatly hindered, suggesting that increasing the amine-polymer content leads to blockage of the porous channels and eventually limits access of carbon dioxide to the total volume of the material. This sets an upper limit to the capacity of the support which is far lower than that for the initial bare material, and shows that amine impregnation cannot be used to improve the MCM-36 support as a capture system.

1. Aruldoss, D., Saigoanker, R., Savarimuthu, J. Das & R., J. Amine-grafted zeolites-mesoporous ceramics: Synthesis and adsorption characteristics. *Ceram. Int.* **40,** 7583–7587 (2014).

2. Choi, S., Drese, J. H., Eisenberger, P. M. & Jones, C. W. Application of amine-tethered solid sorbents for direct CO2 capture from the ambient air. *Environ. Sci. Technol.* **45,** 2420–7 (2011).

3. Choi, S., Drese, J. H. & Jones, C. W. Adsorbent materials for carbon dioxide capture from large anthropogenic point sources. *ChemSusChem* **2,** 796–854 (2009).

4. Hiyoshi, N., Yogo, K. & Yashima, T. Adsorption characteristics of carbon dioxide on organically functionalized SBA-15. *Microporous Mesoporous Mater.* **84,** 357–365 (2005).

5. Kumar, P., Kim, S., Ida, J. & Guliants, V. V. Polyethyleneimine-Modified MCM-48 Membranes□: Effect of Water Vapor and Feed Concentration on N 2 / CO 2 Selectivity. 201–208 (2008).

6. Li, W. *et al.* Structural changes of silica mesocellular foam supported amine-functionalized CO2 adsorbents upon exposure to steam. *ACS Appl. Mater. Interfaces* **2,** 3363–72 (2010).

7. Yang, S.-T., Kim, J.-Y., Kim, J. & Ahn, W.-S. CO2 capture over amine-functionalized MCM-22, MCM-36 and ITQ-2. *Fuel* **97,** 435–442 (2012).

8. Liu, F.-Q. *et al.* Amine-Tethered Adsorbents Based on Three-Dimensional Macroporous Silica for CO2 Capture from Simulated Flue Gas and Air. *ACS Appl. Mater. Interfaces* **6,** 4371–81 (2014).

9. Drese, J. H. *et al.* Synthesis-□□Structure-□□Property Relationships for Hyperbranched Aminosilica CO 2 Adsorbents. *Adv. Funct. Mater.* **19,** 3821–3832 (2009).

10. Drese, J. H. *et al.* Effect of support structure on CO2 adsorption properties of pore-expanded hyperbranched aminosilicas. *Microporous Mesoporous Mater.* **151,** 231–240 (2012).

11. Corma, A., Forn, V., Mart, J. & Pergher, S. B. Delaminated Zeolites□: Combining the Benefits of Zeolites and Mesoporous Materials for Catalytic Uses. *J. Catal.* **63,** 57–63 (1999).

12. Lawton, S. L. *et al.* Zeolite MCM-49□: A Three-Dimensional MCM-22 Analogue Synthesized by in Situ Crystallization. 3788–3798 (1996).

13. Maheshwari, S. *et al.* Influence of layer structure preservation on the catalytic properties of the pillared zeolite MCM-36. *J. Catal.* **272,** 298–308 (2010).

Oxidative Degradation of Aqueous Amine Solvents for CO_2 Capture

Hanbi Liu, Daniel Hatchell, Omkar Namjoshi, Gary T. Rochelle

McKetta Department of Chemical Engineering

In the proposal of using amine solvents to remove CO_2 from coal-fired gas plant post-combustion flue gas, one of the challenges is solvent loss by oxidative and thermal degradation. This study focuses on the oxidative degradation behavior of diamines at absorber conditions. The dmaines used in this study are ethylenediamine (EDA), propylenediamine (PDA), and 1,4-diaminobutane (DAB).

Three main factors were taken into consideration—structural effects of the amines, catalytic effect of metals present at operating conditions, and whether degradation is limited by oxygen mass transfer into the solution or reaction kinetics. Based on previous works by Goff (2005), Sexton (2008) and Voice (2013) focusing on MEA oxidation, it is believed that hindered and substituted amines are more resistant to oxidation. Their studies have also shown that metals like Fe and Cu have synergetic catalytic effects on oxidative degradation. For other metals such as Mn, Goff and Sexton's results suggested that Mn acts as an inhibitor at low concentration while Voice's results suggested the opposite and high (70 °C) and low (55 °C) temperatures.

Diamine solutions were prepared gravimetrically at an initial concentration of 5 m amine loaded initially with 0.4 mol CO_2/mol alkalinity, and an initial metals concentration of 0.4 mM Fe, 0.2 mM Mn, 0.1 mM Ni, and 0.05 mM Cr. Each amine solution was agitated in a stirred reactor maintained at 70 °C, 98 kPa O_2, and 2 kPa CO_2. Samples were taken out from the reactor at scheduled time intervals. Water balance was maintained each time a sample was removed to make up for water loss through evaporation. Cation chromatography was used to analyze for parent amine concentrations and degradation by-product concentrations in each sample. Anion chromatography was used to qualitatively analyze for formate salts and nitrite.

Among the three amines tested, EDA and DAB showed significant degradation over the 2-week experimental period while PDA showed little signs of degradation. The amine concentrations are calculated based on cation chromatography results and plotted against experimental time. Figure 1 shows the PDA concentration vs. experimental time at high and normal metals concentration. PDA does not show significant degradation and the fluctuations are likely noise caused by water balance adjustments.

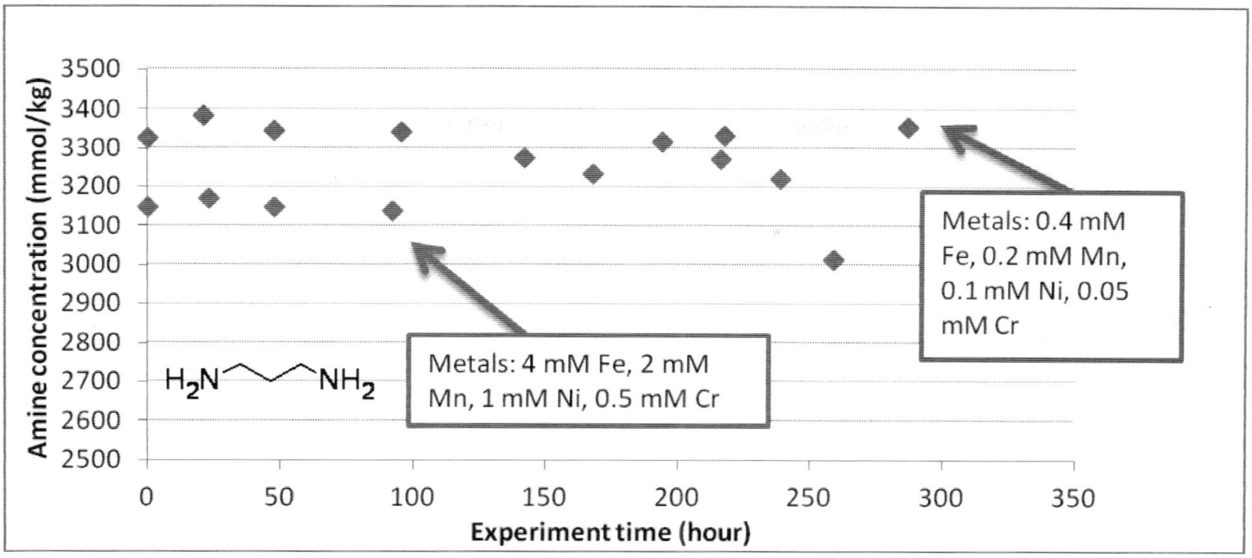

Figure 1: Amine concentration of 5 m PDA as a function of time, showing that PDA is resistant to oxidation with either a high concentration or normal concentration of metals. Conditions: 70 °C, 98 kPa O_2, 2 kPa CO_2, 100 ml/min gas flow, 1400 rpm rotational speed.

Figure 2 shows the degradation of EDA follows 0th order behavior with a rate constant of 9.79 mmol/kg-hr. Figure 3 shows degradation of DAB follows first-order behavior with a rate constant of 0.0104 mmol/hr.

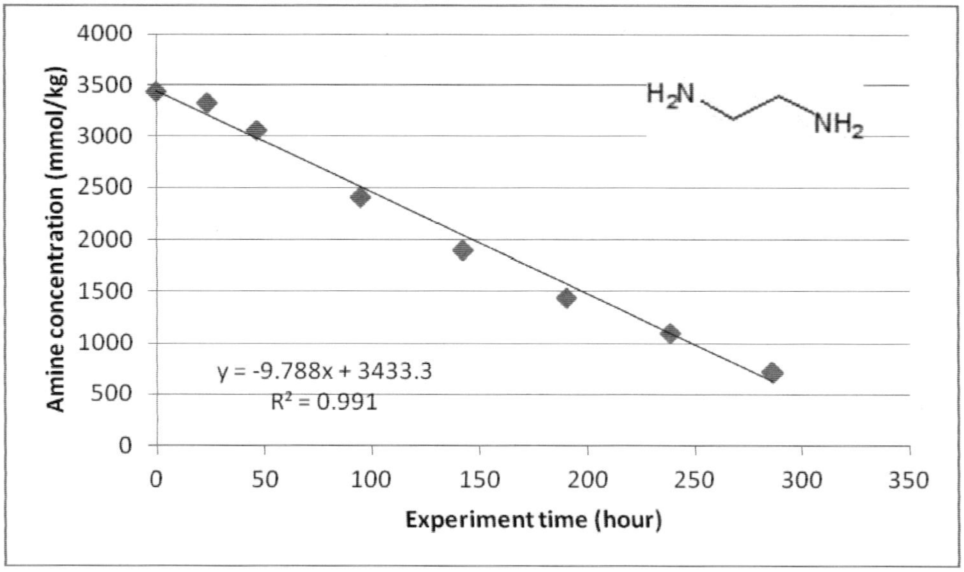

Figure 2: Amine concentration of 5 m EDA as a function of time, showing that EDA oxidative degradation follows 0th order kinetics and is likely mass transfer controlled or catalytically controlled. Conditions: 70 °C, 98 kPa O_2, 2 kPa CO_2, 100 ml/min gas flow, 1400 rpm rotational speed, and metals concentration of 0.4 mM Fe, 0.2 mM Mn, 0.1 mM Ni, 0.05 mM Cr

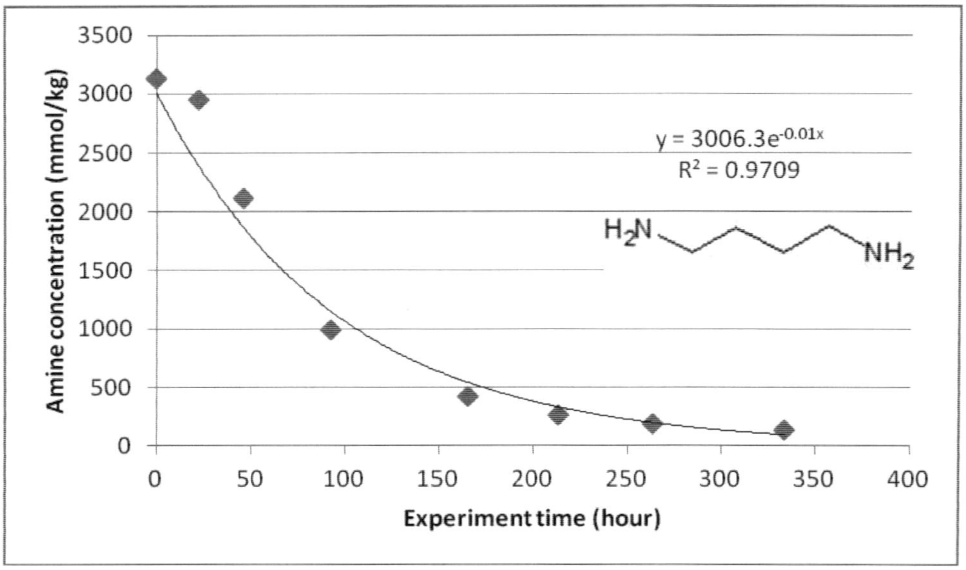

Figure 3: Amine concentration of 5 m putrescine as a function of time, showing that putrescine oxidative degradation follows first-order kinetics. Conditions: 70 °C, 98 kPa O_2, 2 kPa CO_2, 100 ml/min gas flow, 1400 rpm rotational speed, and metals concentration of 0.4 mM Fe, 0.2 mM Mn, 0.1 mM Ni, 0.05 mM Cr

For future work, more experiments with different diamines will be performed to study the structural trend and control mechanism of diamine oxidative degradation. This includes amines such as bisaminoethylether (BAE), Diglycolamine (DGA®), and hindered amines such as Tris(hydroxymethyl)aminomethane (Tris). DGA® is a primary alkanolamine and Tris is a hindered alkanolamine.

The EDA oxidation experiment will be repeated to understand and quantify repeatability of the experimental apparatus. The next step in studying the degradation products of these amines is to quantify the heat stable salt concentration, nitrile concentration, and nitrosamine yield.

398326 3-D Printing for CO2 Capture and Chemical Engineering Design

Abstract:

3-D printing is a form of "additive manufacturing" that allows the fabrication of items directly from digital files. It allows the user to produce virtually any solid object "on demand" and on site, instead of creating it elsewhere. As 3-D printing grows, it has begun to find a number of applications in research laboratories. We have recognized that 3-D printing can have many roles in the design of chemical engineering processes as a means of fabricating parts or perhaps entire unit operations.

Specifically, we foresee many opportunities for 3-D printing as a means of producing novel and advanced components and entire devices for gas treating. Gas treating (i.e. the removal of one or more contaminants from a gaseous mixture) plays a crucial role in many existing and emerging energy-related processes including natural gas sweetening, flue gas desulfurization and pre- and post-combustion CO2 capture. These processes are typically carried out in absorption columns containing trays or packing that provide interfaces for gas-liquid contacting. These trays and packing are traditionally created using conventional manufacturing techniques (i.e. casting and molding). New devices such as membrane contactors are also emerging as alternative mechanisms for achieving separation of gases with potential cost and energy saving benefits derived from having smaller footprints, being of lighter weight and having much larger interfacial areas.

Because all of these devices are produced using conventional manufacturing techniques, the design, cost and ability to optimize gas-liquid contactors may be limited. In this respect, 3-D printing could be a mechanism by which to achieve improvements on existing technologies and more rapidly deploy novel devices. We have begun to design, print, and test new packing materials to be used for CO2 separations in a gas-liquid contactor column. In this presentation we will describe the design and performance of our new packing materials as well as future design plans. We will also present the short- and long-term benefits resultant from our use of 3-D printing in CO2 separations.

394471 Reactive Absorption of CO2 from Industrial Flue Gas Using Aqueous Solution Containing Renewable Prepared Amine

Abstract:

Flue gas from coal-fired power plants, cement manufacturing, refineries, etc., is considered as one of the major sources of CO_2 emissions which contribute to the increasing greenhouse effect. Because of this adverse effect of CO_2 discharge on the global environment, as well as the world's immense dependence on fossil fuels, the development of strategies for the reduction of CO_2 emissions has become increasingly important. With this viewpoint, considerable efforts are underway towards the development of CO_2 capture technologies. Due to its higher efficiency and lower complexity, amine based closed loop chemical absorption/desorption technology (CADT) is preferred industrially as an effective gas sweetening process for separation of CO_2 from flue gas. In this investigation, solvent-1: aqueous solution of renewably prepared ethylaminoethanol (EAE), solvent-2: aqueous blend of EAE and N-methyl-2-pyrolidone (NMP), and solvent-3: aqueous blend of EAE, NMP, and hydroxyl radical scavengers were investigated towards reactive absorption of CO_2 from industrial flue gas. Degradation of all the above mentioned solvents in presence of O_2 and CO_2 was studied at different experimental conditions by using a high-temperature high-pressure subcritical reactor and the samples obtained after performing the degradation reactions were analyzed via gas chromatography (GC). Furthermore, kinetics of reactive absorption of CO_2 in solvent-1, solvent-2, and solvent-3 was also investigated in detail by using stirred cell reactor. A fall in pressure technique was employed for the absorption study and fall in the CO_2 pressure during the absorption experiments was determined by using pressure transducer coupled with a data acquisition system. Effects of various process parameters such as stirring speed, amine concentration, and reaction temperature on absorption kinetics were examined in detail. The results obtained indicate that solvent-1, solvent-2, and solvent-3 possess relatively high solvent stability and CO_2 absorption capacity with faster absorption kinetics as compared to previously investigated amine based solvent systems. The experimental set-up/procedures and results obtained during this investigation will be presented in detail.

395456 Wastewater Treatment Using Microbial Fuel Cells with Peroxide Formation

Abstract:

The breakdown of solids in wastewater is called hydrolysis and may be the rate limiting step of the wastewater sludge treatment process. The microbial fuel cells studied have the capability to produce hydrogen peroxide which can potentially be used to increase the hydrolysis rate. Tests have been conducted on wastewater primary sludge with hydrogen peroxide to determine if it does increase the hydrolysis rate. The tests indicate that it does not dramatically increase the rate, however a microbial fuel cell will be constructed and hydrogen peroxide treated wastewater sludge will be used to determine the positive effects of this method.

Simwatachela, Zambia, Africa Sustainable Water Project

The current project for MSU's Chapter of Engineers Without Borders (EWB) is in Simwatachela, Zambia, Africa. The need for a clean and sustainable source of clean water was presented by the Simwatachela Sustainable Agricultural and Arts Program (SSAAP) to the national EWB organization where the MSU organization then took ownership of the project. The team at MSU has successfully worked with SSAAP to put in place four hand pump water wells and is continuing to work the logistics on four more. The current wells help over four-thousand people. Because of the differences in drilling and logistics a safety manual is currently being made. This project not only helps people directly in Zambia, but also made each team member more aware of the lifestyle many Americans take for granted. Working and coming up with solutions in an environment unlike America has made me more ingenuitive.

TOWARDS STRUCTURE-CYTOTOXICITY CORRELATIONS FOR COMPLEX ENGINEERED NANOMATERIALS

Yutao Gong, Sharlee Mahoney, Thomas Richardson, Kimaya Padgaonkar, Suzanna Hinkle, Ipsita Banerjee and Götz Veser

Department of Chemical & Petroleum Engineering, Swanson School of Engineering, University of Pittsburgh, PA, USA

Email: yug17@pitt.edu

Nanomaterials are widely used across different fields such as electronics, catalysis, and medical treatment, but potential health and environmental effects are triggering concerns due to the lack of standardized, robust tests for nanotoxicity. We aim to develop robust, high-throughput assays for nanotoxicity and combine those with a thorough physiochemical characterization of nanomaterials in order to derive structure-toxicity correlations. In the present work, nickel and silica complex engineered nanomaterials are used, combining a toxic metal (Ni) with a largely non-toxic matrix (porous silica). We use three different configurations: nickel deposited on silica, nickel encapsulated in hollow silica and nickel embedded in non-hollow silica. Here we report on the settling and dissolution characteristics of these three nanomaterials. The settling rate of the nanomaterials was measured using UV-Vis spectroscopy, the agglomeration of the nanomaterials were studied using DLS and ion-coupled plasma-mass spectrometry (ICP-MS) was used to measure the amount of Ni ion dissolution in cell media over 5 days. Our results show that the settling rates for the three nanomaterials were identical within experimental error, indicating that the cells are exposed to a similar number of nanoparticles in the cell experiments. The agglomeration of the nickel nanoparticles was measured both in DMEM media (cell media) and DMEM media supplemented with FBS protein. In pure DMEM, all three types of Ni nanoparticles agglomerated to sizes of ~1,000nm, and only nhNi@SiO2 retained a small fraction of smaller agglomerates. When FBS was added, Ni-SiO_2 formed only agglomerates of 1000 nm, while hNi@SiO2 and nhNi@SiO2 formed also smaller agglomerates. This suggests that proteins play a significant role in nanomaterials agglomeration, likely through the formation of a protein corona, and hence are likely to affect uptake of these materials by the cells. Simultaneous dissolution studies show that Ni-SiO2 has a significantly lower Ni dissolution rate, again suggesting that the protein corona that caps the (external) Ni NP on Ni-SiO2—but not the internal Ni NP in the other two core-shell materials—strongly alters the behavior of these materials in cell culture media.

As a next step we will investigate the intracellular Ni ion concentration (using a fluorescent dye) in order to assess the intracellular Ni^{2+} release and correlate the results with cell viability assays. This will allow us to identify whether the Ni nanoparticles themselves or the intracellular Ni ions cause the nanomaterials toxicity.

Optimization of Low Cost Biodiesel Washing Process from Locally Sourced Materials for sub-Saharan Africa

Max Croft
University of Kentucky Department of Chemical Engineering
Faculty advisor: Dr. Jeffrey Seay

Abstract

o produce biodiesel is commonly carried out

ɔn sophisticated production methods that aren't nderdeveloped regions.

ɜnge since these regions don't have access to processes.

s Cameroon are rich in biological resources.

ited electricity or industrial resources so a low-ɔlogy (AT) based approach is required.

n solving a key challenge in the production of the soap produced as a side product of the ion – in a simple and cost effective method.

available resources for different wash additives ɔnducted to yield the optimal washing process.

Background

- Three major steps required to produce biodiesel: Filtration, Transesterification, and Washing.
- Normal static washing as a means of soap removal does not work well for an AT solution because of the thick emulsion formed.
- This emulsion is due to the interfacial properties of the soap and biodiesel.
- Soap molecules are highly polar and are attracted to water molecules. Therefore pure water is an effective means for removing soap via liquid-liquid extraction.

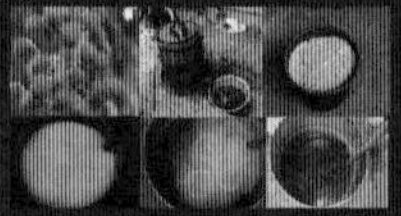

Preliminary Experimental Results

ːs were conducted to determine the effectiveness ɾcerin solutions to wash crude biodiesel.

ːgar and glycerin washes were most effective in ɪtent, while the salt and vinegar washes were ɪng the emulsion formed.

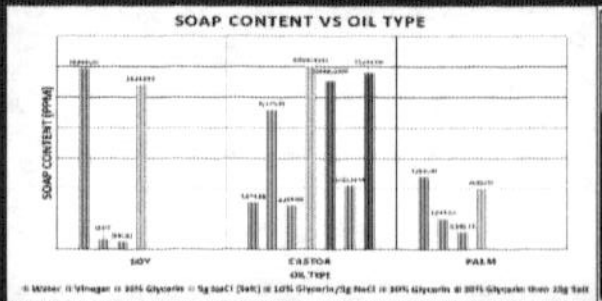

Preliminary Experimental Results

- A Statistical design of experiments was conducted to determine the effectiveness of each washing method.
- Different solvent ratios of vinegar, salt, and glycerin at different temperatures were used to wash the crude biodiesel.
- The 55°C wash solutions were the most effective in reducing the soap content, along with the solutions containing all three wash solvents.
- The results from the washes are shown in the graph below.

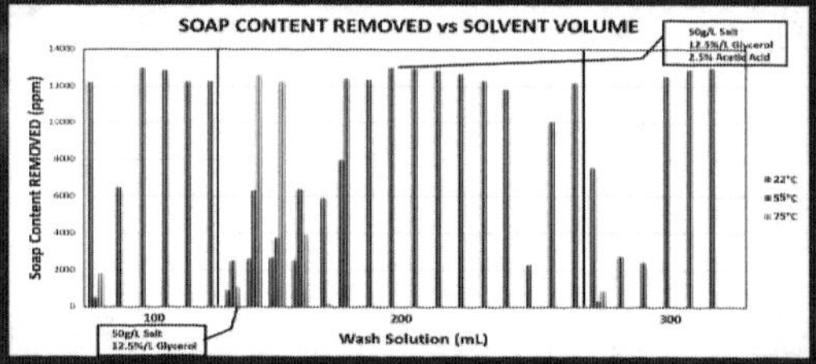

Experimental Methods

- The main focus of this research was to eliminate the emulsion and lower the soap content in the biodiesel by varying: pH, temperature, and the solvent to feed ratio of locally available solvents .
- Vinegar (acetic acid), fermented pineapple and lemon vinegar, and calcium carbonate ($CaCO_3$) were the substances chosen from the result of field research conducted in Cameroon.
- The statistical design of experiments previously conducted was used to design oil and region specific experiments, focusing on simplistic washing methods and solvents.

Results and Discussion

- The experiments yielded promising data, resulting in multiple AT washing techniques.
- Using a 1:1 solvent to feed ratio was the most effective. Signifying, that the wash solution surface area needs to be equal.
- Allowing the biodiesel and wash solution one weeks time to settle, yields a lower soap content in the biodiesel compared to next day testing.
- The results from the wash experiments are shown in the graph below.

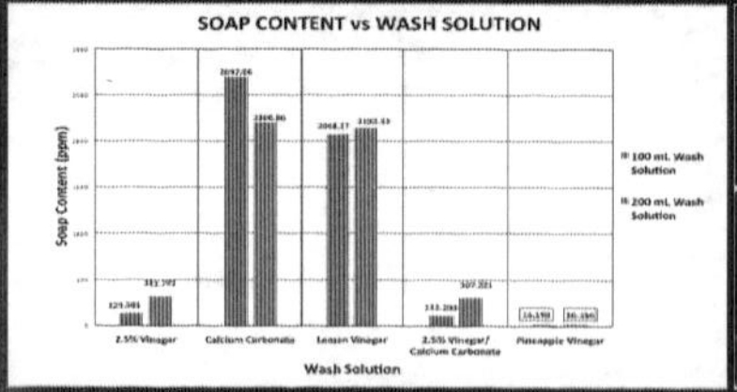

Future Work

- Implementing an AT recycled was process for ACREST.
- Finding another feedstock that could water soa solution.
- Configuring the same process for In fermented

Acknowledgements

- The University of Kentucky.
- The African Center for Renewable Energy and Sustainable Technology (ACREST).

References

- Croft, M., W. Croft and J. Seay (2013). "Developing an Optimal Che Local Feedstock in Rural West Africa", Proceedings of the 2013 Ma Volume.
- Rahayu, Suprihastuti Sri; Mindaryani, Aswati. Optimization of Biodi of the World Congress on Engineering and Computer Science 2007.
- Lkshmi, Ch. Vijaya; Viswanath, K.; Venkateshwar, S.; ... in the production of biodiesel using edible and non-edible ...
- Toledano, Pierre. Phase transitions in complex fluids. 1st ed. World Sci ... Washing by Water Ex

ember 17, 2014 American Institute of Chemical Engineers Annual Student Conference Atlanta,

398411 Thin Film Photovoltaics Response to Heat on Various Substrates

Abstract:

Thin-film photovoltaics are flexible and have the potential to be more cost-effective than crystalline silicon. However, increasing the temperature of a semiconductor reduces the band gap and overall efficiency of the solar panel. A Copper Indium Gallium diSelenide (CIGS) 70 Watt panel was set directly onto an asphalt roof during peak sun hours to test the effects of temperature on electricity production. An Arduino Uno board was programmed to report the temperature of the roof and the temperature of the white platform on which 14 additional CIGS panels were installed. When compared with amorphous silicon (a-Si) solar cells installed on a metal roof near the location, the CIGS panels performed better when the temperature ranged from 18-28 °C and the a-Si were more efficient when the temperature was between 27-36 °C.

Integrating Life Cycle Assessment (LCA) into Design Process

By Armando Maldonado

Abstract: Comparing what is used today to evaluate different design alternatives using a Life Cycle Cost Analysis (LCCA), with the implementation of sustainability metrics for Life Cycle Assessment (LCA) to compare which of the pavement designs impact the environment less. The scope is to integrate an LCA into the design process for decision-making. The interest for us to provide this LCA for pavement design process is because each year the United States places about $100 billion of roadway construction. This level of activity possibly suggests significant sustainability implications in the construction activities for this work. A means to quantify some of this impact is through life-cycle assessment (LCA). The research project provides an overview of the design process used for comparing alternatives with the conventional LCCA tool. Also the LCA metrics was calculate to provide a comparison decision process. With the LCCA done, we have founded the flexible pavement was more cost effective than the rigid pavement. The flexible pavement and rigid pavement were analyzed for a useful life of 30 years and the analysis was performed for a one-mile length. In the LCA metrics it was found that the energy use for asphalt concrete slab was double than that of the PCC-Slab. Also the amount of waste generated, and CO_2 emissions were calculated for both materials. For the future as more sustainability metrics are developed, LCA can be incorporate into pavement design process and this can help us to make more wisely decision at the time of choose which of the alternatives is better. Now see it in a broader way than before.

394473 Thermochemical Storage of Concentrated Solar Radiation Via Two-Step Iron Oxide – Iron Sulfate Cycle

Abstract:

Solar radiation is an essentially inexhaustible energy source that delivers about 100,000 TW to the earth. To harvest the solar radiation and to convert it effectively into storable and transportable chemicals via thermochemical cycles provides a promising path for a future sustainable energy economy. Among the several thermochemical cycles investigated previously, the sulfur-iodine cycle (SI Cycle) and its variation the hybrid sulfur cycle are more appealing as the required operating temperatures are lower as compared to other thermochemical cycles. For both cycles, the most energy consuming step is the dissociation of SO_3 into SO_2 and O_2, which is possible only under catalytic conditions. As sulfation poisoning is a major concern related to such reactions, simply the noble metal catalysts were observed to be active towards the endothermic dissociation of SO_3. Although, the noble metal catalysts are attractive for such reactions, they are less preferable due to the limited availability and high cost.

In this study, a two-step iron oxide – iron sulfate (IO-IS) cycle was thermodynamically investigated towards thermochemical storage of concentrated solar radiation. It is a two-step process in which the first solar step belongs to the endothermic thermal reduction of MSO_4 into MO, SO_2, and O_2. The exothermic step two corresponds to the non-solar oxidation of MO by SO_2 and H_2O producing metal sulfate (MSO_4) and H_2. The MO and SO_2 produced in step 2 are recycled back to step 1 and hence can be used in multiple cycles. The thermodynamic equilibrium compositions during step 1 (solar thermal reduction of $FeSO_4$ under inert atmosphere) and step 2 (oxidation of FeO via water splitting reaction) were determined. The variation of the reaction enthalpy, entropy and Gibbs free energy for the thermal reduction and water splitting steps with respect to the operating temperatures were studied. Furthermore, solar absorption efficiency of the solar reactor, net energy required to operate the IO-IS cycle, solar energy input to the solar reactor, rediation heat losses from the solar reactor, rate of heat rejected to the surrounding from the the water splitting reactor, and maximum theoretical solar energy conversion efficiency of the IO-IS cycle was determined by performing the second law thermodynamic analysis over different solar reactor temperatures and with/without considering the heat recuperation. Findings of this investigation will be presented in detail.

Business Models for Solar Powered Charging Stations to Develop Infrastructure for Electric Vehicles

Jessica Robinson [1,*], Gary Brase [2,†], Wendy Griswold [3,†], Chad Jackson [4,†] and Larry Erickson [3,*]

1

Department of Chemical Engineering, Kansas State University, Manhattan, KS 66502, USA

2

Department of Psychological Sciences, Kansas State University, Manhattan, KS 66502, USA; E-Mail: gbrase@ksu.edu

3

Center for Hazardous Substance Research, Kansas State University, Manhattan, KS 66502, USA; E-Mail: griswold@k-state.edu

4

Department of Management, Kansas State University, Manhattan, KS 66502, USA; E-Mail: cjackson@k-state.edu

†

These authors contributed equally to this work.

*

Authors to whom correspondence should be addressed; E-Mails: jmrobins@live.unc.edu (J.R.); lerick@ksu.edu (L.E.); Tel.: +1-785-532-4313 (L.E.); Fax: +1-785-532-7372 (L.E.).

External Editor: Jack Barkenbus

Received: 3 August 2014; in revised form: 7 October 2014 / Accepted: 14 October 2014 / Published: 23 October 2014

Abstract

: Electric power must become less dependent on fossil fuels and transportation must become more electric to decrease carbon emissions and mitigate climate change. Increasing availability and accessibility of charging stations is predicted to increase purchases of electric vehicles. In order to address the current inadequate charging infrastructure for electric vehicles, major entities must adopt business models for solar powered charging stations (SPCS). These SPCS should be located in parking lots to produce electricity for the grid and provide an integrated infrastructure for charging electric vehicles. Due to the lack of information related to SPCS business models, this manuscript designs several models for major entities including industry, the federal and state government, utilities, universities, and public parking. A literature review of the available relevant business models and case studies of constructed charging stations was completed to support the proposals. In addition, a survey of a university's students, staff, and faculty was conducted to provide consumer research on people's opinion of SPCS

construction and preference of business model aspects. Results showed that 69% of respondents would be more willing to invest in an electric vehicle if there was sufficient charging station infrastructure at the university. Among many recommendations, the business models suggest installing level 1 charging for the majority of entities, and to match entities' current pricing structures for station use. The manuscript discusses the impacts of fossil fuel use, and the benefits of electric car and SPCS use, accommodates for the present gap in available literature on SPCS business models, and provides current consumer data for SPCS and the models proposed.
Keywords:
emissions; kWh; solar panel; smart grid; parking; finance; payback period; electric grid; partnership

1. Introduction

Transportation is a major source of outdoor air pollution. Transportation accounts for one-quarter of greenhouse gas emissions and one-third of US fossil fuel consumption [1]. Carbon dioxide is one of the principal greenhouse gases, which causes global warming and climate change [2]. Eighty-five percent of the high levels of greenhouse gases in the transport sector, including carbon dioxide, can be attributed to vehicle emissions [3]. The atmospheric carbon dioxide concentrations in May 2014 reached 401.88 parts per million, 50 parts per million over the safety limit for carbon dioxide levels [2]. In order to slow the rate of climate change and to protect people's health, the amount of air pollutants must be reduced [4]. Specifically, in order to decrease carbon emissions by 80%, electric power must be less dependent on fossil fuels and transportation must be electric [5,6]. Therefore, the United States and the rest of the world need to shift reliance from gasoline powered vehicles and fossil fuel based electricity generation to electric vehicles and renewable energy.

The most viable current options for making this transition are solar (and wind) power generation and increased infrastructure for electric powered transportation. Both of these are important steps in reducing carbon emissions. These two changes can be combined to maximize their impacts: solar power can be used with charging stations to generate electricity to reduce fossil fuel consumption. Many questions surround the adoption of solar power, of electric cars, and the combination of both of these new technologies together as solar powered charging stations (SPCS). However, Tesla Motors is demonstrating the feasibility of this sustainable development by constructing a network of SPCS to provide charging infrastructure in convenient locations for their electric vehicles across the nation. The purpose of the present research is to provide some initial answers to questions about the viability of solar powered charging of electric vehicles including public reception, physical logistics, business models, and economic viability. There is considerable literature related to this topic [1,2,3,4,5,6,7,8,9,10,11,12,13,14,15,16,17,18,19,20,21,22,23,24,25,26,27,28,29,30,31,32,33,34,35,36,37,38,39,40,41,42,43,44,45,46,47,48,49,50,51,52,53,54,55,56,57,58,59,60,61,62,63,64,65,66]; however, there is a need for review of business models for SPCS, which is provided in this work.

2. Survey

An online survey was conducted from 26 June 2014 to the morning of 7 July. The survey was designed for Kansas State University (K-State) students, staff, and faculty to provide consumer research of peoples' opinions on solar power, electric cars, and the installation of SPCS in campus parking lots. The survey was created on Qualtrics Survey Software and distributed through an advertisement on K-State Today, a daily email with current announcements, to staff and faculty; and through emails to students, staff, and faculty. Participants were entered into the lottery for a chance to win a $50 gift card to the K-State Union. There were a total of 759 respondents comprised of 98 students, 192 faculty, 451 staff, 11 students and staff, five students and faculty, and two staff and faculty. Some respondents are in multiple sub-groups because they share responsibilities of both groups; i.e., someone could be a student at K-State but also work in a staff position. To provide perspective, K-State has about 6600 faculty and staff, and 24,000 students, including undergraduates and graduates [7]. The majority of the respondents were staff members (59%). A large portion of the staff parks on campus and will remain at the university for many years; they will be more impacted by the installation of SPCS in the long-term.

Although these respondents are affiliated with a university, they are also consumers; thus their responses can at least tentatively be extrapolated to a wider consumer base and used to facilitate business model designs. The results of the survey are mentioned throughout the paper and the complete survey questions and results are included in the Appendix.

Table 1 shows how students, faculty, or staff responded to certain survey questions and allows comparison among these different groups.

Table 1. Comparison of survey results for K-State students, faculty, and staff respondents.

Survey Questions	Students	Faculty	Staff
Most frequently drive to campus	43%	85%	90%
Travel 2–5 miles to campus	52%	51%	43%
Own electric cars	0%	3.60%	0.20%
Would like to reduce the cost of commute by using electric vehicles	73%	69%	58%
Are in favor of K-State installing SPCS	53%	82%	75%
Would like to own an electric vehicle	36%	52%	34%
Are more willing to invest in an electric vehicle if K-State has the infrastructure	26%	44%	26%
Believe SPCS on campus will encourage more people to purchase an electric vehicle	37%	47%	36%

Table 1. Comparison of survey results for K-State students, faculty, and staff respondents.

Overall, faculty are the most in favor of installing SPCS, are most willing to invest in electric vehicles, and are most believing that SPCS will encourage more electric vehicle purchases. Faculty's greater interest in SPCS and electric vehicles is likely related to their greater salaries, and possibly more advanced knowledge about electric vehicles and related environmental issues.

3. Values of Electric Vehicles

There are several social, environmental, and economic implications associated with electric vehicles replacing internal combustion vehicles [5]. All-electric vehicles (BEVs) have no tailpipe emissions, which as a result improves air quality and lowers health risks [3]. Electric vehicles also help the nation reduce its reliance on foreign fossil fuels [1]. In addition, electric vehicles have one-tenth the number of parts that gasoline-powered vehicles have [3], thus requiring less maintenance, and saving drivers time and money.

For instance, one study (cited in [3]) found that electric vehicle owners save about 46% in annual maintenance costs. Another economic value of electric vehicles is that energy operational costs are about one-third the fuel cost associated with internal combustion engine vehicles [5].

Table 2 shows price comparisons of electric vehicles to an average gasoline-powered car for varying distances traveled. This table compares the average cost it takes to travel a certain number of miles (one way) for an electric or gasoline powered car. The average US cost of electricity per kWh was estimated to be $0.12 [8]. The average US price for gasoline was estimated as $3.749/gallon [9]. All values used were based on 2014 data [10], but the 2013 mpg value for the average gasoline powered car was used [11].

Table 2. Comparison of travel costs for electric vehicles and gasoline-powered cars.

Distance to Destination (miles)	Tesla	Nissan Leaf	Plug-in Toyota Prius	Ford Focus Electric	Chevrolet Volt	Average Gasoline Powered Car	Ford Focus (gas)
10	$0.42	$0.36	$0.35	$0.38	$0.42	$1.63	$1.21
20	$0.84	$0.72	$1.02	$0.77	$0.84	$3.26	$2.42
30	$1.26	$1.08	$1.77	$1.15	$1.26	$4.89	$3.63
40	$1.68	$1.44	$2.52	$1.54	$1.80	$6.52	$4.84
50	$2.10	$1.80	$3.27	$1.92	$2.81	$8.15	$6.05
60	$2.52	$2.16	$4.02	$2.30	$3.83	$9.78	$7.26
Electrical Efficiency (kWh/100 mi)	35	30	29	32	35	N/A	N/A
Miles per gallon	N/A	N/A	50	N/A	37	23	31
Battery Range (mi)	208	84	11	76	38	N/A	N/A

Table 2. Comparison of travel costs for electric vehicles and gasoline-powered cars.

The average US driver travels 33 miles each day [12]. Using Table 2, if someone drives 30 miles in total to work, five days a week for 50 weeks of the year (250 days), they would spend $1223 on fuel for the average gasoline powered car or $288 on electricity for a 2014 Ford Focus Electric (using US average of $0.12/kWh) [8]. Thus, they would save $935 on work commute annually by driving a 2014 Ford Focus Electric instead of the average gasoline-powered car. If someone drives 100 miles total to work a day for 250 days, they would spend $4075 on fuel for the average gasoline powered car or $960 on electricity for 2014 Ford Focus Electric, saving $3115 for annual work commute fuel costs. The upfront cost of electric vehicles is often a concern of consumers, but the price of electric vehicles is decreasing with technological improvements in the battery and vehicle [1]. The 2014 Ford Focus Electric costs $35,000, but qualifies for a $7500 federal tax credit reducing the price to $27,500; thus, this vehicle is $2500 less expensive than the average 2014 gasoline-powered vehicle [13,14,57].

Four percent of survey respondents travel at least 50 miles one way to work daily. If an owner of the 2014 Ford Focus Electric needed to travel 100 miles total (50 miles one way) to and from work each day, since the battery range for the vehicle is 76 miles, they would not be able to travel the entire distance on a single charge [13]. However, the vehicle owner could make the commute if they charged their vehicle during work. Consumers already have an interest in charging at work; 64% of survey respondents would like to drive an electric vehicle and charge at work and home to decrease the cost of their commute. Thus, there is value in employers adopting a business model for SPCS.

There are three possible charge levels (level 1–3) for electric car charging stations. Level 1 is the slowest and least expensive charging level for electric cars [15], and charges plug-in vehicles at a power level of about 2 kW, which provides 16 kWh in 8 h. Since typical work-days are eight hours long, the electric vehicle could be fully recharged using this simplest charge level while sitting in the parking lot during the work day. When the vehicle owner returned home, they could plug in their vehicle again, if necessary, to recharge for their next destination. Since cars are parked about 95% of the time [16], it is logical to use this sedentary time effectively and charge electric vehicles while owners work or complete other tasks.

4. Solar Powered Charging Stations

Although electric vehicles reduce carbon emissions and reliance on foreign-oil, emissions issues are still present if fossil fuels are used for electricity generation. However, electric vehicles can be net zero emission vehicles if solar power generates the electricity [3]. Electric vehicles charged using solar power emit 96% less mass of pollutants than all-electric vehicles using the grid (with four percent of pollutants remaining from brake and tire wear) [3]. In addition, photovoltaics emit no noise, chemical pollutants or radioactive substances during use, so they are not added nuisances to the surrounding community [17]. Although solar panels have some carbon dioxide emissions associated with their production, the panels become carbon neutral in two years, and have an average lifetime of over 20 years [3]. Reducing carbon emissions using solar power will help the nation reach proposed US federal plans to reduce the nation's 2005 carbon emission levels by 30% from the electric power sector by 2030 [18,59].

4.1. Benefits of SPCS

Physically, SPCS are typically structures several feet tall with solar panels on slanted roofs and with an electric charging station underneath. One social value associated with SPCS is convenience since SPCS can be installed at many diverse locations such as work places, shopping centers, restaurants, rest stops along highways, gyms, parks, and more, allowing people to charge their vehicle while completing daily activities [21]. An infrastructure of many SPCS allows EV owners to charge their EV more frequently, which facilitates driving further before returning home. SPCS also provide shade for the charging vehicle. Shade significantly reduces the internal temperature of the vehicle and protects the car from sun damage, such as warping and cracked interiors [21]. A study in Oklahoma [21] found that internal temperatures of small cars dropped from 172 degrees Fahrenheit when in direct sunlight to 111 degrees Fahrenheit when under shade [53]. This lower internal temperature results in more comfortable conditions for drivers entering their car and reduces risk of heat stroke [21]. In addition, when asked how much survey respondents would like to park in shaded parking spaces on a scale of 1–9, 9 meaning "I would like it very much", the mean score was 6.40 with a standard deviation of 2.45. This average score indicates that the majority of respondents are interested in shaded parking. Another social value of SPCS is shelter since the roof of SPCS protects the vehicle from rain, ice, and snow [21]. This can be beneficial especially in the Midwest of the United States where weather can be unpredictable, and hail and snow are more common.

SPCS can increase the number of green jobs with needs for installation, maintenance, and operation [22]. An estimated number of 4000–5000 jobs will be created per year with distributed photovoltaic power generation [17]. Solar powered charge stations also improve the green image of the establishment that installs the charging stations, and could count towards LEED certification [22]. In addition, SPCS would attract more customers to the establishment [1]. A study showed that 50% of Americans believe it is "very important" that employers are environmentally responsible [1]. Also, that study found that 57% of consumers believe it is important to support products with environmental benefits and 51% will pay more for them, even during tough economic times [1].

SPCS could also serve as an educational tool, especially if installed at a school. The SPCS could raise awareness of solar power and electric vehicles, and could possibly encourage homeowners to install solar panels themselves [21].

Of the K-State survey respondents, 66% were in favor of installing solar powered charging stations for electric cars in campus parking lots. Table 3 shows the percentages for the reasons respondents favor SPCS.

Table 3. Reasons in favor of solar powered charging stations (SPCS) in K-State campus parking lots.

Reasons	**Percentage**
It would be good for people with electric vehicles to charge their car	86%
It would be good for air quality and reducing air pollution	75%
It would encourage more electric vehicles	65%
It would fit with the 2025 K-State Sustainability Plan	64%
It would help the image of K-State	60%
It would provide more shaded parking	40%

Table 3. Reasons in favor of solar powered charging stations (SPCS) in K-State campus parking lots.

As Table 3 shows, most people are in favor of SPCS because they believe it would help those with electric vehicles charge their car, and it would be good for air quality and reducing air pollution. Over half the respondents also believe SPCS would encourage more electric vehicles, would fit with the University's strategic vision plan, and would help the image of the University.

4.2. Design and Price Aspects

The ideal SPCS would be 27 m^2 of solar panel area on an overhead structure with a power output of about 3–4 kW and a level 1 or level 2 electric charge station underneath. Level 1 charge stations are the slowest of three charge levels, with a power level of about 2 kW. Level 1 is also the least expensive charge level, making it the most economical option. The slower level 1 charging rate should not be a concern for those parking for a significant time. For example, for a typical workday or school day employees or university students, faculty, and staff are parked for about eight hours, which is sufficient time to charge the car battery. In contrast, for customers who are parked on average for one hour, like a grocery shopper for example [40], level 2 charging is likely the best option for owners of BEVs [15]. Although this level of charging is slightly more expensive to install, the faster charging rate is more logical for the time spent parked.

A south-facing solar panel in Manhattan, Kansas with a DC to AC derate factor of 0.819 and 39.1 degree tilt, with an area sufficient to cover a SPCS of 27 m^2, is estimated to produce an average of 16 kWh per day and 5840 kWh annually. The solar power produced is fed into the grid and electricity is drawn from the grid when an electric vehicle is charging. This design guarantees that an electric car can always be charged and that the generated solar power is always being utilized. SPCS provide many benefits for the installer and customers, but they ultimately are financial investments. Installing a 4 kW solar panel system at $4.00 per watt costs about $16,000 or $11,200 with a federal tax credit of 30% of the net cost at installation; this is an estimate since solar panel costs vary depending on manufacturer, panel type, and installation costs.

The total costs for SPCS depend on the charging station hardware equipment, installation and maintenance costs, and cost of electricity [1]. Charging station hardware equipment for SPCS, such as the communications, charge level, and customer interface, can vary from $500–$7000 depending on the complexity of the system [1]. The $500 equipment is the simplest equipment option where there is a cable and electric connection while the $7000 equipment is the most complex option where there is a pre-built charging station that allows smart metering and collection of customer information [1]. Level 1 charge stations tend to be less expensive than level 2 charge stations. Installation and maintenance prices can range from about $165–$400 on average [1]. Using the estimated values, the SPCS may range from about $11,865–$18,600 per SPCS; but the station costs likely average closer to $18,600 based on previous SPCS installments. When simply considering savings from generated solar power (using the average U.S electricity cost of $0.12 per kWh) [8] and excluding additional revenue for an entity, such as parking permits, the payback period for SPCS is estimated to be about 17 years for the lowest estimated cost and about 26.5 years for the highest estimated cost. Since electricity costs vary from one entity to another and with time of day, these estimated payback periods are also subject to change. Since solar panels produce power during peak hours when demand is greater, they have increased value where time of use (TOU) prices are used. Although the estimated payback periods of 17–26.5 years seem relatively long, other federal incentives or benefits, or partnerships could decrease the cost estimate of SPCS. In addition, SPCS increase a company's customer base because of its green image [22], which would increase the company's sales, and provide another location for advertisements [1]. Therefore, the total payback period for SPCS is likely shorter than the estimated range of 17–26.5 years.

4.3. Financing SPCS

There are multiple ways to finance the installation of SPCS including federal tax incentives or grants, partnerships with a utility or business, donations, or purchase power agreements. A power purchase agreement is where a private company pays for the installation and maintenance of the SPCS for an entity, and receives payments from that entity for the power and services received [23]. Once the contract expires, the university or entity can buy the solar charging station system at a reduced price or renew the contract [23]. Power purchase agreements are currently allowed in 22 US states, Washington D.C., and Puerto Rico, but banned by law in NC, GA, FL, IA, OK, KY and are unclear or unknown in the remaining states [23].

Table 4 shows the percentage of respondents who were in favor of certain financing options for SPCS.

Table 4. Finance Options for Solar Powered Charging Stations.

Finance Option	Percentage
Federal grant	73%
Partnership with electric charging station company	57%
Income from user permits	55%
Power Purchase Agreement	51%
Partnership with Westar Energy Utility	51%
Donations to K-State Foundation	50%
Tax incentives	31%
University funds	22%

Table 4. Finance Options for Solar Powered Charging Stations.

The results in Table 4 suggest that respondents are open to many different finance options, which is a positive result since installers of SPCS may likely use several financial avenues to fund an installation. Also, the actual financing of SPCS that have been installed includes several of the options in Table 4. Goldin et al. have reviewed the financial aspects of several SPCS [3]. Federal grants have been received and used to help finance many of the early installations [3]. Electric charging station companies have been a partner in many installations [3,61,62]. Income from users of the charge stations is common; in some cases the charge is for parking; in other cases it is for the electricity; in other cases there is a fixed charge similar to a user permit [3,61,63]. For some installations, the SPCS are installed and financed by a power company [46,47,64]. Tax incentives have reduced the cost of the installation of SPCS [3]. Examples of SPCS financing include the Tesla Supercharger network which is financed by Tesla and provides free rapid charging to Tesla owners [30,41]. The sales of new Tesla vehicles provide the income to pay for the network. Similar to the Tesla Superchargers, in the Netherlands the Fastned network is being installed by a corporation that has many investors [63]. The investments and revenue from those who use the stations to charge their vehicles are financing the construction of the charging network [63]. In addition, Stanford University proposed a scenario of installing SPCS on campus and using a power purchase agreement to fund the purchase and installation of solar panels [24].

Another important aspect to consider for SPCS is the parking price structure. The three most common pricing structures that universities, businesses, utilities, and cities have used for charge stations include free charging, metered charging, or a flat rate. Since free parking is already provided at many businesses or around town, these entities often choose to provide free charging to encourage customers to use the stations and to have a competitive edge over other businesses [22]. Cities or companies that already have a pricing structure, that cannot financially support free charging, or that want to avoid scrutiny of customer preferential treatment often choose to provide metered pricing [22]. Metered parking is best when drivers are not parked for many hours and when level 2 charging is used. These hourly charge rates encourage vehicle turnover and prevent the exploitation of a charge station, Metered pricing must be reasonable to compete with other businesses [22].

Entities usually have flat parking rates when customers or employees are parked for many hours a day, multiple times a week. Level 1 charging is often the most logical charge level to use for these situations. Since level 1 charging takes several hours and the cars are often parked for about eight hours a day (typical workday), 250 days of the year, it is more reasonable for the user to pay a flat rate rather than pay an hourly price every day. This option is also more convenient for the driver since they do not have to move their car during the workday as they would with metered charging. The survey proposed two permit price structures for use of SPCS. One price structure option was for all drivers to pay the same price, but slightly higher than usual, for their parking permit, and all drivers have access to the SPCS. However, 77% preferred the staggered price structure instead, in which, drivers who park in open lots without using the SPCS have the least expensive permit, drivers who park under the SPCS for simply use of the shade and shelter have a slightly more expensive permit, and drivers who park under the SPCS to charge their car have the most expensive parking permits. The average amount survey participants chose to pay more for their annual permit to park in the shade was $25. Since shade is typically a free commodity, the fact that these respondents are willing to pay extra for it indicates that consumers value this benefit.

5. Electric Charge Station Infrastructure

Consumers hesitate to purchase electric vehicles because of high costs, concerns of battery capacity, and lack of charge station infrastructure [26]. Consumers fear that the electric vehicle would need to be charged before they reached their destination or located a charging station, experiencing "range anxiety" [3]. With recent technological advancements, battery capacity and large costs are no longer major barriers to electric vehicle purchase; however, lack of charging station infrastructure continues to be a major concern [3,26].

To provide perspective, as of May 2014, there are about 200,000 plug-in EVs sold in the US and 20,000 publicly available electric charging stations located throughout the nation [27]. As a result, there are about 10 electric cars per public charging station. Plug-in electric vehicles need to be charged every day or every other day. Thus, there is a need for a greater number of SPCS to satisfy the growing demand. In addition, several hours to charge an electric vehicle is a relatively large amount of time to allot in one's schedule; therefore, vehicle owners must be able to charge their car while completing other tasks or at home. More SPCS need to be available at work sites, shopping centers, restaurants, and other places where drivers spend time in addition to their homes.

Electric car charging cannot follow the same model as gas stations. Instead, electric car charging needs to be incorporated within existing parking lots so that drivers can multi-task and run errands, eat, or work as their vehicle is charging. Electric car charge stations need to be located along highways to facilitate drivers making long trips. Tesla Motors has started building Superchargers along major highways at places like restaurants and motels that allow drivers to eat or sleep while recharging their car before continuing their drive [30]. Charge stations can be built at restaurants and rest areas, which have restrooms, vending machines, and maps, and provide additional amenities such as wireless Internet connection for people to use their electronic devices and multi-task as their car charges. Infrastructure must also be developed in locations around town for those completing everyday tasks such as working, eating, buying groceries, or spending time at the gym or library. As the number of electric vehicle owners increases, places that lack charging stations could lose business.

With technological improvements, the time for electric cars to charge will decrease. However, more strategically placed solar powered electric charging stations are still needed to increase the locations that are available and decrease the ratio of electric cars to charging stations [5]. If there is greater car charging infrastructure, a positive feedback loop will be initiated and more drivers will purchase electric vehicles [5]. Currently, 1% of survey respondents own an electric vehicle, but 31% are more willing to invest in an electric vehicle if the university has sufficient charging infrastructure. If there happens to not be many electric cars in the area yet, installers of the SPCS could still feed solar power into the electrical grid, and customers or employees with any car could use the charging stations for shade and shelter [5].

NRG Energy Inc. has installed hundreds of eVgo charging stations in California, Washington DC, and Texas with plans for more charging station installations. Coulomb Technologies has installed over 4600 stations in Texas, Florida, New York, Michigan, and other US regions, and more stations in 13 other countries [31,32]. These companies and others have begun building the necessary electric car charging station infrastructure and have played a key role in the growth of electric vehicles. Thirty-one percent more electric vehicles have been purchased in the first five months of 2014 compared to the first five months of 2013 [27]. As the number of electric vehicles increase, the number of solar powered electric charging stations must increase to support the greater charge demand and to avoid overloading the grid [3]. If there is a 30% increase in plug-in hybrid electric vehicles, there may be a 1.5 time increase in peak load electricity consumption in a residential area [19,20]; SPCS can help meet this peak load.

For entities that decide to install SPCS, an aspect to consider is where to best locate them. In the survey, specific K-State parking lots and locations were proposed to respondents including west of old stadium/alumni center, the Federal National Bio and Agro-Defense Facility on campus, north of Call Hall, a location near the building they frequent most, the Bill Snyder Family Football Stadium, and a central location on campus.

Descriptions of the K-State locations with pros and cons were included in the survey questions and are used in Table 5; these can be applied to parking lots with similar characteristics at other universities, businesses, or entities. Respondents scored the proposed locations a number between 1 and 9 where 1 meant not a good location at all and 9 meant an excellent location.

When observing the results, it is evident that the location with the highest score was not a central location on campus, which may have been expected, but rather a large lot with many available parking spots that is located far from main buildings. Some survey respondents expressed concerns about SPCS reducing the number of available parking spots, making it more difficult to find a place to park; therefore, this large lot may have the highest score because it has open spaces available most of the day. In addition, the second highest scoring location is a new federal building planned to be built on campus. It is possible respondents favored this location because the future parking lot could be constructed with extra parking spots for SPCS. The federal building lot may also rank second because it would support the proposed new U.S. federal rule to reduce greenhouse gas emissions [18,59], which was mentioned in the survey question. The survey respondents' preferences and hesitations associated with SPCS installations indicate the concerns the majority of non-EV owners may feel with the construction of SPCS. These anxieties should be considered when designing an entities' SPCS business model. When looking specifically at the preferred SPCS locations of respondents who own electric cars or would like to own an electric car, the most favored locations were large lots with many available spots and multiple locations so there is a station near the building they most frequent.

Table 5. Preferred Locations for SPCS.

Location	Mean Score	Standard Deviation
Large lot, often with available spots for parking, but quite far from main buildings	5.48	2.57
Federal building on campus. SPCS would mainly be for federal employees but would support the new federal emissions rule *	4.75	2.68
A certain lot with a central location on campus, but not many buildings relatively close to the lot	4.62	2.49
Location near building respondent frequents most	4.54	2.77
Football Stadium at far end of campus	4.12	2.73
Central location on campus in general	3.97	2.46

* The proposed new US federal emissions rule involves reducing the nation's 2005 carbon emission levels by 30% from the electric power sector by 2030 [18

Table 5. Preferred Locations for SPCS.

The reduction of greenhouse gas emissions and the improvement of urban air quality can be accomplished through the electrification of transportation, increased SPCS infrastructure, and greater renewable energy generation. This can be achieved with present technology, but the transition is expected to benefit from continuing advances in solar energy, battery technology, and SPCS developments. Additional choices in electric vehicles and plug-in hybrid vehicles are expected as more individuals select these options. As the world moves forward with new emission guidelines, reducing their countries reliance on fossil fuels, and upholding other environmental policies, it is important countries continue building electric car charging infrastructure that incorporates solar power.

6. Business Models

A business model for SPCS describes the rationale associated with providing value in social, economic, and environmental contexts [65,66]. For commercial companies, a business model is a plan to generate revenue and make a profit from operations [65]. For SPCS, the social value of conveniently located SPCS, the environmental value of green solar energy, and the economic aspects are important parts of the business model to be addressed [65,66]. The payback period, the length of time required for the income associated with an investment to equal the capital invested, is often used to evaluate alternatives.

The survey results can be used to design business models for various situations in which SPCS may be viable to install, as well as identify parameters outside of which the installation of SPCS are not yet economically recommended. In addition, the survey results can indicate the opinions or hesitations the general public may have with aspects of SPCS installation and logistics. The preliminary models developed here include models for non-profit organizations (using universities as an example), industries like single businesses and shopping centers, utilities, public parking, and federal and state governments. Because of length of time and frequency, workplace charging has been given more attention. The U.S. Department of Energy has a popular Workplace Charge Challenge program with over 60 participating companies and organizations [56].

6.1. University Non-Profit Organization Model Parameters

Parking services at large non-profit organizations, such as universities, are often self-supported. The income acquired from the parking permits and citations is used to develop and maintain the university parking lots. The majority of university students, faculty, and staff park in open parking lot spaces, as the survey shows, in order to save money and park in more convenient locations. Installing solar powered charging stations in these lots can provide some of the benefits a parking garage provides, such as shade and protection from weather, but at a more convenient location and comparable price to parking garages. Although the amount of electric vehicle ownership at universities is still relatively low overall, the survey indicates that there is a growing interest in owning an electric vehicle and support for the installation of SPCS. The purchases of electric vehicles are expected to increase with the greater availability of electric car charging stations [5,41].

SPCS have been estimated to be about $11,865–$18,600 each. SPCS seem to be relatively expensive, but there are values associated with these stations in addition to power generation and charging such as shade, protection from rain, ice, and snow, the improved image of the university, and reduced carbon dioxide emissions that make the price seem more reasonable [21]. Student, staff, and faculty do not need to own an electric car to use the shade and shelter of the SPCS. When considering the value of these benefits, the charging stations have the potential to be successful.

When considering methods to finance SPCS, there are multiple options; one popular option many installers use is tax incentives. In order to qualify for certain renewable energy credits or tax incentives, the university may need to form partnerships with other companies [5]. Federal tax incentives include the Solar Investment Tax Credit, which provides a 30% tax credit for solar systems, valid through 2016 [33]; the Renewable Energy Property Tax Exemption, which exempts renewable energy equipment from property taxes [60]; the Electric Vehicle (EV) Infrastructure tax credit, which is a 30% credit of the charge station cost with a max of $1000 for consumers and $30,000 for businesses [34]; and Business Energy Investment Tax Credit (ITC) which equals 30% of expenditures [34]. The Database of State Incentives for Renewables and Efficiency is a helpful database that lists incentives for all US states, and should be further perused for updated tax incentives [35].

A partnership with an electric charging station company may also be a viable option for financing SPCS for a non-profit organization. For example, ChargePoint, an electric charging station company, donated a charging station to K-State, which charges an hourly rate for use of the station. The company benefits from receiving 10% of the monthly revenue generated from use of the station and from customers subscribing to the ChargePoint infrastructure [36]. On the other hand, K-State benefits from receiving the charging station for free and 90% of the income generated from station use, while improving their image. A similar scenario could be followed at other universities.

Another possible finance option is a partnership with a business. In this case, the business purchases the SPCS and pays for the installation, but in return receives free advertising on the campus charge station for a certain number of years. Spending about $11,865–18,600 per station in return for multiple years of free advertising on a college campus can be an attractive deal for businesses. This partnership allows the university to spend no money on the charge stations, but still improve their image, generate revenue from station use, and reduce their electricity bill with generated solar power. In addition, it allows businesses to reach a large advertising audience for multiple years. The university could request Tesla Motors, Nissan, or Chevrolet to donate a SPCS for example and in exchange the car company would receive free advertising. Partnerships with businesses or electric charging station companies (or both) are win-win situations.

Most universities sell annual parking permits for students, faculty, and staff to park on campus. University permit parking prices vary depending on classification as student, staff, or faculty and the location of the parking for faculty and staff. For example, K-State students pay $170 for an open parking lot spot while K-State faculty and staff pay $150–$190 for an open parking lot spot, $410 for a preferred garage spot, and $610 for a reserved garage spot [37]. The addition of solar powered charging stations on campus will alter the pricing structure for university parking. Based on survey results, the best pricing structure is staggered pricing where it is least expensive to park in open lots without using the SPCS, slightly more expensive to park under the SPCS to simply use the shade and shelter, and most expensive to park under the SPCS to charge an electric vehicle. The average amount survey participants chose to pay to park in the shade in addition to their existing parking permit price was $25. Therefore, for the new pricing structure the shaded parking spot prices would be $25 more expensive than the average spots, and the parking permit prices for using the charging station would be about $300 more expensive. The additional $300 is actually a reasonable pricing amount because it includes not only shade and shelter (amenities which parking garages provide), but also allows unlimited charging for electric cars. When comparing the new permit price to the K-State case, it is evident the SPCS charging permit is actually between the $410 and $610 K-State garage permit prices. Thus, the additional $300 for unlimited annual charging appears to be a reasonable price that provides several benefits and is comparable to current university permit prices.

As already mentioned, the estimated payback period for SPCS is slightly below or equivalent to the typical 25-year warranty for solar panels [38]. It is important to remember that the payback period would decrease with the addition of tax incentives, partnerships, and user parking permits, which 55% of respondents were in favor of. Also, the payback period could decrease further if the university or non-profit entity slightly increases the permit prices annually.

Besides price and payback period, another logistical aspect to consider regarding SPCS is location. According to the survey results, the majority of respondents prefer SPCS to be placed in parking lots that are typically located far from main campus buildings but have an abundant amount of parking spaces. Most respondents seemed to value available parking spots higher than conveniently located parking. Respondents who own or would like to own an electric car preferred locations for SPCS in large lots with many available spots or in multiple locations. The future plans for the university should also be considered when planning locations for SPCS. For example in the K-State case, the University has a 2025 Master Plan that plans to move surface parking from around campus locations to the Bill Snyder Family Football Stadium parking lots at the edge of campus where shuttles will take people to campus [39]. Therefore, in anticipation of this future plan of campus parking relocation, it is logical for K-State to build some SPCS at the Bill Snyder Stadium lots.

There is also value for other universities to build SPCS at their football stadium parking. Often many alumni will attend university football games and may travel many miles, especially in the Midwest of the US, to attend the games. Thus, it would be convenient to have a charge station infrastructure to allow commuters with electric vehicles to charge their car during the game, especially if the commuter could not return home on the remaining battery charge otherwise. Tailgating and football games often last for about 6 h or more collectively, which is sufficient time to charge many of the car batteries. Another benefit includes allowing alumni or others to use the charge stations for social reasons as they tailgate before a football game. The SPCS would provide shade and shelter for the tailgating party. In addition, level 1 SPCS would allow tailgaters to plug in their television, grills, or other electronics. The SPCS would be an additional tailgate amenity, making alumni's football game experience more enjoyable. The SPCS in the football parking lots would demonstrate to alumni and other game-goers the sustainability progress the university is making and thus further improve its image. In addition, the SPCS could generate more revenue for the university if tailgaters or electric vehicle owners were charged for use of the station. Extra income from tailgaters would help pay for the cost of the SPCS installation and maintenance, further reducing the payback period.

6.2. Industry

A number of employers now have workplace charging for their employees and customers [56]. Many benefits associated with SPCS related specifically to businesses include increased clientele base, credit towards LEED certification, shelter, convenience, and a car charging service for customers and employees [5,22]. As Table 3 demonstrates, customers value that SPCS encourage more EV purchases, improve air quality and reduce air pollution, improve the image of the establishment, and provide more shaded parking.

Customers typically spend an average of one hour at a store [40]; therefore, SPCS at these businesses should be capable of both level 1 and level 2 charging. Level 1 charging is needed because some electric cars only have capacity for this charge level, some employees or customers may be parked for a significant time in the lot, and because level 1 is the best level to use with solar power. Level 2 charging is needed because it will allow customers to almost or completely charge their vehicle in the time they are shopping at the store [15]. Since a SPCS generates about 4 kW, solar power cannot provide all the electricity required for a 7 kW level 2 charge station. Thus, the SPCS must be able to access electricity from the grid when necessary.

Since the average industry electricity cost is less than the average residential electricity cost (although exact electricity costs vary by region and negotiated agreements) [8], the estimated payback periods may be larger for industry compared to residential. Despite the larger estimated payback period, businesses may continue to invest in SPCS due to the associated benefits such as improved green image, shade, convenience, and wider clientele base [5,22]. In other cases, charge stations are installed without solar panels.

Due to the relatively high initial cost for SPCS, the economics of industrial-based SPCS do not seem particularly attractive at this point. However, some businesses are proving this perception incorrect. For example, Tesla Motors is working to install electric vehicle Superchargers across the US and other countries as well. Superchargers are 120 kW chargers that have the capability of charging 50% of a Tesla car battery in 20 min [30]. Each station also has a half-megawatt-hour battery pack [30]. Tesla is also working to have every charge station have grid storage capabilities and solar power [41]. There are already 102 stations in North America with complete East and West coast coverage and coast-to-coast travel in the US, as well as 44 stations and nine stations located in Europe and Asia, respectively [30]. The chargers are located near other amenities like restaurants, shopping centers, and cafes, allowing users to complete another task while their car is recharging [30]. Each Supercharger costs $150,000 to install, but use of the station is free for Tesla car owners [41]. The solar panels, covering six parking spaces, add an additional $150,000 to installation costs [41]. These high upfront costs may seem prohibitive, but Tesla Motors' business model is faring well. The provision of free, abundant, quick charge stations is reducing customers' anxieties, like range and prevalence of stations, and encouraging customers to purchase Tesla's electric vehicles [41]. Tesla is also able to use the half-megawatt-hour battery pack to avoid utility "demand charge" fees that occur if a Model S Tesla plugs into the grid, causing high increases in electricity use [41]. The battery pack may be used in the future for grid services, like grid stabilization, which would generate additional revenue for the company [41]. Superchargers help support Tesla's business model and play a large role in increasing the number of electric vehicle purchases. Despite the high initial costs for the Supercharger infrastructure, Tesla is demonstrating SPCS to be a feasible investment.

6.2.1. Single Business

The typical payback period for businesses to invest in an item or technology is 5–7 years [55]. The larger estimated payback period for SPCS is significantly greater. One option for a business to install SPCS in an economically viable way is to engage in a partnership with another entity to share the cost.

NRG eVgo has partnered with Walgreens and is installing 120 electric car charge stations across areas in Texas [42]. Walgreens is serving as the host site for the stations while NRG eVgo pays for the installation and maintenance, and offers subscription plans of monthly flat rates for customers [42]. Walgreens is able to benefit from having the electric car charging stations by increasing their customer base and improving their image without paying for the charge stations. Establishing partnerships with another entity reduces the initial cost associated with SPCS.

According to the present survey, 57% of respondents were in favor of the business partnership finance option. The host business and SPCS owners both benefit from this partnership. The host company receives the benefits of SPCS, such as improved image, greater customer base, and shaded parking for free while the partnering company is able to market their name, improve their own image, and increase their profits from customers' charging. In this case, the two partners usually determine the pay structure for the stations, which can be free charging, a membership flat fee, or an hourly rate. It is best to model the new pricing structure based on the current parking price structure at the host business; this provides consistency and reduces customer accusations of preferential treatment [22] (pp. 28–31). If parking at the host business previously was free, it should continue to be free for customers to use the charge stations in order to compete with other businesses hosting charge stations [22] (pp. 28–31). If parking at the host business previously cost money, the business should use the same pricing structure (hourly rate or flat fee) for the SPCS fees, although the prices may be higher.

Another viable form of partnering for the installation and use of SPCS at a reduced cost, if allowed in that state, is power purchase agreements (PPA), which 51% of survey respondents favor. PPA reduce the initial investment and maintenance cost of SPCS for host businesses since the partnering business owns the stations and thus funds the project [23]. The host business would pay a contracted fee for the SPCS to the partnering station owner for a certain number of years, usually around 25 years, until the contract ends. Once the contract ends, the host business could purchase the SPCS from the owners at a reduced cost or renew the contract. This agreement allows host businesses to avoid the initial upfront cost of the SPCS installations, and allows the partnering business to benefit from the financial incentives and market their company name [24]. Partnerships can allow multiple entities to benefit from the installation of SPCS.

6.2.2. Shopping Center

A shopping center can have 10 or more businesses in a plaza with shared parking spaces; if the 10 businesses share the cost of the SPCS, this has a benefit similar to sharing the cost of the parking area. SPCS would improve the plaza's image and draw more customers to the entire plaza, benefiting all the businesses.

Parking offered in shopping centers is usually free; therefore, the SPCS should also be a free amenity. In order to pay for the SPCS installation and maintenance, and the electricity consumed that is not already offset by the generated solar power, the businesses could use store income [5]. Another possible option is for the government, with the community's approval, to engage in an agreement with several shopping centers around town, allocating a percentage of sales tax to cover SPCS installation and maintenance costs. Using sales tax allows customers to indirectly pay for the SPCS as they purchase food or items at the shopping center stores. Those who do not have EVs can benefit from shaded parking provided by the SPCS.

If not all the businesses in the shopping center are interested in financing SPCS, the businesses who are interested could still split the cost or engage in a partnership or PPA with another entity. The participating businesses could require customers to use a loyalty card or a code, available only at their stores, to use at the SPCS; this way, the other businesses not paying for the SPCS do not receive the benefits of improved image or increased clientele for free [22].

Since the solar panels on SPCS are designed to feed electricity straight into the grid, it does not matter what time of day electric vehicle owners charge their car. Therefore, it is not an issue that many customers typically shop at shopping centers in early to late evening when there is little or no solar energy generated; the SPCS will be just as beneficial and will provide the needed infrastructure.

6.3. Utilities

Utilities often deal with a reputation of being anti-environment due to their continued fossil fuel use. Utilities can improve their image with the installation of SPCS. Most customers would be more inclined to support a utility if it is advancing environmental initiatives, like electric vehicles and SPCS [1]. Regulated utilities have the obligation to provide power at a profitable rate and to provide power for electric vehicles at public locations. SPCS could generate more revenue for utilities while providing necessary charging infrastructure, and counting toward the Renewable Energy Portfolio if present in that US state. The Renewable Energy Portfolio Standard (RPS) is a mandate designed to increase electricity generation from renewable sources like solar, wind, and biomass [43]. There is no national RPS for the US, but 31 states currently have an RPS and seven other states have renewable or alternative energy goals [44]. This standard has been effective in encouraging electricity producers to increase the percentage of their electricity generation on renewable sources rather than fossil fuels [43]. In addition, SPCS would help utilities meet the proposed new US federal rule to reduce the nation's 2005 carbon emission levels by 30% from the electric power sector by 2030 [18,59]. Another benefit is that the utility could sell electricity when the utility's SPCS are being used in the evening or at night (when there is little to no solar power generated).

Utilities have the opportunity to partner with businesses or universities, supplying SPCS for an entity that serves as the stations' host. According to the survey, 51% of respondents are in favor of a partnership with a utility to fund SPCS. Austin Energy, located in Texas and the nation's eighth largest public owned utility, has engaged in partnerships with Wal-Mart, Sam's Club, H-E-B, and Denny's to install SPCS at locations around Austin [45,46]. By January 2013, Austin Energy's electric car charging sessions collectively helped save 7363 gallons of gasoline, and the electric charging, which is totally powered by renewable energy, helped prevent 1391 pounds of greenhouse gases from being emitted into the atmosphere [47]. Other utilities can also have a positive impact on the environment and market this fact with the installation of SPCS. In addition, utilities can use electric vehicle batteries as grid stabilizers, similar to Tesla's Supercharger battery, which is helpful for the utility's operations [48]. Benefits for SPCS host businesses include increased clientele base, improved image, and recognition for the support of renewable energy and encouragement for more electric vehicle purchases [22]. It is still being debated in some states whether utilities can own and operate charging stations, and whether utilities should be the only ones to own public charging stations [49]. Most states have yet to rule on this issue, but those who have, such as Washington and Oregon, have permitted utilities to own public stations [49].

In order to pay for electricity consumed, Austin Energy customers engage in a six-month subscription for unlimited charging within the network or pay an hourly fee for use of the SPCS [46]. For other utilities, subscriptions, hourly priced charging, or a price per kWh of electricity used are also viable finance methods for level 1 and level 2 SPCS. The SPCS charge level would depend on the utility's partner and customer needs.

Unlike other entities, utilities have the ability to charge a certain price per kWh of electricity used if sanctioned by the utility regulatory agency that approves prices. For example, if approved, the utility agency could charge $0.15 per kWh of electricity delivered at a charge station, which would be $0.05 per mile for an electrical vehicle that averages 3 miles per kWh. With this charging structure, the payback period for SPCS is 13.5–21 years. This $0.15 per kWh charge rate is a reasonable price for utilities that results in a profit, but it is slightly high for the consumer. However, the idea for these higher priced SPCS is to have these stations available in many locations so customers can use them if they need to recharge. With an appropriately regulated price per kWh charge rate, it is profitable for the utility to sell power from the grid at the charge station.

Another finance option for these utility-owned SPCS is metered charging where customers pay an hourly rate for use of the station. This allows for vehicle turnover and encourages customers to be more conscious of their electricity consumption [25,58]. To streamline the payment process, customers could simply enter their utility account information into the station and add the hourly charging fee to their utility bill [22].

Subscriptions are another option that ensure the utility a set amount of revenue each month and allow the customers to charge their vehicle however often they want that month.

Charging a fee to use a charging station located at host businesses with general free parking risks low use of the station due to the competition of free charging from other entities [22]. Customers would likely not be pleased to pay to park and charge their vehicle, when parking is usually free, and might go elsewhere to charge their vehicle. For example, survey respondents were asked how much they would be willing to spend for shaded parking, which is usually a free commodity, and 33% were not willing to spend any money. Thus, customers may choose to go to free charge station parking locations. Therefore, it would be in the utility's and host business's best interests to provide free charging if free parking is already offered. The electricity provided at the charge stations can be free to the EV owner, but paid for by the business associated with the parking lot. Time limits could be placed on the charging to encourage turnover and prevent exploitation of the free charging. In order to ensure that only customers use the charge stations, businesses could require customers to use a loyalty card or code, available only at their store [22]. Another finance option includes charging businesses to place advertisements on the charging stations and using this additional revenue to help cover free charging costs. There are several finance options that utility-owned SPCS could employ. A partnership with a utility and the businesses they serve can be an effective option for increasing the SPCS infrastructure.

6.4. Public Parking Model

In cities, particularly in downtown areas, businesses often do not own parking lots, but rather the city does. Public parking is provided along streets and in lots, where there is free or metered parking, and in parking garages, where there is often an hourly rate. Parking availability is viewed as an incentive for downtown investment opportunities [50]. Independent business owners in these urban areas believe that accommodating vehicle users allows the businesses to compete with suburban shopping centers [50]. Customers of these businesses are from immediate and surrounding communities, and often travel downtown for errands, dining, or entertainment purposes. The local economy is improved when other city dwellers also travel to the city and spend money. However, electric vehicle owners with significant commutes may be hesitant to travel the long distance to the city if charge stations are not provided. The commuter may be unable to return home without charging their car battery beforehand. Therefore, it is wise for cities to install SPCS for public parking to encourage economic activity from these commuters as well as from other electric vehicle owners and environmentalists.

In areas with free public parking, which are often less convenient or less frequented places, free charging for SPCS should be provided. Free charging will encourage electric vehicle owners to travel to this city or area over another, and will entice them to shop at local businesses. Charging for SPCS spots adjacent to free spots, conversely, could discourage electric vehicle owners from using the stations and may cause them to charge elsewhere.

For parking spots with metered parking, SPCS should be provided with hourly rate charging. Charging an hourly rate will encourage more vehicle turnover, allowing more drivers to use the well-located SPCS spaces.

Both level 1 and level 2 charging options should be provided. Level 1 should be provided since some electric vehicles can only handle this level charging and some customers spend the majority of their day in town. Level 2 should also be provided since some customers only stay in the city for an hour or two and thus need a faster charging rate. Level 2 charging should also be used for metered parking since drivers will be parked in these spots for shorter times and will need a faster charging rate.

Cities can use a portion of sales tax to help pay for the installation, maintenance, and electricity usage of the SPCS as long as the community and government approves. Customers who are using the SPCS are purchasing items at the local businesses while parked, so the customers would be indirectly paying for the charge stations with sales tax. In addition, since commuters or tourists must pay sales tax as well, using this tax to fund SPCS seems logical.

Other options for financing SPCS include federal tax incentives, federal grants, or a partnership with a utility or company. For example, the city of Saint Paul, Minnesota partnered with the utility Xcel Energy and installed two SPCS. The stations were paid for partly by Xcel Energy and the City of St. Paul, and primarily by the US Department of Energy federal stimulus grant [51]. This city and utility partnership has worked well for St. Paul and these entities are continuing to work with businesses and non-profit entities to develop electric car charging infrastructure in Minnesota [51].

SPCS have an estimated 17–26.5 year payback period when simply considering revenue received from generated solar power (although this varies by utility and region). This payback period decreases, however, when factoring in federal incentives, grants, or partnerships the city has formed. In addition, revenue from sales tax and hourly charge rates for charging, if implemented by the city, can also decrease the payback period further. To further decrease the payback period, the revenue rate for SPCS can also be increased annually. Cities may still choose to invest in SPCS no matter the payback period because of the benefits associated with them such as infrastructure for EVs, shade, convenience, and improved image [5,22].

6.5. Federal and State Government Model

Created under the Obama administration, a new federal plan has been proposed that calls for the reduction of 30% of the US's 2005 carbon emissions from the electric power sector by 2030 [18,59]. The US federal and state governments have the opportunity to lead the nation by example and work towards achieving this plan through the installation of SPCS in federal and state employee parking lots. Other country governments also have the potential to lead by example by integrating SPCS throughout their government's existing physical infrastructure.

Since employees typically work 8 h days, level 1 charging (the less expensive installation option) is sufficient for the charging infrastructure. Use of the charge station should be free and can be considered an employee benefit. US federal employees can currently apply for the Transit Benefit Program, an employer provided benefit that encourages employees to use mass transit for their work commute by providing a monthly stipend that covers commute costs [52]. Mass transit is encouraged because it decreases emissions and the nation's reliance on fossil fuels. Providing free SPCS charging is equivalent to providing a mass transit stipend. Free charging will improve the image of the federal agency, support the proposed new US federal plan or other country's plan to reduce emissions, and encourage purchases of electric vehicles. Therefore, federal and state government agencies should install SPCS in employee parking lots and provide free charging as an employee benefit.

7. Conclusions

In order to decrease carbon emissions, electric power must be less dependent on fossil fuels and transportation must be electric. Due to electric vehicles' current high purchase costs and high costs of battery capacity, the infrastructure for charging stations must be increased to encourage more EV purchases. A significant SPCS infrastructure generates electricity for the grid and provides locations for public charging. In addition, the charging infrastructure must be integrated with the existing physical infrastructure, such as shopping centers and work places, to allow drivers to charge their vehicle while completing other tasks. Non-profit organizations, businesses, utilities, cities, the government, and other entities have the ability to facilitate this transition to an electric charge station infrastructure. There are key aspects these or similar entities should consider when designing their business models. One includes determining the best fit SPCS charging levels (level 1 or 2 or 3) for the consumers, which depend on the amount of time parked at the facility. Level 1 charging should be provided for vehicles parked for a full work day or about eight hours or more. Level 2 charging should be provided for vehicles parked about 1–2 h. Also, several financial avenues should be pursued to reduce the initial cost of the SPCS. The most widely used options include federal and state tax incentives, or a partnership with another business or charging company. The pricing for use of the SPCS largely depends on their current pricing structure for parking. Entities with free parking should primarily provide free charging, whereas those with metered parking should charge an hourly rate. Maintaining a similar price structure provides consistency.

Implementing SPCS into an entity's parking provide benefits such as improved image, increased clientele base, credit towards LEED certification, as well as convenience and shade for customers [22]. In addition, benefits of constructing SPCS extend beyond the entity and associated consumers; it provides a greater charging infrastructure for electric vehicles and helps shift the nation from a focus on fossil fuels to alternative fuel sources.

Acknowledgments

Thank you to the National Science Foundation for partially funding this research with the REU EEC-1166549 grant and thank you to Black and Veatch for their funding as well.

Author Contributions

Jessica Robinson: The main author for this publication who wrote the survey questions, conducted the survey, recorded and analyzed the responses, conducted the literature review, made calculations such as payback period, and wrote the manuscript.

Gary Brase: Gary Brase assisted with the Institutional Review Board Process, editing the survey questions, and with the survey process. He advised ways to word survey questions and ways to distribute the survey. He also assisted with editing the manuscript and added a paragraph when necessary.

Wendy Griswold: Wendy Griswold assisted with the Institutional Review Board process, editing the survey questions, and editing the manuscript.

Chad Jackson: Discussed and provided guidance for business models, and edited the manuscript.

Larry Erickson: Larry Erickson served as a research mentor to Jessica Robinson. He assisted with the planning process for the project, the Institutional Review Board, editing the survey and manuscript, and the survey process. He helped distribute the survey and the survey prize gift card. He added a paragraph when necessary to the manuscript and helped with some of the calculations. He also provided guidance and advice during the duration of the project.

Appendix

Survey Questions

Section 1:

- Are you a student, faculty or staff?
    - Student—**14%**
    - Faculty—**27%**
    - Staff—**61%**
- What is your age?
    - 18–22—**10%**
    - 23–30—**12%**
    - 31–40—**17%**
    - 41–50—**19%**
    - 51–60—**31%**
    - 61–70—**9%**
    - 71 or more—**1%**
- How do you get to campus most frequently?
    - Walk—**10%**
    - Bike—**2%**
    - Drive—**83%**
    - Carpool—**4%**
    - Public transportation—**0%**
    - Other—**1%**
- How many miles do you travel one way to get to campus?
    - 0–1 miles—**13%**
    - 2–5 miles—**44%**
    - 6–10 miles—**15%**
    - 11–20 miles—**10%**
    - 21–30 miles—**7%**
    - 31–40 miles—**5%**
    - 41–50 miles—**2%**
    - 51–60 miles—**2%**
    - 61 miles or more—**2%**
- Do you own or have use of a car?

    - Yes—**100%**
    - No—**0%**
- Do you have an electric car?
    - Yes—**1%**
    - No—**99%**
- What type of electric car do you drive?
    - All electric plug-in—**0%**
    - Plug-in vehicle that also uses gasoline—**25%**
    - Gasoline vehicle that uses some battery, but is not a plug-in vehicle—**75%**
- Do you park a vehicle on campus or plan to in the next 9 months?
    - Yes—**93%**
    - No—**7%**
- Do you typically park a car or motorcycle on campus?
    - Car—**99%**
    - Motorcycle—**1%**
- What type of permit do you have?
    - A permit to park in open parking lots—**78%**
    - A permit to park in a reserved space in an open lot—**4%**
    - A Reserved permit to park in the garage—**4%**
    - A Preferred permit to park in the garage—**8%**
    - I do not have a permit—**6%**
- Please help us understand why you park in the garage. You may select all answers that apply to you.
    - The garage is a convenient location for me to park—**86%**
    - The price seems reasonable—**23%**
    - I value the ability to park in the shade—**57%**
    - I value the shelter from rain, ice, and snow—**77%**
    - I want the protection for my car, so it stays looking nice—**31%**
    - I want the interior temperature of my car to not get too hot or too cold—**38%**
    - I want to help pay for the garage—**1%**
    - It is easier to find an open parking space in the garage—**47%**
    - Other—**22%**
- Why did you decide to not pay the additional cost to park in the garage? You may select all answers that apply to you.
    - The garage is not a convenient location for me—**65%**
    - The price does not seem reasonable—**48%**
    - I don't care much about the ability to park in the shade—**14%**
    - I don't care much about shelter from rain, ice, and snow—**11%**
    - There is congestion entering and leaving at some times of the day—**26%**
    - I don't care much about the interior temperature of my car getting too hot or too cold—**12%**
    - It is not my responsibility to help pay for the garage—**12%**
    - It is just as easy to park elsewhere as it is to park in the garage—**29%**
    - Other—**11%**
- In general, which do you value more?

- Parking close to my destination, but having a more expensive permit—**50%**
- Parking farther from my destination, but having a less expensive permit—**50%**

Section 2:

14.

Independent of the electric car charging aspect, how much would you like to park in shaded parking spaces, rather than spaces exposed directly to the sun, rain, and snow? (Scale of 1 to 9 where 1 = I would not like to at all, 9 = I would like it very much)

Mean = 6.40 Standard deviation = 2.45

15.

Think about how much you currently pay for a campus parking permit (or how much you would pay once you get one). How much more would you be willing to spend annually for shaded parking? (Scale of $0 to $200)

[To provide perspective for this question, several K-State parking permit prices are listed: annual parking lot permits for faculty and staff range from $150–$190, student parking is $170, garage preferred parking is $410, and garage reserved parking is $610.]

- **Mean = $25.52 Standard deviation = 39.33**

16.

Suppose you have an electric vehicle. You can reduce your cost to travel to and from campus (compared to a gasoline powered vehicle) by charging your vehicle on campus and at home. Would you like to be able to do this?

- Yes—**64%**
- No—**36%**

If shaded electric charging stations were added to the parking areas of the university it would cost some money to build. There are two different ways that these costs could be put into the parking permits.

17.

Which parking permit system would you prefer?

- Parking permit prices vary depending on where the driver is allowed to park. People could pay the same amount as now, but have fewer parking spots to choose from. Other people would pay more for the ability to park in shaded spots or to use the solar electric charging stations.—**77%**
- Everyone pays the same parking permit price, which is a little bit more, and everyone can park in the shaded solar electric charging spaces and charge their car if they wish.—**23%**

18.

Which parking permit system do you think is <u>most fair</u>?

- Parking permit prices vary depending on where the driver is allowed to park. People could pay the same amount as now, but have fewer parking spots to choose from. Other people would pay more for the ability to park in shaded spots or to use the solar electric charging stations.—**76%**
- Everyone pays the same parking permit price, which is a little bit more, and everyone can park in the shaded solar electric charging spaces and charge their car if they wish.—**24%**

19.

What is your general feeling or opinion towards solar energy? (Scale of 1 to 9 where 1 = very negative opinion, 9 = very positive opinion)

Mean = 6.48 Standard deviation = 2.39

20.

Are you in favor of K-State installing some solar powered charging stations for electric cars in parking lot(s)?

- Yes—**66%**
- No—**34%**

21.

Please mark all of the reasons why you favor having solar powered charging stations in K-State parking lots:

- It would help the image of K-state—**60%**
- It would be good for people who need to charge their electric vehicles—**85%**
- It would fit with the 2025 K-State Sustainability Plan—**64%**
- It would be good for air quality/reducing pollution—**75%**
- It would encourage more electric vehicles—**65%**

- It would provide more shaded parking—**40%**
- Other—**6%**

22.

Please mark all of the reasons why you are not in favor of having solar powered charging stations in K-State parking lots:

- Presently the demand for charging stations is small—**78%**
- It would make parking more expensive—**78%**
- It would be a waste of money and time—**50%**
- It would be visually distracting—**12%**
- Other—**21%**

23.

Parking spots with solar powered electrical charging stations are shaded. How much would you like to park at and use a shaded solar powered charging station while at the university? (Scale of 1 to 9 where 1 = I would not like it at all, 9 = I would like it very much)

Mean = 3.91 Standard deviation = 2.77

24.

Would you like to own an electric vehicle?

- Yes—**40%**
- No—**60%**

25.

Would you be more willing to invest in an electric vehicle if K-State had sufficient infrastructure for charging stations?

- Yes—**31%**
- No—**69%**

26.

Do you think installing solar powered charging stations on campus will encourage more people to purchase an electric vehicle?

- Yes—**40%**
- No—**60%**

27.

How should the solar powered charging stations be paid for? Select as many as you think apply.

University funds

- Federal grant—**22%**
- Donations to K-State Foundation—**73%**
- Income from user permits—**50%**
- Power purchase agreement. [A private company pays for the installation and maintenance of the system, and sells electricity to the university at a specified rate. Once the contract expires, the university can buy the solar charging station system or renew the contract.]—**54%**
- Partnership with Westar Energy utility—**51%**
- Partnership with electric charging station company—**58%**
- Tax incentives—**31%**

Section 3:

If Kansas State University were to install solar powered charging stations, there are several different possible locations. Each location has advantages and disadvantages. For each of the following possible locations, please rate how good of a location it would be for installing solar powered charging stations. There are also multiple ways to fund the installation of shaded solar powered charging station parking. For the following questions, please indicate which payment methods you think are best. (Scale of 1 to 9 where 1 = Not at all a good location, 9 = an excellent location)

- 28 Bill Snyder Family Football Stadium parking lot. These parking spots are far from the center of campus, but could be used for shade while tailgating before football games. Do you consider this a good location for some solar powered charging stations?

  Mean = 4.14 Standard deviation = 2.74

- 29 A central location on campus. These parking spots are convenient to many buildings, but there are limited spaces and they are rarely open to park in.

  Mean = 3.98 Standard deviation = 2.46

- 30 Multiple locations, so there is one near the building you frequent most. This would involve more expense because of the multiple locations, but the location would be most convenient.

  Mean = 4.93 Standard deviation = 2.77

- 31 West of the old stadium/alumni center. This lot is large and often has many available spots for parking, but the lot is quite far from the destinations of many students, staff, and faculty.

  Mean = 5.48 Standard deviation = 2.56

- 32 North of Call Hall. This has a central location on campus, but there are not main buildings relatively close to the lot.

  Mean = 4.62 Standard deviation = 2.48

- 33 K-State will be the location of a new federal building for the National Bio and Agro-Defense Facility. Mainly federal employees will work in this building, but solar powered charging stations here would support Obama's new plan to reduce the nation's carbon emissions from electricity 30% below 2005 levels by 2030.

  Mean = 4.74 Standard deviation = 2.68

- 34 What is your opinion on the environmental sustainability of K-state?

  (Scale of 1 to 9 where 1 = A lot of need for improvement, 5 = some need for improvement, 9 = No need for improvement)

  For this question, "environmental sustainability" is defined as sustained efforts to reduce one's carbon footprint, amount of pollution, consumption of resources such as water, and to engage in environmentally responsible activities such as recycling, composting, etc.

  Mean = 4.91 Standard deviation = 2.10

Conflicts of Interest

The authors declare no conflict of interest.

References

1. May, J.W.; Matila, M. Plugging in: A Stakeholder's Guide for Public Electric-Vehicle Charging Infrastructure; Rocky Mountain Institute: Snowmass, CO, USA, 2009. [Google Scholar]
2. What the World Needs to Watch. Available online: http://co2now.org (accessed on 24 June 2014).
3. Goldin, E.; Erickson, L.; Natarajan, B.; Pahwa, A. Solar Powered Charge Stations for Electric Vehicles. Environ. Progress Sustain. Energy **2014**, 33, 1298–1308. [Google Scholar]
4. TST Issue Brief: Sustainable Transport. Available online: http://sustainabledevelopment.un.org/content/documents/2634Issues%20Brief%20on%20Sustainable%20Transport_FINAL_21_Nov.pdf (accessed on 24 June 2014).
5. Erickson, L.E.; Boguski, T.; Babcock, M.W.; Leven, B.A.; Pahwa, A.; Brase, G.L.; Griswold, W.; Kramer, K.; Miller, R.D. Community Infrastructure for the Electrification of Transportation. Available online: http://www.smartgrowth.org/nationalconversation/papers/Erickson_SGN%20Natl_Conv_Paper_Electrification_of_Transport.6-22-2012.pdf (accessed on 10 June 2014).
6. Williams, J.H.; DeBenedictis, A.; Ghanadan, R.; Mahone, A.; Moore, J.; Morrow, W.R., III; Price, S.; Torn, M.S. The technology path to deep greenhouse gas emissions cuts by 2050: The pivotal role of electricity. Science **2012**, 335, 53–59. [Google Scholar]
7. About K-State. Available online: http://www.k-state.edu/about/ (accessed on 10 July 2014).
8. U.S. Department of Energy. Electric Power Monthly with Data for November 2013. Retrieved from U.S. Energy Information Administration Website. Available

online: http://www.eia.gov/electricity/monthly/current_year/january2014.pdf (accessed on 13 July 2014).
9. Weekly Retail Gasoline and Diesel Prices. U.S. Energy Information. Available online: http://www.eia.gov/dnav/pet/pet_pri_gnd_dcus_nus_w.htm (accessed on 10 June 2014).
10. Find and Compare Cars. Available online: http://www.fueleconomy.gov/feg/findacar.shtml (accessed on 13 June 2014).
11. Fuel Economy Guide. US Department of Energy. Available online: http://www.fueleconomy.gov/feg/pdfs/guides/FEG2013.pdf (accessed on 22 June 2014).
12. Chang, D.; Erstad, D.; Lin, E.; Rice, A.F.; Goh, C.T.; Tsao, A.A.; Snyder, J. Financial Viability of Non-Residential Electric Vehicle Charging Stations; Luskin Center for Innovation: Los Angeles, CA, USA, 2012. [Google Scholar]
13. Focus Electric. Available online: http://www.ford.com/cars/focus/trim/electric/ (accessed on 16 July 2014).
14. Nickel. Average Price of a New Car? Available online: http://www.forbes.com/sites/moneybuilder/2012/05/10/average-price-of-a-new-car/ (accessed on 10 May 2012).
15. Installing a Charging Station at a Commercial or Employee Parking Lot [Pamphlet]. Available online: http://www.psrc.org./3967/Draft_Fact_Sheet.Commercial.051110.GD.pdf (accessed on 24 July 2014).
16. Barter, P. "Cars are Parked 95% of the Time." Let's Check! Available online: http://www.reinventingparking.org/2013/02/cars-are-parked-95-of-time-lets-check.html (accessed on 10 July 2014).
17. Tsoutsos, T.; Frantzeskaki, N.; Gekas, V. Environmental impacts from the solar energy technologies. Energy Policy **2005**, 33, 289–296. [Google Scholar] [CrossRef]
18. Parkinson, G. Obama Sends a Shock through Abbott's Energy Ideals. Available online: http://reneweconomy.com.au/2014/run-abbott-run-53988 (accessed on 3 June 2014).
19. Mets, K.; Verschueren, T.; Haerick, W.; Develder, C.; de Turck, F. Optimizing Smart Energy Control Strategies for Plug-in Hybrid Electric Vehicle Charging. Network Operations and Management Symposium Workshops (NOMS Wksps), 2010 IEEE/IFIP, Available online: http://dx.doi.org/10.1109/IEVC.2012.6183168 (accessed on 3 July 2014).
20. Cutler, S.A.; Schmalberger, B.; Rivers, C. An Intelligent Solar Ecosystem with Electric Vehicles. Electric Vehicle Conference (IEVC), 2012 IEEE International, Available online: http://dx.doi.org/10.1109/IEVC.2012.6183168 (accessed on 17 June 2014).
21. Erickson, L. Third Report on the Installation of Shaded Parking Lot Structures with Integrated Photovoltaic Power Systems at Kansas State University. Available online: https://www.engg.ksu.edu/chsr/files/chsr/sustainability/docs/ParkingLot3rdReport2011.pdf (accessed on 5 June 2014).

22. Morton, J. Plug into EV Charging Stations. Buildings **2012**, 106, 28–31. [Google Scholar]
23. Third-Party Solar Financing. Available online: http://apps3.eere.energy.gov/greenpower/onsite/solar_financing.shtml (accessed on 9 June 2014).
24. Corcoran, B.; Golden, D.P.; Larson, K.; Schneider, S. Evs with Pvs: Analysis of Electric Vehicle Integration at Stanford University Using Solar Pv Panels. Available online: http://web.stanford.edu/~bethanyc/Papers/AASHE_EVs%20with%20PVs_FINAL.pdf (accessed on 11 June 2014).
25. Why Parking Meters should be Outlawed. Available online: http://www.taphilo.com/tom/parkingmeters.shtml (accessed on 3 June 2014).
26. Masoum, A.S.; Deilami, S.; Moses, P.S.; Abu-Siada, A. Impacts of Battery Charging Rates of Plug-In Electric Vehicle on Smart Grid Distribution Systems. Smart Grid Technologies Conference Europe (ISGT Europe), 2010 IEEE PES, 1,6,11–13. Available online: http://www.cricte2004.eletrica.ufpr.br/anais/IEE_ISGT_2010/2047502.pdf (accessed on 17 July 2014).
27. Kelly, R. Plug-in Vehicle Sales Roundup. Plug in America. Available online: https://org.salsalabs.com/o/2711/t/0/blastContent.jsp?email_blast_KEY=1300492 (accessed on 25 June 2014).
28. Gas Station Statistics. Available online: http://www.statisticbrain.com/gas-station-statistics/ (accessed on 24 June 2014).
29. Number of vehicles registered in the United States from 1990 to 2012 (in 1000 s). Available online: http://www.statista.com/statistics/183505/number-of-vehicles-in-the-united-states-since-1990/ (accessed on 24 June 2014).
30. Road Trips Made Easy. Available online: http://www.teslamotors.com/supercharger (accessed on 16 July 2014).
31. Sign Up for an NRG eVgo Plan. Available online: http://www.nrgevgo.com/charging-plans-by-market/ (accessed on 2 July 2014).
32. Garthwaite, J. Coulomb to Deploy 4600 Electric Car Charge Spots Thanks to DOE. Available online: http://gigaom.com/2010/06/02/coulomb-to-deploy-4600-electric-car-charge-spots-thanks-to-doe (accessed on 17 June 2014).
33. Solar Investment Tax Credit (ITC). Available online: http://www.seia.org/policy/finance-tax/solar-investment-tax-credit (accessed on 13 July 2014).
34. Financial Incentives. Retrieved from Database of State Incentives for Renewables and Efficiency. http://www.dsireusa.org/incentives/allsummaries.cfm?State=us&re=1&ee=1 (accessed on 13 July 2014).
35. DSIRE. Retrieved from Database of State Incentives for Renewables and Efficiency. http://www.dsireusa.org/solar/ (accessed on 13 July 2014).
36. Davis, T. Charge Your Ride: University Installs Electric Vehicle Charging Station. Available online: http://www.kstate.edu/media/newsreleases/jan13/station13013.html (accessed on 1 July 2014).

37. Parking Permit Prices. Available online: http://www.k-state.edu/parking/permits/permitprices.html (accessed on 11 July 2014).
38. Maehlum, M.A. The Real Lifespan of Solar Panels. Available online: http://energyinformative.org/lifespan-solar-panels/ (accessed on 13 July 2014).
39. Campus Master Plan Update Task Force. 2025 Campus Master Plan. Available online: http://www.k-state.edu/planning/master_plans/2025_plan/ (accessed on 18 July 2014).
40. U.S. Department of Labor. American Time Use Survey- 2013 Results; U.S. Department of Labor: Washington, DC, USA, 2014. [Google Scholar]
41. Herron, D. The Business Model for Solar-Powered Electric Car Charging. Available online: http://www.plugincars.com/solar-powered-electric-car-charging-business-model-127710.html (accessed on 15 July 2014).
42. NRG's eVgo Charging Stations Hope to Catch the Eyes of Texas. Chatham: Newstex. Available online: http://search.proquest.com.er.lib.k-state.edu/docview/861143581?accountid=11789 (accessed on 10 July 2014).
43. Most States Have Renewable Portfolio Standards. Available online: http://www.eia.gov/todayinenergy/detail.cfm?id=4850 (accessed on 13 July 2014).
44. Renewable and Alternative Energy Portfolio Standards. Available online: http://www.c2es.org/node/9340 (accessed on 13 July 2014).
45. About. Available online: http://goo.gl/hbIC4i (accessed on 13 July 2014).
46. HEB, Walmart and Sam's Club to Host Electric Vehicle Charging Stations. Available online: http://www.greenretaildecisions.com/news/2011/08/11/heb-walmart-and-sams-club-to-host-electric-vehicle-charging-stations (accessed on 13 July 2014).
47. Austin Energy surpasses 10,000TH plug-in Charge at Public Stations. US Fed News Service, Including US State News. Available online: http://search.proquest.com.er.lib.k-state.edu/docview/1282750492?accountid=11789 (accessed on 10 July 2014).
48. Andersen, P.H.; Mathews, J.A.; Rask, M. Integrating private transport into renewable energy policy: The strategy of creating intelligent recharging grids for electric vehicles. Energy Policy **2009**, 37, 2487–2486. [Google Scholar] [CrossRef]
49. Vonderschmitt, K. State Utilities Law and Electric Vehicle Charging Stations. Available online: http://knowledgecenter.csg.org/kc/content/state-utilities-law-and-electric-vehicle-charging-stations (accessed on 21 July 2014).
50. Mavridis, M. Parking Lot Aesthetic and the City Image: A Practicum Document. (Order No. MQ90482, the University of Manitoba (Canada)). ProQuest Dissertations and Theses. Available online: http://search.proquest.com.er.lib.k-state.edu/docview/305089741?accountid=11789. (305089741) (accessed on 23 July 2014).
51. Mayor Coleman Unveils New Solar-Powered Electric Car Charging Stations in Como Park, First of Their Kind in Minnesota. Available online: http://mccollum.house.gov/press-release/mayor-coleman-unveils-new-solar-powered-electric-car-charging-stations-como-park-first (accessed on 21 July 2014).

52. U.S. Department of Transportation. FAQs. Available online: http://transerve.dot.gov/faq.html (accessed on 25 July 2014).
53. Surpure, J.S. Heat-Related Illness and the Automobile. Ann. Emerg. Med. **1982**, 11, 263–266. [Google Scholar] [CrossRef]
54. Gas Pump Advertising. Available online: http://www.allovermedia.com/our-solutions/gas-pump/ (accessed on 29 July 2014).
55. Levy, J.I.; Dilwali, K.M. Economic incentives for sustainable resource consumption at a large university. J. Sustain. High. Educ. **2000**, 1, 252–266. [Google Scholar]
56. Vehicle Technologies Office: EV Everywhere Workplace Charging Challenge. Available online: http://energy.gov/eere/vehicles/vehicle-technologies-office-ev-everywhere-workplace-charging-challenge (accessed on 21 September 2014).
57. Retail Light-Vehicle Sales Strengthening in March. Available online: http://www.jdpower.com/press-releases/2014-march-automotive-forecast (accessed on 21 September 2014).
58. Conrad, E.; Aberg, W.J.P. Validity of parking meter ordinances. Va. Law Rev. **1943**, 29, 617–624. [Google Scholar] [CrossRef]
59. U.S. EPA. Carbon Pollution Emission Guidelines for Existing Stationary Sources: Electric Utility Generating. Available online: https://www.federalregister.gov/ (accessed on 4 October 2014).
60. Johnson, R.M. Smart Grid: Carbon and Economic Implications for Colorado. (Order No. 3403932, University of Colorado at Boulder). ProQuest Dissertations and Theses, 153. Available online: http://search.proquest.com.er.lib.k-state.edu/docview/507901067?accountid=11789. (507901067) (accessed on 4 October 2014).
61. Uda, R. UCSD Installs First Electric Vehicle Charging Stations. The Guardian, Available online: http://ucsdguardian.org/2012/10/8/ucsd-installs-first-electric-vehicle-charging-stations/ (accessed on 6 October 2014).
62. Electric Vehicles. Available online: http://www.sullivansolarpower.com/solar-power/electric-vehicles (accessed on 3 October 2014).
63. Fast Charging Along the Highway. Available online: http://www.fastned.nl/en (accessed on 7 October 2014).
64. Use of Solar Powered Electric Vehicle Charging Stations at San Diego Zoo on the Rise [Blog post]. Available online: http://cleantechsandiego.wordpress.com/2014/08/26/use-of-solar-powered-electric-vehicle-charging-stations-at-san-diego-zoo-on-the-rise/ (accessed on 7 October 2014).
65. Mariotti, S.; Glackin, C. Entrepreneurship & Small Business Management; Prentice Hall: Upper Saddle River, NJ, USA, 2012. [Google Scholar]
66. Teece, D.J. Business models, business strategy, and innovation. Long Range Plan. **2010**, 43, 172–194. [Google Scholar] [CrossRef]

Solar Thermochemical Energy Storage

Joanna Julien, Jeremy Grunewald, Kelvin Randhir, Nathan Rhodes, Conrad Cole, Nick AuYeung, Like Li, Renwei Mei, David Hahn

UNIVERSITY of FLORIDA

Abstract: Thermochemical energy storage is the next step in creating a self-sustaining society because it allows for energy supply to meet the electricity demand. Chemical bonds provide much higher storage capacities than the conventional energy storage methods; renewable storage schemes with greater energy storage density will potentially have a faster path to economic viability.

A potential thermochemical storage cycle lies in the carbonation/decomposition of $SrO/SrCO_3$. It offers the prospect of capturing thermal energy and releasing it at temperatures above 1200°C. One of the fundamental aspects of the project depends on the amount of surface area of the physical structure. To obtain an optimal amount of surface area, the current project involves creating a matrix through the mixing of heat treated SrO and decomposition of sacrificial carbon. The temperatures at which the SrO is being heat treated varied along with the size of the particles being used; also the ratio and size of the carbon particles are being varied to find the optimal structure with the most surface area.

Objective: To find the variable/s that most affects the energy density. It is hypothesized that an optimized amount of surface area will maximize the energy density.

Procedure:

1. A matrix was created

	Parameter	Level	Selected Values
A	Heat treatment temperature (T_H)	-	1235 °C
		+	1400 °C
B	SrO particle size (S)	-	25 μm - 38 μm
		+	106 μm - 125 μm
C	Graphite particle size (S_g)	-	25 μm - 38 μm
		+	106 μm – 125 μm
D	Mass ratio of graphite to SrO (r)	-	0.2
		+	0.7
E	Structure formation temperature (T_s)	-	1100°C
		+	1300°C

2. Heat treat the SrO

3. Crush and sieve the materials to the appropriate sizes

4. Mix the SrO and graphite to the appropriate ratios

5. Evaluate the reactivity of the samples using a thermo gravimetric analyzer (TGA)

$$SrCO_3 \leftrightarrow SrO + CO_{2(g)}$$

Reversible | **No catalysis** | **High temperature (~1200°C) Safe**

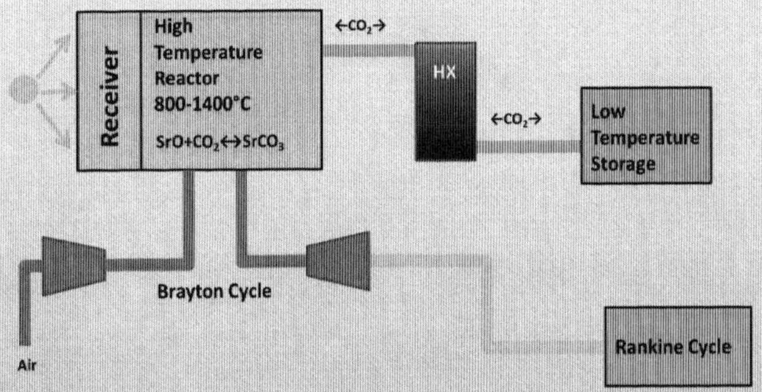

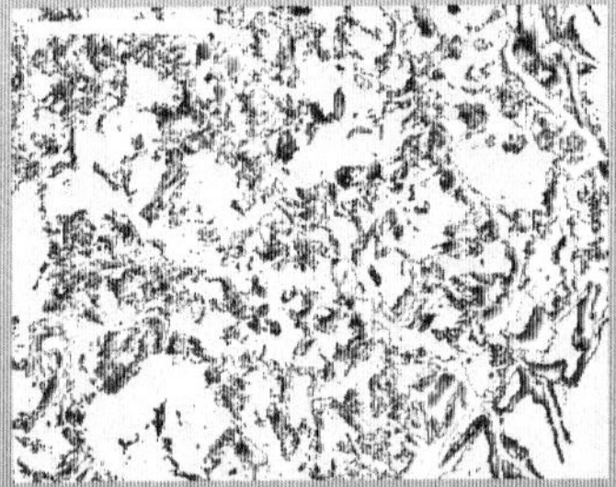

*SEM image of porous SrO/SrCO3 sample 8.

Energy Density of Samples

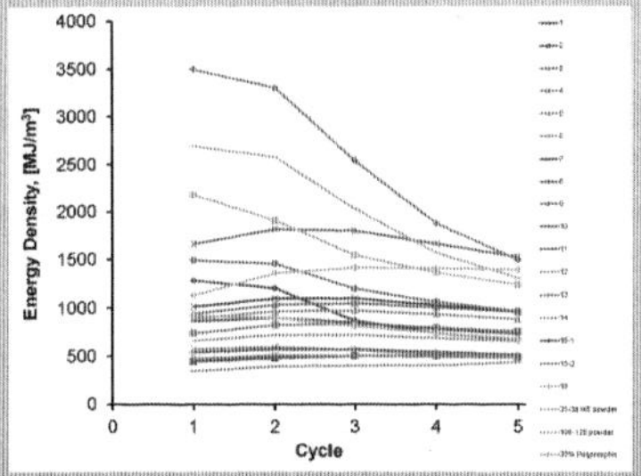

Z-value of the Parameters

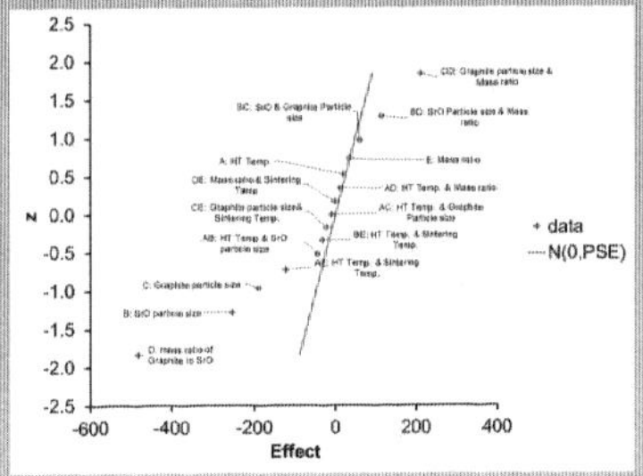

Correlation of results: Lenth's Method

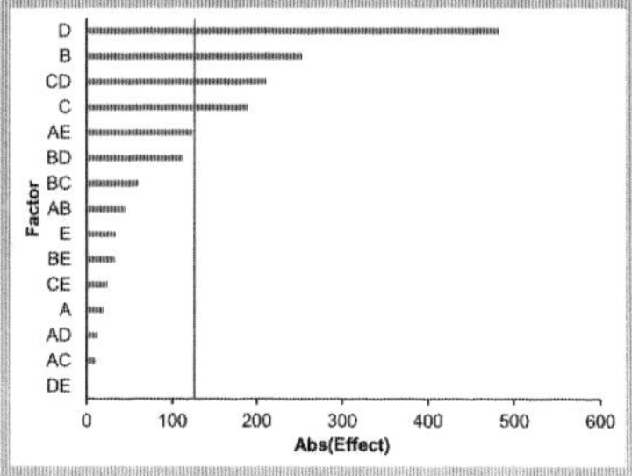

*A baseline of 95 percent.

Conclusion:

With 5 cycles done with the TGA, the assumption is that the energy density will be stable with continued cycles for each sample. The sample that showed the most promise was sample 15 in terms of energy density. The energy density of 15 started off around 3500 MJ/m^3 in the first cycle and tapered off to just above 1500 MJ/m^3, about 7% higher than the base powder. When comparing the effects of the parameters alone and together, the data shows that both the size of the SrO particles and the graphite particles create a large effect on the energy density. Surface area appears to have an affect on energy density. The next question should be what stabilizes the energy density ?

Future Work

Investigation into carbonation at elevated pressure

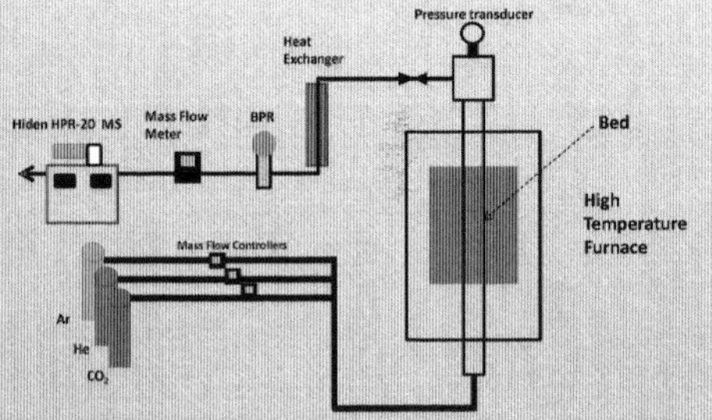

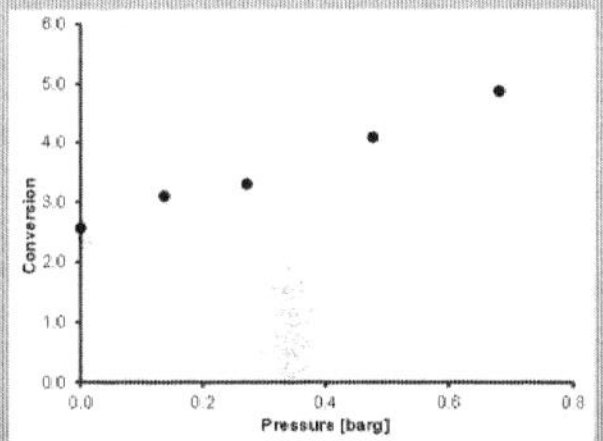

Preliminary data showing dynamic cycling between 1100 and 1300°C at 10°C/min. pCO_2 = 0.33 bar

Acknowledgements
We are very appreciative of funding from Hiden Analytical for participation in this poster session.
This work was funded by the U.S. Department of Energy Sunshot Initiative as part of the ELEMENTS program, award DE-EE0006534.

398479 An Overview of the Efficiency of Human Waste Pyrolysis Using Concentrated Solar Power

Abstract:

As sanitation poses a major issue for much of the developing world, a sustainable solution must be implemented to handle these issues. The Sol-Char Sanitation approach takes concentrated solar energy to disinfect, dry, and pyrolyze waste to create an easily usable solid fuel. This is done through parabolic sunlight concentrators which concentrate the light into fiber optic cables. These cables transmit the light to the reactor which absorbs the light. Using a CCD camera, flux maps were obtained showing relative intensity of the concentrated solar power along the optical train. These images showed the majority of the light was being captured at each inlet along the optical track, while also showing some oddities in the optics such as slight size differences and a relatively elliptical focused image at the reflected focal point. The average optical efficiency was found to be 0.28 +/- 0.01 through various on-sun power testing, while the reactor efficiency was found to be 0.72 +/- 0.02 from an on-sun water boiling test.

398491 Optimizing Energy Efficiency Under Uncertainty for Appropriate Technology Based Renewable Energy Processes in Developing Regions

Abstract:

The University of Kentucky Appropriate Technology and Sustainability (UKATS) Research Team has been engaged in designing a low-cost, sustainable processor for producing biodiesel and biochar in Sub-Saharan Africa. The team has constructed a multifunctional, non-automated (equipment) processor from simple, locally sourced materials based on appropriate technology fundamentals. The processor was tested and implemented during a research trip to Bangang Village, Cameroon, proving to be very productive and suitable for making desired biodiesel and biochar. To assess the total energy efficiency of the processor under sustainability applications, computational modeling has been conducted. The model was used to analyze the processor design under uncertainty due to non-standard constructional materials and parameters present in rural, developing regions. In return, its results describe the optimum energy efficiency reached by the processor when possible variations in its constructional design occurred. This research in Bangang village is performed in collaboration with the NGO, African Center for Renewable Energy and Sustainable Technology (ACREST).

Towards Understanding Nanoparticle Toxicity

Kimaya Padgaonkar[1,2], Sharlee Mahoney[1], Thomas Richardson[1], Ipsita Banerjee[1,2,3], Götz Veser[1,4]

Department of Chemical Engineering[1] & Department of Biongineering[2], Swanson School of Engineering; McGowan Institute for Regenerative Medicine[3], Mascaro Center for Sustainable Innovation[4]

Material properties at the nano-scale level vary greatly from the macroscopic components due to their small size, motivating a rapid growth in nanomaterials development. Functional nanomaterials have already found application in many consumer products such as apparel, cosmetics, and electronics. However, nanomaterials have the potential to exhibit different properties than shown at the macro-scale in various categories including toxicity[1]. Nanomaterials have shown increasing evidence of elevated toxicity and with that the potential to harm people and the environment. Hence there is an urgent need for a more thorough understanding of the effect of 'nano-sizing' on toxicity[1-2]. This project hypothesizes that supporting or embedding metal nanoparticles on/in a nontoxic support will create less toxic nanomaterials. The goal is to create nanomaterials that are "safer by design" through correlations between structure and toxicity.

Our complex engineered nanomaterials (CENs) combine nickel and silica, i.e. a known toxin (Ni) with a non-toxic matrix (amorphous SiO_2). Due to their versatility, modifiable capabilities, and nontoxic properties, porous silica nanomaterials are used in a wide range of applications such as biomedical imaging, catalysis, and drug delivery. Therefore, we hypothesized that embedding nickel in porous silica will reduce the toxicity of nickel by modifying transport and interaction of the embedded nanoparticles with its environment while still providing full access to the metal nanoparticle surface. In this manner, the functionality of the embedded nanoparticle can be maintained while toxicity can be reduced.

Three nickel and silica nanostructures, nhNi@SiO_2 (nickel embedded in silica), hNi@SiO_2 (nickel encapsulated in a hollow core in silica), and Ni-SiO_2 (nickel deposited on silica) were synthesized. Pure SiO_2 is used as a negative control, and $NiCl_2$ as a positive control. The latter is chosen both due to its established toxicity and to "represent" nickel ions that dissolve from the nanoparticles. We are employing an in vitro model to test the effects of the synthesized nanomaterials. 3T3 mouse-skin derived fibroblasts are well-suited and well established for nanotoxicity studies because they are inexpensive, robust, and proliferate quickly. 3T3 fibroblasts were exposed to varying nickel concentrations over 24 hours before being analyzed for viability via qualitative and quantitative analysis.

Qualitative analysis included using microscopy and live/dead imaging, while MTS assays were used to quantitatively determine cell viability. As expected, exposure to pure SiO_2 did not indicate any toxicity, while $NiCl_2$ showed increasing toxicity with increasing concentrations of nickel exposure. Our Ni/SiO_2 showed generally similar toxicity trends to $NiCl_2$. Among the three CENs, nhNi@SiO_2 was found to be the least toxic compared to hNi@SiO_2 and Ni-SiO_2. The MTS assay used for this analysis determines the metabolic activity of cells, which is directly correlated to cell count. Hence there is a possibility of CENs affecting metabolism along with affecting cell viability. In order to eliminate this artifact, we directly measured the cell number by DNA-staining and quantitative fluorescence imaging using Li-cor Odyssey imaging platform. Comparing Li-cor to MTS we confirmed that the NP toxicity was indeed from a reduction in cell number, not a reduction in cell metabolism.

We have successfully used 3T3 cells to study cell viability after exposure to our nanomaterials. As expected, pure silica showed no toxicity, while $NiCl_2$ salt and our NPs exhibited dose-dependent toxicity. Confirming that the porous SiO_2 in indeed nontoxic, suggests that the nanomaterial toxicity seen is associated with the nickel nanoparticles. However, since we did not observe a significant mitigation of toxicity through our CENs, we were not about to validate our original hypothesis. Future work includes the study of nanomaterial uptake, and continuing analysis on the nanoparticle toxicity mechanism and persistency.

References

[1] Nel, A.; Xia, T.; Mädler, L.; Li, N., Toxic potential of materials at the nanolevel. Science 2006, 311 (5761), 622-627.
[2] Oberdoester, G., Nanotoxicology: An Emerging Discipline Evolving from Studies of Ultrafine Particles (vol 113, pg 823, 2005). Environmental Health Perspectives 2010, 118 (9), A380-A380.

Determining an Appropriate Solvent for Extracting Oil from Spent Coffee Grounds for Producing Biodiesel

Devan Buckhalter, Chemical Engineering, University of Kentucky, Paducah, KY

Abstract Text:
Determining an Appropriate Solvent for Extracting Oil from Spent Coffee Grounds for Producing Biodiesel

Biodiesel is a renewable fuel made by the transesterification of vegetable oil. However, the process for making biodiesel can be sustainable by utilizing used oil as a feedstock. This contribution will present the results of research to evaluate different solvents to be used for extracting oil from spent coffee grounds. The motivation for this project comes from the overwhelming volume of spent coffee grounds being disposed of each year. Since it is a waste product, oil extracted from spent coffee grounds is a sustainable option for producing biodiesel. It is estimated that 340 million gallons of biodiesel could be produced each year from the waste of coffee grounds.

Experiments were conducted using a Soxhlet extractor with the following solvents; a 50:50 ratio Hexane to Isopropanol, 100% Methanol, a 50:50 ratio Methanol to Isopropanol, and a 50:50 ratio Ethanol to Isopropanol. The experimental tests yielded 3 – 12% oil depending on the solvents used. The solvent with the lowest yield was the 100% Methanol, while the solvent with the highest yield was the 50:50 ratio of Hexane to Isopropanol. In addition to its use as a biodiesel feedstock, the extruded coffee grounds can be repurposed as fuel pellets or garden fertilizers. These experiments were funded in part by American Reusable Energy, LLC.

Nanoparticle Tracer Method: Exploring the viability, possibility and stability of CO_2 geological and simultaneously enhancing shale gas recovery

*Nan Yang**

Affiliation: State Key Laboratory of Chemical Engineering, East China University of Science and Technology

Abstract

Understanding flow through physical heterogeneities in the subsurface and particularly how heterogeneities lead to preferential (e.g., much faster than average) flow is useful in diverse fields such as enhanced oil recovery, radioactive waste treatment and groundwater contaminant migration. Fractures are a principal factor controlling both the amount of gas that can be stored in shale and the risk associated with that storage. Sequestering CO_2 in shale depends on the diffusion of CO_2 into stagnant areas adjacent to the fractures into which the CO_2 is injected. According to our knowledge, the chemical tracer has a long enough transit through the system to diffuse into the matrix, but a nanoparticle tracer does not. So the nanoparticle tracer arrives first and the chemical tracer later, and the separation measures the degree of preferential flow.

The experiment dispersed the nanoparticles we synthesized in a supercritical CO_2 fluid analogue, and passed them, together with an inert chemical tracer, through a stainless cylindrical column we processed. The cylindrical column has a central channel of large diameter beads surrounded by an annulus filled with smaller diameter beads. For the particle synthesis, a $scCO_2$-phylic polymer was grafted to the particles and fluorophore molecules were added to make the polymer corona fluorescent. Thanks to the great attention paid to the polymer grafting process, mono-dispersed batches of SiO_2 and Fe_3O_4 cored nanoparticles were successfully polymerized and produced using our method. The nanoparticles display satisfying properties such as showing little tendency to agglomerate, having good stability, highly fluorescent and small (~40nm) and uniform in size. Ultraviolet absorption spectroscopy is used to measure the concentration of the Trifluorotoluene (TFT) chemical tracer and the concentration of the nanoparticle tracers are measured mainly by the fluorescence of fluorophore molecules imbedded in the polymer coronae.

In both cases the nanoparticles arrived before the chemical tracer. This reflects the fact that the chemical tracer diffuses more in to the slit adjacent to the core flow channel than do the nanoparticles. Models of the process have been developed to design field tests assessing the viability of sequestrating CO_2 in shale. In the paper, the synthesis of the particles, the laboratory tests we have performed and the modeling will be discussed in detail.

Key words: preferential flow, CO_2 sequestration, nanoparticle tracer, diffusion

Reference:

[1] Subramanian, S. K., Li, Y., Cathles, L. M., Assessing Preferential Flow by Simultaneously Injecting Nanoparticles and Chemical Tracers. Water Resources Research, 2012.

Supporting Information

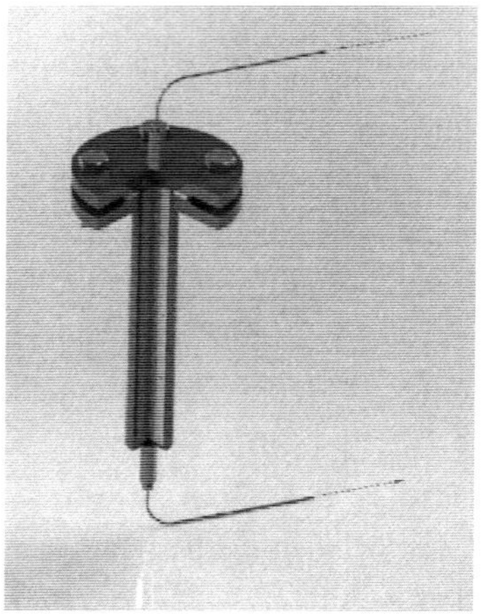

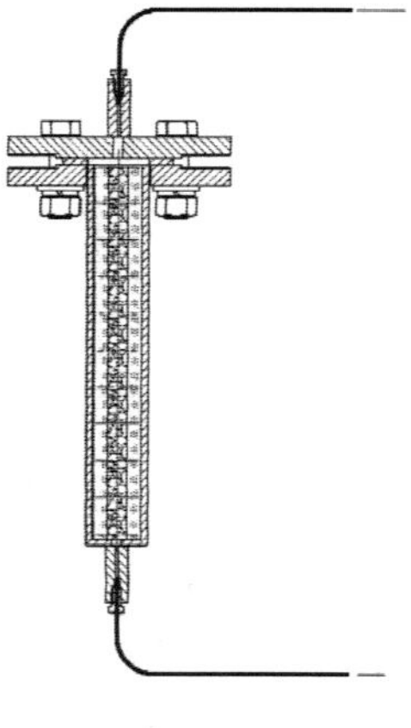

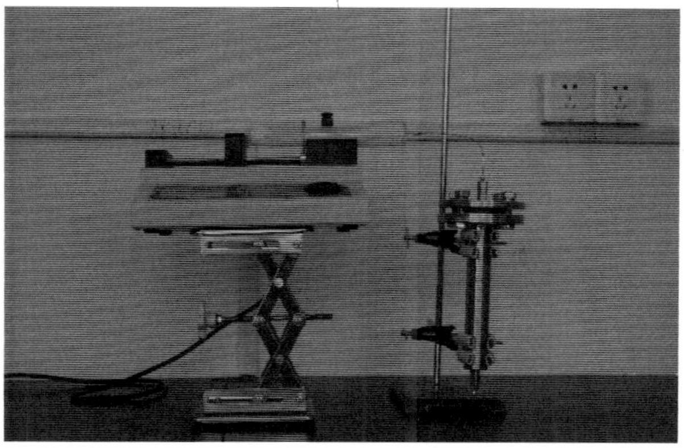

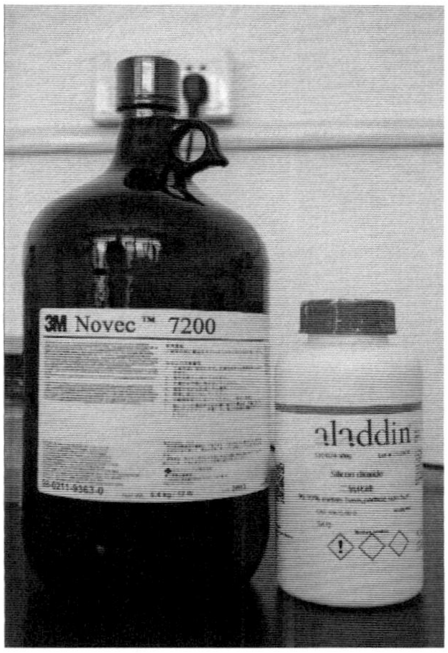
3M Novec ™ 7200
aladdin

Photobioreactors are sustainable devices used for the growth of microalgae, wherein the photosynthetic algae produce their own biomass by utilizing light energy. The tubular types of photobioreactors are most widely used for their ease of use at a laboratory scale. They generally face problems during scale-up. Mixing is an important aspect in airlift tubular photobioreactor systems.

This study presents a computational modelling of microalgae culture using an airlift tubular photobioreactor. The objective of this work is to study the mixing and light intensity within the tubular photobioreactor at three different inclinations 5°, 10° and 15° based on ground level. The COMSOL Multiphysics Version 4.4 software is used to simulate this system. The fluid-flow module will be used to carry out a stationary as well as a time-dependent study. The model will be simulated for the cultivation of *Spirulina platensis,* which has numerous uses: as an antioxidant, a food supplement, an anti- cancer dose and as an animal feed.

The best inclination will be reported at which the most desirable mixing and light intensity is achieved, for maximum biomass productivity. At this inclination, a scale-up will be done and affects on mixing and light intensity will be studied.

Key-Words: - microalgae, Spirulina Platensis, tubular photobioreactor, modelling, mixing, light intensity, scale-up

A Proposed Alternative Measure for Climate Change Potential

F. DeGroff
CCSF

Abstract

Carbon dioxide equivalents (CO_2e) is the current benchmark standard for expressing greenhouse gases equivalents and climate change potential, often the basis for carbon fees.

CO_2e is a one-dimensional scalar measure that does not account for carbon source or ultimate disposition, nor for either spatial or temporal shifts in carbon emissions. A carbon standard vector, however, would take both source and ultimate sink into consideration when calculating climate change potential.

We propose that climate change potential be measured using a logarithm temporal scale, named "carbon quality", abbreviated 'cq', of the form:

$$cq_{(carbon)} = f(\log 10\ [\text{mean carbon residence time}]_{(carbon)}).$$

The carbon quality vector measures when organic carbon from any carbon source is converted to CO_2 or to any lower energy organic carbon in any carbon reservoir, eliminating incentives for consumers to avoid carbon fees through temporal shifts in carbon emissions and spatial shifts via cross-border movement of embodied carbon.

Objectives

The purpose of this poster is to explore an alternative, comprehensive measure of climate change potential to the traditional measure of carbon dioxide equivalents.

Background

Carbon dioxide equivalents is the current benchmark standard for expressing climate change potential, expressed as CO_2e. According to the U.S. EPA,

> "The international standard practice is to express greenhouse gases in carbon dioxide (CO_2) equivalents, or CO_2e. Emissions of gases other than CO_2 are translated into CO_2 equivalents using global warming potentials."

The CO_2e standard is a proxy measure of climate change. The use of this standard, in one form or another as the basis for determining carbon fees, is limited by the one-dimensional, single-point nature of such a measure. The CO_2e standard makes no accommodation for the origin or source of the carbon, nor for alternative dispositions, and thus cannot account for either spatial or temporal shifts in carbon emissions.

Because the CO_2e standard is a scalar, one-dimensional measure of climate change, it has no directional component; it simply measures the equivalent quantity of CO_2 that is emitted into the atmosphere, with no regard for the source of carbon from which the emissions are derived.

> "Vector quantities have two characteristics, a magnitude and a direction. Scalar quantities have only a magnitude. When comparing two vector quantities of the same type, you have to compare both the magnitude and the direction. For scalars, you only have to compare the magnitude. When doing any mathematical operation on a vector quantity (like adding, subtracting, multiplying) you have to consider both the magnitude and the direction. This makes dealing with vector quantities a little more complicated than scalars." (Benson, 2014).

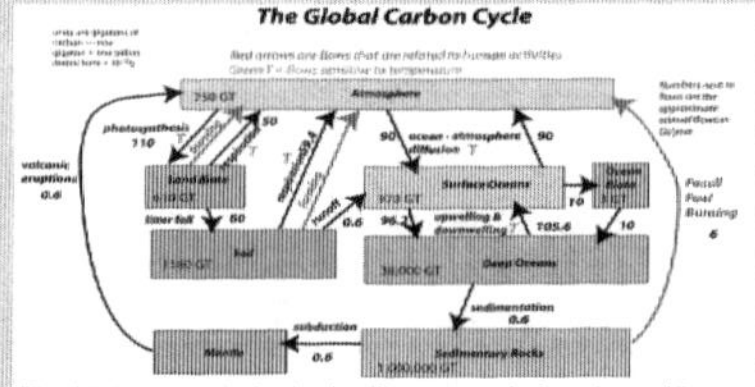

The global carbon cycle showing the different reservoirs for carbon and the exchanges between reservoirs in GT per year

Presumably, the underlying assumption for the CO_2e standard is that fossil fuels are the source of carbon. Such an assumption may be a reasonable approximation in a world where spatial or temporal shifts in carbon emissions (often referred to as 'leakage') are negligible, but such is not the case. Spatial shifts in carbon emissions to areas with no or lower carbon fees are already a significant portion of worldwide global emissions, and we expect that temporal shifts in carbon emissions, such as the conversion of carbon dioxide to biomass, will emerge and expand when and where carbon fees provide the incentive to do so.

Leakage

Carbon leakage is defined as:

> "the increase in CO_2 emissions outside the countries taking domestic mitigation action divided by the reduction in the emissions of these countries." (IPCC 2007)

We have an expanded view of carbon leakage which includes not only spatial displacement of carbon emission, as outlined by the IPCC, but temporal displacement of carbon emissions including emissions of organic carbon intended to avoid taxes on inorganic atmospheric carbon, as well as any method of carbon capture and sequestration intended to avoid taxes on emissions of carbon dioxide.

Numerous studies have highlighted the expanding scope of the spatial displacement form of carbon leakage, including:

> **"Embodied carbon flows are large and growing**
>
> Approximately 25% of all CO_2 emissions from human activities "flow" (i.e. are imported or exported) from one country to another." (Carbon Trust 2011)

and

> **"Embodied carbon imports are significant for many developed economies**
>
> Major developed economies are typically net importers of embodied carbon emissions. UK consumption emissions are 34% higher than production emissions: Germany (29%), Japan (19%) and the USA (13%) are also significant net importers of embodied emissions. For some economies with very carbon-efficient production processes, the relative importance of imported carbon is even greater. The high levels of net imports in France (43%) and Sweden (61%) reflect in part the low-carbon intensity of their energy systems." (Carbon Trust 2011)

There is even evidence of outsourcing of carbon emissions within countries such as China:

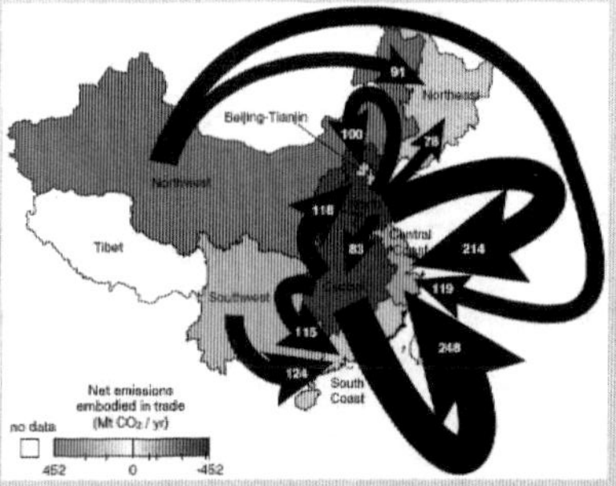

"... wealthy coastal provinces in China are now "outsourcing" a big chunk of their carbon pollution to the poorer inland provinces." (Plumer)

And there is evidence of increasing research into expanded forms of temporal displacement of carbon emissions such as the conversion of carbon dioxide emissions into algae:

Scripps and SoCalGas System to Use Algae to Capture Carbon Dioxide

July 30, 2012 | By Mario Aguilera
Science and Engineering | SIO

Algae could absorb pollution emissions at the source

A new system will explore using algae to capture carbon.

Scripps Institution of Oceanography at UC San Diego (Scripps) and Southern California Gas Co. (SoCalGas) have entered into an agreement focusing on the design of an innovative system in which algae consume carbon dioxide (CO_2) emissions from natural gas combustion and cost-effectively convert it into valuable byproducts such as biomethane, biodiesel and animal feed.

For several years, researchers at Scripps, a member organization of the San Diego Center for Algae Biotechnology, and a number of commercial companies around San Diego and elsewhere, have been studying how algae can most efficiently be developed into a clean, renewable biofuel to one day replace non-renewable fossil fuels. Taking this research a step further, Scripps' researchers hope to leverage algae's natural ability to absorb CO_2 in the environment and convert it into oil rich biomass or biomethane or refined into fossil fuel replacements. After extracting the oils for biodiesel, the remaining biomass can be sold as a safe, protein-rich animal feed.

Reducing climate change without accounting for embodied carbon will be challenging as higher carbon prices increase the incentive to avoid taxes.

Proposal

We have no doubt that as the science and knowledge base of climate change increases, the need for a more robust measure and metric of climate change potential will become self evident. Advances in the methodology and data collection of climate change, along with improved understanding of the linkages between ecosystem services and human well-being, could well form the basis for an expanded, multi-dimensional metric for climate change potential in the future.

In the meantime, to develop a proforma accounting of the effect of spatial or temporal shifts in carbon emissions, we propose a carbon vector scale or metric as a proxy of climate change, one that, in contrast to the CO_2e standard, measures the change in residence time as carbon is moved or shifted between carbon reservoirs due to anthropogenic activity.

To best approximate the climate change effect, we propose that such a shift be measured using a logarithm scale, and the unit name be carbon quality, abbreviated as 'cq'. The general form of the equation is:

$$cq_{(carbon)} = f(\log 10\ [\text{mean carbon residence time}]_{(carbon)})$$

The carbon quality vector will measure the climate change when carbon, in any organic form from any carbon source, is converted to CO_2, or is converted to any lower energy organic carbon, in any carbon reservoir. This measure eliminates the incentive for carbon consumers to avoid carbon fees through temporal shifts in carbon emissions, and can be used to measure cross-border movement of embodied carbon, thus eliminating the incentive for carbon consumers to avoid carbon fees through spatial shifts in carbon emissions.

We view our proposed carbon metric as possibly the first element in what will become a multi-dimensional carbon metric vector and, by extension, of climate change potential (as measured by metrics such as ecosystem services) that will be needed to account for the myriad of factors that affect climate change.

For maximum economic effectiveness, any carbon fee based on anthropogenic changes in carbon quality must be assessed on the actor involved at the same time and place as the pertinent activity. Any delay or displacement of any carbon fee will only contribute to a reduced intensity of the price signal that is intended to influence the actor's behavior.

Such immediate placement of carbon fees based on changes in carbon quality will ensure the capture of the broadest, most comprehensive, and equitable measure of the changes in climate potential due to anthropogenic activity. Economists have long argued that capturing the effect of upstream activity is important to implement an efficient and effective carbon management strategy.

Benefits

The most pronounced benefit of having such a carbon metric is the ability to now account for the climate change potential in everyday commerce, including domestic and international trade, all financial transactions, all investment decisions, and consumer activity for any and all changes in carbon quality. The carbon metric can be used to rank all entities and institutions, ranging from individuals and companies to multinational corporations and entire nations.

With proper placement of appropriate carbon fees based on changes in carbon quality, such upstream fees will follow the embodied carbon throughout the production cycle, allowing for the greatest economic efficiency.

By properly accounting for the climate change potential in international trade, the carbon metric will permit the use of existing accounting systems to implement a worldwide system whereby individual countries choose a carbon management strategy that is best suited to their factors of production, balanced by carbon fees on imports and credits on exports that reflect domestic policies on carbon management.

Metric-based carbon fees and credits would stimulate and nurture longer-term investment in research and development as well as capital for the production of lower-carbon goods and services. Such an approach would allow those countries that wish to pursue a more aggressive carbon management to do so at a lower risk from industries in higher carbon-intensive economies. The lower risk would allow nascent low-carbon industries to reach the economies of scale often necessary to compete on an economic level now occupied by higher carbon-intensive industries and economies.

Another advantage of the carbon metrics themselves is that they allow for two degrees of freedom which are missing from the current worldwide negotiations to reduce carbon intensity. First, the carbon metrics can be developed independently by the scientific community separate from the political posture of individual carbon-intensive entities that wish to maintain their market share in energy and feedstocks. Second, the carbon metrics allow for countries to pursue low-carbon strategies without a complete worldwide agreement to lower carbon intensity.

Implementing such flexibility would allow early adopter countries to pursue a greater reduction of carbon intensity than would be called for in any one-size-fits-all approach to mitigate climate change, while providing trade incentives via carbon parity adjustments for those countries that would not otherwise pursue such a reduction strategy.

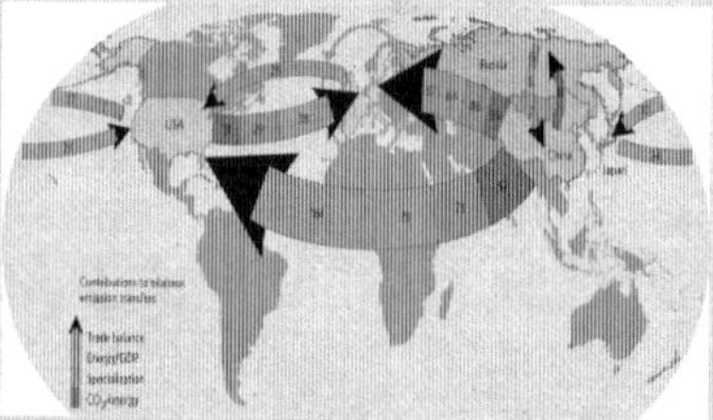

Conclusions

Given the importance of carbon to those countries with high carbon avidity, our view is that efforts to lower carbon intensity via any Cap-and-Trade scheme will prove unworkable and unenforceable worldwide. Combined with the dilution effect from both spatial and temporal leakage of carbon emissions, attempts to reach a statutory worldwide agreement that does not provide for the varying factors of production among countries to reduce carbon intensity will likely be fruitless.

Contact

Email: F. DeGroff at fdegroff@mail.ccsf.edu

398834 Rethinking the Traditional Wall- Thermodynamic and Hygrothermal Modeling of a Textile Wall Assembly Systems for Healthier Energy Efficient Structures

Abstract:

Techstyle Haus is a collaboration between Brown University, the Rhode Island School of Design, and the University of Applied Sciences in Erfurt, Germany, that aimed to build a rapidly-deployed, net-energy-zero house for the 2014 European Solar Decathlon Competition completely out of textiles. Our goal is to design the first textile passive house, in order to explore new ways of sustainable design strategies and inspire others to innovate. Passive House is the world's most rigorous energy efficiency standard for residential construction. Our design process used the fundamentals of thermodynamics and complex understanding and manipulation of construction material properties to achieve a new type of energy efficient construction design. We find that ultimately this construction system design process can be applied to create more efficient, comfortable, residential structures.

Starting with architectural CAD models, we designed a system to break complex curved surface into a tessellated surface of planes that could be input into existing modeling software. This project employed WUFI-Passive, a program that combines a static energy balance model specific to passive house construction with a dynamic modeling software, the WUFI Plus engine. This pair of WUFI products allows us to dynamically simulate the entire building for comfort and energy performance as well as hygrothermal analysis. Through an iterative process, working with architects, engineers, and the WUFI Passive software, Techstyle Haus evolved to its final design. Through the use of our thermodynamic project model, we were able to ensure thermal comfort, indoor air quality, and building assembly durability. Ultimately our prototype design uses approximately 90% less energy than the average American residence for maintaining comfort conditions.

396243 Carbonation of Thermally Activated Serpentine Minerals: Potential for Portland Cement Replacemnt

Abstract:

By thermally activating and carbonating serpentine minerals, a viable Portland cement substitute which produces water as a byproduct and has negative carbon footprint can theoretically be produced. The reaction kinetics of activation have been thoroughly studied by varying parameters including time, mass, powder bed thickness and temperature of the serpentine samples for optimization. Carbonation testing has been done by placing the activated material in a CO_2 -rich environment to produce the Portland cement substitute. The first 'concrete' pellet that showed significant carbonation was produced and is being studied. The activation and carbonation processes will continue to be optimized.

397691 Tumbleweed Mitigation through Biochar Production

Abstract:

Russian Thistle (*Salsola tragus*), which turns into "tumbleweed" when it dries, is an invasive species and poses a threat to surrounding native vegetation and ecosystems. Though originating from the Russian steppe, tumbleweed has adapted to the American southwest readily. This pest is almost impossible to kill; the seeds are viable for 2-3 years and only need a few hours of moisture to germinate. Each plant, when mature, can grow to be the size of a small car and can release up to 200,000 seeds. In addition to competing with native plants, tumbleweed creates fire hazards, dramatically reduces crop yields, promotes wind erosion of soil, and causes allergies when flowering. There have been numerous control measures, including biocontrols, tested but these have yet to give satisfactory results. Pyrolysis, a thermochemical process that transforms biomass through heating under limited-oxygen conditions, can be used to make tumbleweed into biochar and renewable energy. Biochar has the potential to sequester carbon, enrich soils, and increase water retention.

367597 General Thermodynamic Modeling of Gasification of Diverse Biomasses

Abstract:

A thermodynamic equilibrium approach is used in a gasification process in order to create a model and program, via MATLAB, that calculates the composition of the product gas and the gasification temperature. The model is flexible, taking in diverse biomass feedstock, as well as a variety of gasification agents, in order to apply to a wide variety of gasification conditions. Through equilibrium reactions and mass balances, the syngas composition is calculated. Then, energy balance computes the gasification temperature using an iterative method. The model and program were created and then verified through convergence and validated with experimental data from other papers. Finally, a parametric study was completed in order to examine the effects on syngas composition caused by gasification temperature, biomass moisture content, amount of air input, and carbon content of the feedstock. The model will have more dimensions added, including the kinetics of the gasification process, and be used to optimize the process of converting municipal waste to energy and syngas, which is a prime example of converting waste to wealth.

AUTHOR INDEX

AUTHOR INDEX

AUTHOR INDEX

AUTHOR INDEX

AUTHOR INDEX

AUTHOR INDEX

AUTHOR INDEX

AUTHOR INDEX